Eukaryotic Transposable Elements as Mutagenic Agents

The organizers gratefully acknowledge the support of the International Commission for Protection against Environmental Mutagens and Carcinogens (ICPEMC). Banbury Report 30 constitutes ICPEMC Meeting Report No. 4.

Banbury Report Series

Banbury Report 1: Assessing Chemical Mutagens
Banbury Report 2: Mammalian Cell Mutagenesis
Banbury Report 3: A Safe Cigarette?
Banbury Report 4: Cancer Incidence in Defined Populations
Banbury Report 5: Ethylene Dichloride: A Potential Health Risk?
Banbury Report 6: Product Labeling and Health Risks
Banbury Report 7: Gastrointestinal Cancer: Endogenous Factors
Banbury Report 8: Hormones and Breast Cancer
Banbury Report 9: Quantification of Occupational Cancer
Banbury Report 10: Patenting of Life Forms
Banbury Report 11: Environmental Factors in Human Growth and Development
Banbury Report 12: Nitrosamines and Human Cancer
Banbury Report 13: Indicators of Genotoxic Exposure
Banbury Report 14: Recombinant DNA Applications to Human Disease
Banbury Report 15: Biological Aspects of Alzheimer's Disease
Banbury Report 16: Genetic Variability in Responses to Chemical Exposure
Banbury Report 17: Coffee and Health
Banbury Report 18: Biological Mechanisms of Dioxin Action
Banbury Report 19: Risk Quantitation and Regulatory Policy
Banbury Report 20: Genetic Manipulation of the Early Mammalian Embryo
Banbury Report 21: Viral Etiology of Cervical Cancer
Banbury Report 22: Genetically Altered Viruses and the Environment
Banbury Report 23: Mechanisms in Tobacco Carcinogenesis
Banbury Report 24: Antibiotic Resistance Genes: Ecology, Transfer, and Expression
Banbury Report 25: Nongenotoxic Mechanisms in Carcinogenesis
Banbury Report 26: Developmental Toxicology: Mechanisms and Risk
Banbury Report 27: Molecular Neuropathology of Aging
Banbury Report 28: Mammalian Cell Mutagenesis
Banbury Report 29: Therapeutic Peptides and Proteins: Assessing the New Technologies
Banbury Report 30: Eukaryotic Transposable Elements as Mutagenic Agents

Eukaryotic Transposable Elements as Mutagenic Agents

Edited by

MICHAEL E. LAMBERT
Cold Spring Harbor Laboratory

JOHN F. McDONALD
University of Georgia

I. BERNARD WEINSTEIN
Institute of Cancer Research
Columbia University College
of Physicians & Surgeons

COLD SPRING HARBOR LABORATORY
1988

Banbury Report 30: Eukaryotic Transposable Elements as Mutagenic Agents

© 1988 by Cold Spring Harbor Laboratory
All rights reserved
Printed in the United States of America
Cover and book design by Emily Harste

Library of Congress Cataloging-in-Publication Data

Eukaryotic transposable elements as mutagenic agents.

(Banbury report, ISSN 0198-0068 ; 30)
Based on a conference held at the Cold Spring
Harbor Laboratory, Apr. 21, 1987, sponsored by Abbott
Laboratories and others.
Includes index.
 1. Insertion elements, DNA—Congresses.   2. Mutagens—
Congresses.   I. Lambert, Michael E.   II. McDonald,
John F. (John Francis), 1947–   .   III. Weinstein,
I. Bernard.   IV. Cold Spring Harbor Laboratory.
V. Abbott Laboratories.   VI. Series.   [DNLM:
1. Cells.   2. Cytology—congresses.   3. DNA Insertion
Elements—congresses.   4. Gene Expression Regulation—
congresses.   5. Mutation—congresses.   W3 BA19 v.30 /
QH 450.2 E873 1988]
QH462.I48E95   1987   575.2′92   87-21850
ISBN 0-87969-230-8

Authorization to photocopy items for internal or personal use, or the internal or personal use of specific clients, is granted by Cold Spring Harbor Laboratory for libraries and other users registered with the Copyright Clearance Center (CCC) Transactional Reporting Service, provided that the base fee of $1.00 per article is paid directly to CCC, 27 Congress St., Salem, MA 01970. [0-87969-230-8/88 $1.00 + .00]. This consent does not extend to other kinds of copying, such as copying for general distribution, for advertising or promotional purposes, for creating new collective works, or for resale.

All Cold Spring Harbor Laboratory publications may be ordered directly from Cold Spring Harbor Laboratory, Box 100, Cold Spring Harbor, New York 11724. (Phone: 1-800-843-4388) In New York State (516) 367-8325.

BANBURY PROGRAM SPONSORS

Corporate Sponsors

Abbott Laboratories
American Cyanamid Company
Amersham International plc
Becton Dickinson and Company
Cetus Corporation
Ciba-Geigy Corporation
CPC International Inc.
E.I. du Pont de Nemours & Company
Eli Lilly and Company
Genentech, Inc.
Genetics Institute
Hoffmann-La Roche Inc.
Monsanto Company
Pall Corporation
Pfizer Inc.
Schering-Plough Corporation
Smith Kline & French Laboratories
Tambrands Inc.
The Upjohn Company
Wyeth Laboratories

Core Supporters

The Bristol-Myers Fund, Inc.
The Dow Chemical Company
Exxon Corporation
International Business Machines Corporation
The Procter & Gamble Company
Rockwell International Corporation Trust
Texas Philanthropic Foundation Inc.

Special Program Support

James S. McDonnell Foundation
Alfred P. Sloan Foundation

Participants

Sid Aaron, The Upjohn Company, Kalamazoo, Michigan

Garth R. Anderson, Department of Molecular and Cellular Biology, Roswell Park Memorial Institute, Buffalo, New York

Paul M. Bingham, Department of Biochemistry, State University of New York, Stony Brook

Jef D. Boeke, Department of Molecular Biology and Genetics, Johns Hopkins University School of Medicine, Baltimore, Maryland

John Cairns, Department of Cancer Biology, Harvard School of Public Health, Boston, Massachusetts

Robert Callahan, Laboratory of Tumor Immunology and Biology, National Cancer Institute, Bethesda, Maryland

Vicki L. Chandler, Institute of Molecular Biology, University of Oregon, Eugene

Victor G. Corces, Department of Biology, Johns Hopkins University, Baltimore, Maryland

Nina Fedoroff, Department of Embryology, Carnegie Institution of Washington, Baltimore, Maryland

David J. Finnegan, Department of Molecular Biology, University of Edinburgh, Scotland

Elisabeth Gateff, Institut für Genetik, Johannes Gutenberg-Universitat, Mainz, Federal Republic of Germany

Mel M. Green, Department of Genetics, University of California, Davis

Warner C. Green, Howard Hughes Medical Institute, Duke University Medical Center, Durham, North Carolina

Margaret Kidwell, Department of Ecology and Evolutionary Biology, University of Arizona, Tucson

Robert Krauss, Institute of Cancer Research, Columbia University, New York

Edward L. Kuff, Laboratory of Biochemistry, National Cancer Institute, Bethesda, Maryland

Michael E. Lambert, Cold Spring Harbor Laboratory, New York

John F. McDonald, Department of Genetics, University of Georgia, Athens

Charlotte Paquin, Department of Biological Sciences, University of Cincinnati, Ohio

Denise Roberts, Department of Biochemistry and Molecular Biology, Harvard University, Cambridge, Massachusetts

Rodney Rothstein, Department of Genetics and Development, Columbia University, New York, New York

Krishnaswamy Sankaranarayanan, Department of Radiation Genetics and Chemical Mutagenesis, State University of Leiden, The Netherlands

James A. Shapiro, Department of Biochemistry and Molecular Biology, University of Chicago, Illinois

Maxine F. Singer, Laboratory of Biochemistry, National Cancer Institute, Bethesda, Maryland

Frits H. Sobels, Department of Radiation Genetics and Chemical Mutagenesis, State University of Leiden, The Netherlands

Marjorie Strobel, Mammalian Genetics Laboratory, N.C.I.-Frederick Cancer Research Facility, Maryland

I. Bernard Weinstein, Institute of Cancer Research, Columbia University College of Physicians & Surgeons, New York

Michael C. Wilson, Department of Molecular Biology, Research Institute of Scripps Clinic, La Jolla, California

Fred Winston, Department of Genetics, Harvard University School of Medicine, Boston, Massachusetts

Evelyn M. Witkin, Waksman Institute of Microbiology, Rutgers—The State University, Piscataway, New Jersey

Harald zur Hausen, German Cancer Research Center, Heidelberg, Federal Republic of Germany

Row 1: I.B. Weinstein; R. Rothstein, M. Lambert; J. McDonald
Row 2: J. Boeke, V. Chandler; M. Strobel, C. Paquin
Row 3: J. Cairns, B. McClintock; M. Green, V. Corces
Row 4: D. Finnegan; R. Sadorsky, E. Gateff

Preface

Recently, there have been major advances in our understanding of the molecular mechanisms underlying retrotransposition and a growing appreciation of the importance of repetitive DNA sequences and transposable elements in shaping eukaryotic genome structure. This work has focused attention both on the potential of these genetic elements to act as mutagenic agents and on the regulation of such mutagenic processes at the cellular level. Many different experimental approaches in yeast, *Drosophila,* and mammalian systems, are now in use to characterize the range of mutational events mediated by such genetic elements and to study the interplay of both host genes and various forms of environmental and physiological stress, including DNA damage, on these processes. Despite this progress, there have been few efforts to bring together interested investigators for the explicit purpose of reviewing and assessing this field. To create further interest in this field, a conference was recently held within the setting of the Banbury Center, devoted exclusively to an in-depth evaluation of current work in this area. The conference brought together experts in the areas of eukaryotic retroviruses and transposable elements, cancer biology, DNA damage and repair, and human risk assessment. Particular emphasis was placed not only on description of the types of mutations caused by rearrangement of these elements, but also on the involvement of DNA damage, "genomic stress" and host genes in the modulation of viral latency, alteration in viral gene expression, and mediation of transpositional processes. This current report continues a Banbury Center tradition of publishing critical reviews in emerging areas in molecular biology central to human risk assessment and cancer biology.

The funding for this meeting was generously provided by grants from the National Cancer Institute, the Environmental Protection Agency, and the International Commission for Protection Against Environmental Mutagens and Carcinogens.

We would like to thank James D. Watson, Director of Cold Spring Harbor Laboratory, for making the Banbury Center available for this meeting. We are also grateful to Steve Prentis, the director of the Banbury Center at the time this meeting was planned and to Bea Toliver and Katya Davey. We would also like to express our appreciation to the Publications Department, including Nancy Ford, the Managing Director, Ralph Battey, and Inez Sialiano.

M.E. Lambert
J.F. McDonald
I.B. Weinstein

Contents

Participants, vii

Preface, xi

SECTION 1: OVERVIEW OF PROKARYOTIC TRANSPOSABLE ELEMENTS

What Transposable Elements Do in Bacteria / James A. Shapiro   3

The Role of the Bacterial Host in the Mechanism and Regulation of Tn*10* Tranposition / Denise Roberts, Donald Morisato, and Nancy Kleckner   17

SOS Mutagenesis and Replisome Reactivation in Bacteria: Options Acquired by Transposition? / Evelyn M. Witkin and Vivien Roegner-Maniscalco   29

SECTION 2: MUTATIONAL EFFECTS OF TRANSPOSABLE ELEMENT INSERTIONS

Mobile DNA Elements and Spontaneous Gene Mutation / Melvin M. Green   41

Molecular Genetic Studies of the Mouse Dilute Locus: Analysis of Two Dilute Alleles and Dilute-suppressor / Majorie C. Strobel, Peter K. Seperack, Karen J. Moore, Neal G. Copeland, and Nancy A. Jenkins   51

Regulation of the Maize Suppressor-Mutator Element / Nina Fedoroff, Patrick Masson, and Jody Banks   63

The Functional Potential of the Human LINE-1 Family of Interspersed Repeats / Maxine F. Singer, Jacek Skowronski, Thomas G. Fanning, and Skorn Mongkolsuk   71

Factors Affecting Retrotransposition of Intracisternal A-particle Proviral Elements / Edward L. Kuff   79

Two Families of Human Endogenous Retroviral Genomes / Robert Callahan — 91

SECTION 3: INDUCERS/REGULATORS OF TRANSPOSABLE ELEMENT EXPRESSION AND TRANSPOSITION: HOST EFFECTS

Concerted Transposition in *Drosophila melanogaster* / Nickolai A. Tchurikov, Tatiana I. Gerasimova, Sofia G. Georgieva, Lev J. Mizrokhi, Pavel G. Georgiev, and Yurii V. Ilyin — 103

Transposons as Probes of Regulatory Machinery / Paul M. Bingham, Tze-Bin Chou, and Zuzana Zachar — 115

Molecular Basis of Transposable Element-induced Mutations in *Drosophila melanogaster* / Pamela K. Geyer, Melvin M. Green, and Victor G. Corces — 123

Regulation of Expression and Transposition of Murine Endogenous Retroviral Elements / Michael C. Wilson, Paul F. Policastro, and Merete Fredholm — 131

Genes That Affect Ty-mediated Gene Expression in Yeast / Fred Winston — 145

The Genetic Control of Recombination between Repeated Sequences / Rodney J. Rothstein, W. Lane Arthur, John W. Wallis, Jr., Hans Ronne, and Barbara J. Thomas — 155

Regulation of Yeast Ty Element Transposition / Jef D. Boeke, Daniel Eichinger, and Gerald R. Fink — 169

SECTION 4: INDUCERS/REGULATORS OF TRANSPOSABLE ELEMENT EXPRESSION AND TRANSPOSITION: GENETIC MUTATOR SYSTEMS

Regulatory Aspects of the Expression of P-M Hybrid Dysgenesis in *Drosophila* / Margaret Gale Kidwell — 183

Mutation Induction by *MR(P)* and Its Modification by Various Conditions / Frits H. Sobels and Jan C.J. Eeken — 195

The Structure of Mutations Induced by I-R Hybrid
  Dysgenesis in *Drosophila melanogaster* / David J.
  Finnegan, Isabelle Busseau, Michael Lynch, Diana H.
  Fawcett, Claire K. Lister, Alain Pélisson, Helen M.
  Sang, and Alain Bucheton   209

SECTION 5: INDUCERS/REGULATORS OF TRANSPOSABLE ELEMENT EXPRESSION AND TRANSPOSITION: GENOMIC STRESS AND ENVIRONMENTAL EFFECTS

Evidence of Host-mediated Regulation of Retroviral
  Element Expression at the Post-transcriptional Level /
  John F. McDonald, Dennis J. Strand, Mark R. Brown,
  Susan M. Paskewitz, Amy K. Csink, and Susan H.
  Voss   219

Effect of Temperature on Ty Transposition / Charlotte
  Elder Paquin and Valerie Moroz Williamson   235

Effects of DNA Damage on Transcription and
  Transposition of Ty Retrotransposons of Yeast /
  Kevin McEntee and Victoria A. Bradshaw   245

Inducible Cellular Responses to DNA Damage / Michael E.
  Lambert and James I. Garrels   255

Induction of VL30 Element Expression as a Response to
  Anoxic Stress / Garth R. Anderson, Daniel L. Stoler,
  Joseph P. Scott, and Becky K. Farkas   265

DNA Modification of *Mu* Elements in Maize / Vicki Lynn
  Chandler   275

SECTION 6: INDUCERS/REGULATORS OF RETROVIRAL ELEMENT EXPRESSION

A Possible Role of Retrotransposons in Carcinogenesis / I.
  Bernard Weinstein, Wendy-L. Hsiao, Ling-Ling Hsieh,
  Tommaso A. Dragani, Carl Peraino, and Martin
  Begemann   289

Retrovirus-like Particles, Reoviruses, and c-*src* Expression in Malignant Tumors of *Drosophila melanogaster* / Elisabeth Gateff   299

HTLV-I, HIV, and Human T Cell Growth / Warner C. Greene, Mitchell Dukovich, Yuji Wano, and Miriam Siekevitz   309

Mobile Genetic Elements, Spontaneous Mutations, and the Assessment of Genetic Radiation Hazards in Man / Krishnaswamy Sankaranarayanan   319

Author Index   337

Subject Index   339

Overview of Prokaryotic Transposable Elements

What Transposable Elements Do in Bacteria

JAMES A. SHAPIRO
Department of Biochemistry and Molecular Biology
University of Chicago
Chicago, Illinois 60637

OVERVIEW

Transposable elements are well known as agents of genome restructuring and mutagenesis in bacteria. In addition to the widely recognized phenomena of sequence disruption and transcriptional blocks, their influence on genome functioning comprises multiple positive and negative effects. Some examples of these are detailed: Tn*1000* can act at a distance as a transcriptional silencer of the *ebgA* promoter; IS*2* utilizes several kinds of sequence rearrangements to create new promoters; IS*1* and IS*5* can act as transcriptional enhancers of the existing but poorly expressed β-glucoside (*bgl*) promoter. Besides having a wide repertoire of mechanisms for altering genome function, the activities of transposable elements display considerable responsiveness to changes in bacterial physiology. Some examples of such responsiveness are detailed: Insertions of Tn*3*, IS*1*, and IS*5* are highly sensitive to the topology of target DNA; phage Mu can mediate the fusion of two different protein coding sequences, and this fusion occurs in a highly regulated manner in response to selection for hybrid protein function. A Mu*lac* element can block plasmid inheritance, and this inhibitory activity can be programmed to occur at specific times in colony development. These examples of genomic modulation and physiological responsiveness point to useful roles for transposable elements in facilitating bacterial survival and proliferation.

INTRODUCTION

Transposable elements are agents of genetic change built into the genomes of all organisms. In bacterial species, they can move through the genome and promote rearrangements of other DNA sequences. As they change the structure of the bacterial genome, they display a variety of mechanisms for altering how the information stored in DNA molecules is expressed and inherited. The genomic restructuring activities of bacterial transposable elements are intricate, well-regulated biochemical systems and can be seen to operate in response to changes in the conditions under which bacteria must survive and reproduce. The evidence leading to the first of these generalizations, about the role of bacterial transposable elements as agents of genome restructuring, is well known. The relevant data have been summarized in the last five years in a number of reviews which discuss the structure, biology, biochemistry, and genetic behavior of bacterial transposable elements in considerable detail (Heffron 1983; Iida et al. 1983; Kleckner 1983; Toussaint and Resibois 1983; Grindley and Reed 1985; Shapiro 1985a; Mizuuchi and Craigie 1986). Since other data relating to the multiple effects of transposable elements on genome functioning are less well known, this article will present a few exemplary cases to illustrate how sophisticated and responsive bacterial transposable elements can be.

RESULTS

Consequences of Rearrangements Involving Transposable Elements

Blocks to Expression. It is obvious that a deleted or disrupted coding sequence will not be properly expressed. Thus, many genomic rearrangements lead to loss of function. However, transposable elements influence the operation of the bacterial genome by a variety of effects that go far beyond destroying coding sequences. One of the first clues that transposable elements were a new source of mutation came from the observation that they often had dramatic effects in blocking transcription of intact cistrons, and it is now known that they carry various classes of transcription termination signals. Thus, transposable elements can uncouple a cistron or an intact operon from its previous transcription signals and open the way for the generation of new controls.

Expresson of intact cistrons can be blocked by mechanisms other than transcription termination. An intriguing case where a transposable element acts at a distance as a transcriptional "silencer" was discovered by Stokes and Hall (1984). They were trying to identify the location of the evolved β-galactosidase (*ebgA*) operon in a cloned 9.6-kb DNA fragment by mutagenesis with Tn*1000*. To their surprise, they found that all inserts of Tn*1000* into the plasmid carrying their fragment inactivated *ebgA* expression and led to a Lac$^-$ phenotype. Further investigation showed that only some of these inserts completely destroyed β-galactosidase expression; these were within the *ebgA* operon itself. The other inserts were in both orientations outside *ebgA* (including some in vector plasmid sequences), and these reduced β-galactosidase expression by a factor of >200 without completely eliminating enyzme activity. Certain explanations for Tn*1000* silencing of *ebgA* could be excluded because of further observations on the role played by plasmid organization: (1) the silencing effect was not seen with Tn*1000* inserts into a plasmid carrying the 9.6-kb fragment in the alternative orientation with respect to the vector, and (2) the Tn*1000* silencing effect reappeared in this second plasmid when a 0.8-kb piece (located >3 kb from *ebgA*) was removed from the end of the insert. Thus, overall plasmid organization, rather than the relative positions of individual transcription signals in Tn*1000* and *ebgA*, was responsible for the inhibitory effect. Although there is not yet a detailed explanation for this Tn*1000* silencer effect, it appears to have something to do with the relationships between primary sequence organization, plasmid topology, and functioning of the *ebgA* promoter. The idea that Tn*1000* insertion alters plasmid topology is very plausible because this element contains both the substrate sequence and coding sequence for a site-specific recombinase/topoisomerase activity (Grindley and Reed 1985). Moreover, this interpretation of Tn*1000* silencing is consistent with the *bgl* system described below, where other transposable element insertions also appear to have significant effects on promoter function through DNA topology.

Activation of Expression. Transposable elements can activate unexpressed sequences in bacteria by a variety of mechanisms, some of which are understood in molecular detail. These activation events apply to sequences that are nonfunctional for several reasons: They were inactivated by mutations (sometimes insertions); they were introduced into bacteria from other organisms without functional promoters by genetic engineering methods; or they exist in a normally "cryptic" state in the bacterial genome.

The case of *galOP*::IS2 insertion mutations illustrates the first kind of activation situation (see Iida et al. 1983, for the most complete summary of the rather extensive data). These mutations lead to a Gal$^-$ phenotype because transcription of the intact *galETK* cistrons from the *gal* promoter region is almost completely eliminated by a ρ-dependent transcription termination signal in the IS2 insert. Nonetheless, the mutant bacteria can revert to a Gal$^+$ phenotype by several mechanisms. At a very low frequency, IS2 excises to leave a stable, normally regulated *gal* operon. However, most revertants are unstable and have abnormally regulated *gal* expression. These kinds of revertants result from an amazing variety of rearrangements that use the ends of the IS2 insert to create new promoters. The rearrangements include (1) duplications of terminal IS2 sequences (at least three kinds of duplications have been found), (2) duplications plus deletions of terminal IS2 sequences, (3) insertions of either IS2 or IS3 into the resident IS2, and (4) a combination of insertion of a second IS2 and deletion of sequences from the original IS2.

The impression derived from the *galOP*::IS2 results is that IS2 has a wide repertoire of mechanisms for creating novel promoters. This impression is reinforced by analyzing many other situations where IS2 inserts next to silent cistrons and activates their transcription through new promoters. The transcriptionally activated coding sequences include the cloned yeast *his3* locus (Brennan and Struhl 1980), promoter-damaged *arg* cistrons (Glansdorff et al. 1981), and the cryptic chromosomal *ampC* β-lactamase locus found in some *E. coli* strains (Jaurin and Normark 1983).

A very different mechanism by which IS elements can activate expression has been studied in the aromatic β-glucoside (*bgl*) operon of *E. coli* (Reynolds et al. 1981, 1985). This cryptic but inducible operon is normally present in an inactive state in the chromosome of *E. coli* K-12 and other *E. coli* isolates (Schaefler and Malamy 1969). When cultures are plated on medium containing an aromatic β-glucoside such as salicin or arbutin as sole carbon source, Bgl^+ colonies invariably appear within a few days of plating. These mutants display inducible expression of a phospho-β-glucosidase and a β-glucoside-specific transport protein encoded by the *bglB* and *bglC* cistrons. In each case, the genetic event that activated expression to give rise to these colonies was insertion of either an IS1 or an IS5 element upstream of the *bgl* promoter. Different insertions lead to different levels of expression, but in all cases the inducible mRNA starts at the same position as that used at a very low level in the Bgl^- parent strain (Reynolds et al. 1985). Thus, the IS1 and IS5 inserts do not create new promoters but activate the existing, yet silent, *bgl* promoter. The mechanism of transcriptional activation appears to have something to do with changes in local DNA topology, because two other kinds of mutational events can also give increased *bgl* expression: (1) mutagen-induced changes in a CAP-cAMP binding site upstream of the *bgl* promoter (Reynolds et al. 1984) and (2) independently selected mutations that alter DNA gyrase activity (DiNardo et al. 1982). Both altered gyrase activity and altered binding of activator protein can be expected to change the structure of the DNA helix in the *bgl* promoter region.

The parallel between the role of IS1 and IS5 elements in stimulating the *bgl* promoter and eukaryotic enhancers is intriguing. In addition to the obvious mechanistic problems of how promoter stimulation occurs, two significant biological questions come out of the *bgl* activation results: (1) How is a cryptic operon like *bgl* maintained in the genome in the absence of function? (2) Why does the *bgl* selection only yield IS1 and IS5 insertions when other kinds of activating mutations have been found? In other words, is there something special about *bgl* that specifically utilizes IS1 and IS5 for activation so that insertion is actually a dedicated regulatory mechanism?

Other instances of IS5 insertional activation of cloned sequences have been reported, but it is not yet known whether the stimulation mechanisms involved in these cases resemble the *bgl* system or create new promoters like IS2 activations.

Responsiveness of Transposable Elements

Sensitivity to DNA Structure. A great deal of evidence is accumulating from in vitro studies about the roles of higher-order DNA-protein complexes in specific recombination systems, such as bacteriophage λ insertion and excision (Echols 1986) and phage Mu transposition (Mizuuchi and Craigie 1986; Surette et al. 1987). Thus, it is to be expected that the movement of transposable elements will show a high degree of sensitivity to alterations in DNA structure. There are two cases where this sensitivity has been documented by comparing the magnitude of insertion into particular genomic targets.

One case involves the transposon Tn*3*. Tn*3* and its mutants utilize a replicative insertion mechanism that produces cointegrate molecules when both the donor element and the target site are present on plasmids. Although replicon fusion normally proceeds independently of whether the donor molecule is or is not replicating, a block in the ability of the target molecule to replicate reduces the frequency of fusions by over four orders of magnitude (Muster et al. 1983).

The second case involves IS*1* and IS*5*. In the course of characterizing recombinant plasmids carrying an immunoglobulin κ-light-chain cDNA fragment, Amster et al. (1982) were surprised to observe that 12% of the molecules had acquired IS*1* or IS*5* insertions. If the cDNA fragment was inserted into the same vector in the opposite orientation, then no inserts were found. Both recombinant plasmids exerted comparable inhibitory effects on bacterial growth, but only one DNA structure acquired inserts that relieved the inhibition. These results indicated that there must be some topological difference between the plasmids which influenced their effectiveness as transposition recipients, and a difference in superhelical densities was in fact detected by electrophoresis on agarose gels containing chloroquine phosphate (Amster and Zamir 1986).

From these results, it is apparent that the choice of targets by at least three different transposable elements, Tn*3*, IS*1*, and IS*5*, is linked to DNA replication and DNA topology. Such higher-order structural effects may explain some of the complex relationships that have been observed between primary sequence organization of target molecules and the specificity of transposable element insertion.

Sensitivity to Physiological Changes. Detailed studies of the kinetics of genomic rearrangements mediated by phage Mu and its derivatives reveal even more striking instances of physiological specificity in the activity of transposable elements. One case involves the use of a Mu prophage as a region of portable DNA homology to construct strains for the selection of bacteria that synthesize hybrid proteins (Casadaban 1976). In this in vivo genetic engineering method (Fig. 1), a Mu prophage insertion is selected in the coding sequence for the protein that is to contribute the upstream amino terminal portion of the hybrid protein. The appropriately located prophage is then used to direct the downstream insertion by homologous recombination of a circular molecule that carries the coding sequence for another protein

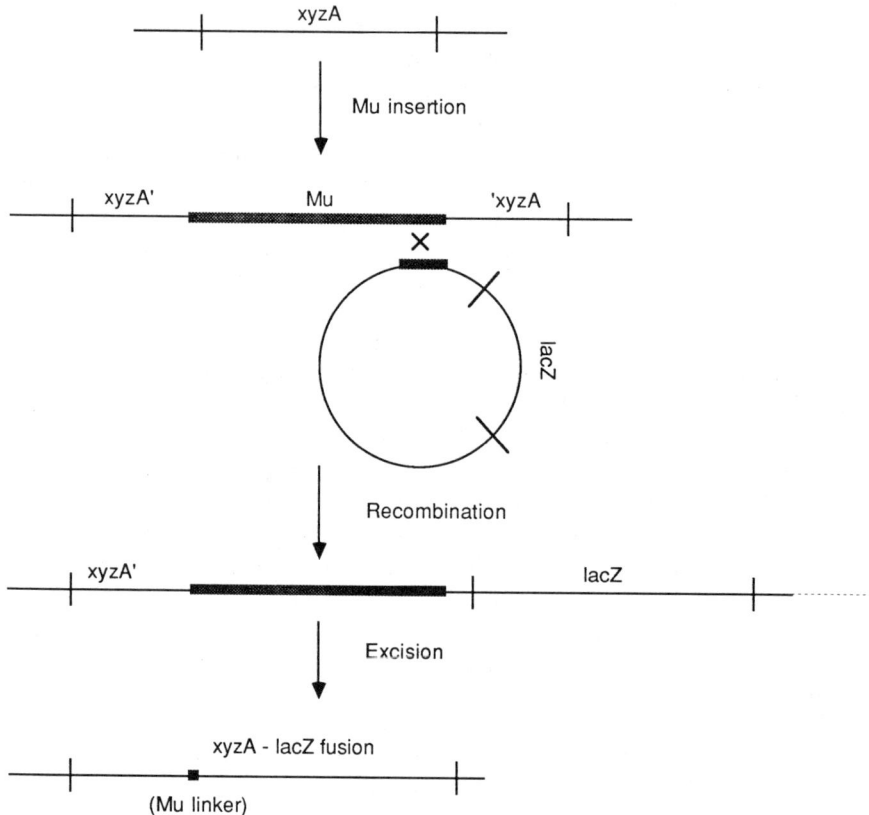

Figure 1

The Casadaban (1976) technique for isolating protein fusion strains. This method depends on Mu as a region of portable DNA homology to direct the insertion of *lacZ* downstream of any other protein-coding sequence in *E. coli*. The homologous recombination substrates are the end of the Mu prophage inserted into the chosen cistron and a terminal fragment of another Mu prophage upstream of the *lacZ* cistron in a λ-bacteriophage. The reciprocal recombination event integrates the λ-DNA and produces the appropriate prefusion structure shown with all sequences in the correct orientation. A Mu-dependent excision event removes all blocks to transcription and translation between the beginning of the *xyzA* cistron and all but the first 20 or so codons of the *lacZ* cistron. A small number of Mu-derived nucleotides constitute the "Mu linker," which is often found in the hybrid-coding sequence. (For further details, see Casadaban 1976; Shapiro 1984a, 1987b).

(generally the *Escherichia coli* β-galactosidase). Thus, a region of DNA is constructed, where appropriate excision of Mu prophage sequences to remove blocks to transcription and translation can fuse the two coding regions and create a novel cistron encoding a hybrid protein. Two observations tell us that Mu plays an active role in the process of creating these hybrid coding sequences: (1) Mu nucleotides are often found as linkers between the two parental cistrons (Fig. 2) and (2) repression of Mu functions by a second Muc$^+$ prophage blocks fusion formation (Shapiro 1984a).

```
Mu S end:
TGAAGCGGCG CACGAAAAAC GCGAAAGCGT TCACGATAAA TGCGAAAACT TTAGCTTTCG
ACTTCGCCGC GTGCTTTTTG CGCTTTCGCA AGTGCTATTT ACGCTTTGA AATCGAAAGC
<--------- ---------- ---------- ---------- ---------- ----------

lamB fusion 42-12:
leu lys arg arg thr lys asn ala lys ala phe gly val
CTG AAG CGG CGC ACG AAA AAC GCG AAA GCG TTT GGC GTT
GAC TTC GCC GCG TGC TTT TTG CGC TTT CGC AAA CCG CAA
BB< --- --- --- --- --- --- --- --- --- --Z ZZZ ZZZ

lamB fusion 61-4:
ile glu ala ala his gly lys arg glu ser leu
ATT GAA GCG GCG CAC GAA AAA CGC GAA AGC CTT
TAA CTT CGC CGC GTG CTT TTT GCG CTT TCG GAA
BB< --- --- --- --- --- --- --- --- --- ZZZ

malF fusion 11-1:
leu thr pro leu his gln leu
CTG ACG CCG CTT CAC CAA CTT
GAC TGC GGC GAA GTG GTT GAA
FFF F-- --- --- ->Z ZZZ ZZZ

malF fusion 14-4:
thr thr gln     lys ala phe phe val arg arg phe thr gly val thr
ACC ACG CAG NCG AAA GCG TTT TTC GTG CGC CGC TTC ACT GGC GTT ACC
TGG TGC GTC NGC TTT CGC AAA AAG CAC GCG GCG AAG TGA CCG CAA TGG
FFF FFF F    -- ---/--- --- --- --- --- --- --- >ZZ ZZZ ZZZ ZZZ

malF fusion 53-1:
val gly ala pro leu his gln leu
GTG GGT GCG CCG CTT CAC CAA CTT
CAC CCA CGC GGC GAA GTG GTT GAA
FFF F-- --- --- --- ->Z ZZZ ZZZ

malF fusion 6-3:
asp ala leu ser arg phe ser cys ala ala ser thr gln leu
GAC GCG CTT TCG CGT TTT TCG TGC GCC GCT TCA ACC CAA CTT
CTG CGC GAA AGC GCA AAA AGC ACG CGG CGA AGT TGG GTT GAA
FFF F-- --- --- --- --- --- --- --- --> ZZZ ZZZ ZZZ

ompF fusion 62-1:
thr ala phe ala phe phe val arg arg phe thr gly val thr
TAC GCT TTC GCG TTT TTC GTG CGC CGC TTC ACT GGC GTT ACC
ATG CGA AAG CGC AAA AAG CAC GCG GCG AAG TGA CCG CAA TGG
F-- --- --- --- --- --- --- --- --- --- >ZZ ZZZ ZZZ ZZZ
```

Figure 2
The fusion sequences of some hybrid cistrons containing Mu linkers. The dashed lines below the DNA sequence indicate Mu-derived base pairs, and the arrowhead indicates the last Mu base pair (which was next to the upstream cistron in the preexcision strain). Similarly, Z indicates a base pair derived from *lacZ*, and B or F indicates a base pair derived from either the *lamB*, *malF*, or *ompF* upstream cistrons. Note that the Mu nucleotides are inverted from their original orientation in the last 5 fusions (Reprinted, with permission, from Shapiro 1987b).

By following the emergence of fusion colonies from a strain constructed as shown in Figure 1, it was possible to demonstrate that fusion formation did not occur stochastically with a fixed (low) probability but rather at a surprisingly high frequency in a temporally specific way in response to the appropriate selective environment (Shapiro 1984a). No fusions were formed when the bacteria were grown under nonselective conditions (none in 3×10^{10} bacteria tested). Nonetheless, fusion colonies appeared almost synchronously in large numbers within three weeks of plating under selective conditions. In the particular situation that was examined, up to 39% of all plated colony-forming units produced fusions. It was further observed that the formation of fusions by a subpopulation of the plated bacteria was accelerated when the medium was slightly enriched, so that the transition from growth to nongrowth conditions occurred on the selection plates. This last result indicated a further element of physiological responsiveness to the Mu-mediated fusion process.

Specific genomic responses to physiological change are not limited to mutant selection regimes. On agar medium, such changes occur in a regular way as colonies develop, and the results of bacterial differentiation can be visualized by examining colonies histochemically (Shapiro 1984b,c; 1986) or microscopically (Shapiro 1985b, 1987a). Differentiated bacteria are organized in colonies in two basic ways: as outwardly directed *sectors* of cells that share a common ancestry and as concentric rings of cells that share common positions in the developmental history of the colony. Sectors can display radically different types of growth control from the rest of the colony, sometimes giving rise to neoplastic growth with a tumorous appearance (Fig. 3). Concentric rings are composed of groups of cells that may differ with respect to biochemical activity, cellular morphology, and pattern of multicellular alignment. This concentric differentiation can be seen clearly in scanning electron micrographs of the edge of a 68-hour-old *E. coli* colony (Fig. 4).

Obviously, cells that differ morphologically will also differ biochemically, and there is no reason to think that DNA metabolism will not be affected by overall physiological regulation. A close look at sectoring events reveals instances where particular genomic alterations can occur in a regular manner during colony development. Another way that transposable elements can modify genome functioning is to interfere with plasmid inheritance. For example, expression of Mu functions from a Mud*lac* element inserted in plasmid R388 can block inheritance of that replicon (Shapiro 1984b; Shapiro and Brinkley 1984). Plasmid loss due to such interference during growth on agar is illustrated in Figure 5. When R388::Mud*lac* plasmids were lost from some of the bacteria that gave rise to outwardly expanding clones, white sectors appeared in colonies growing on medium containing the chromogenic β-galactosidase substrate commonly known as X-gal. The sectors were quite

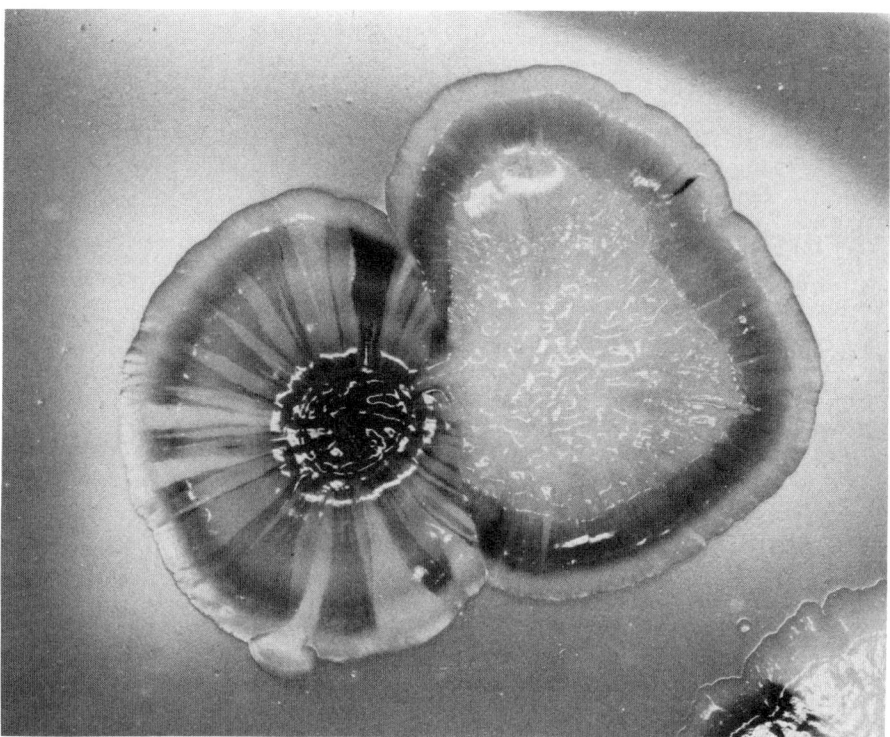

Figure 3
A sectored *E. coli* colony on β-galactosidase indicator medium. This colony grew from a small inoculum containing about 10^4 bacteria. The inoculation site corresponded to the circular indentation at the center of the colony. The agar contained the chromogenic β-galactosidase substrate, commonly known as X-gal, which turns blue upon hydrolysis. Thus, pigmentation indicates enzyme expression from the *lac* sequences that were present in a Mud*lac* element in the genome. The various sectors emerging from the inoculation site showed different enzyme levels. One sector produced a very large neoplasm with distinct growth control and a novel pattern of enzyme expression. (Reprinted, with permission, from Shapiro 1985c.)

distinct because they had an intensely pigmented border that formed where plasmid-free and plasmid-containing cells grew together. It can be seen from Figure 5 that several very large sectors formed when a number of adjacent progenitor bacteria underwent plasmid loss at about the same time during colony growth. In other words, it appears that the system controlling plasmid maintenance in these different progenitor bacteria responded in the same way to a specific physiological change occuring at a particular period during colony development. The positions of the responding bacteria indicate that they were themselves descendants of a clonal population that had originated earlier in colony development. Presumably, some heritable event occurred in

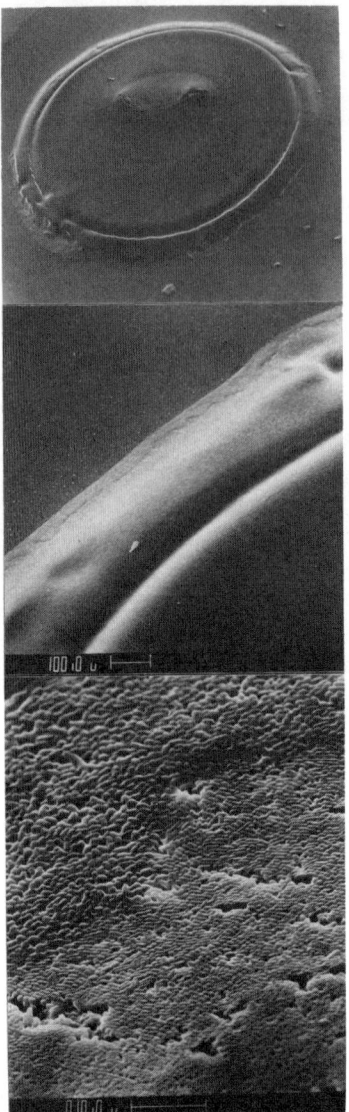

Figure 4
Concentric organization in an *E. coli* colony visualized by scanning electron microscopy. The top panel shows a low-magnification picture of a 68-hr-old colony prepared for electron microscopy. A small emerging sector can be seen at 7 o'clock, but the predominant morphological pattern of this colony is concentric, displaying several grooves and rings that are constant across any radius. The middle panel shows a higher magnification view of the colony edge at about 11 o'clock on the circumference (note where the dimples in the outer mound correspond in the top two panels). Here there are at least four concentric zones visible. The bottom panel shows a high magnification view of the boundary between the outermost group of cells, which are large, ovoid, and irregularly arranged, and the concentric zone just inside, where the cells are a smaller, more bacillary, and arranged in regular palisades.

the ancestral bacteria of these clones which conditioned the genome to lose the R388::Mud*lac* plasmid at a particular subsequent stage of colony development.

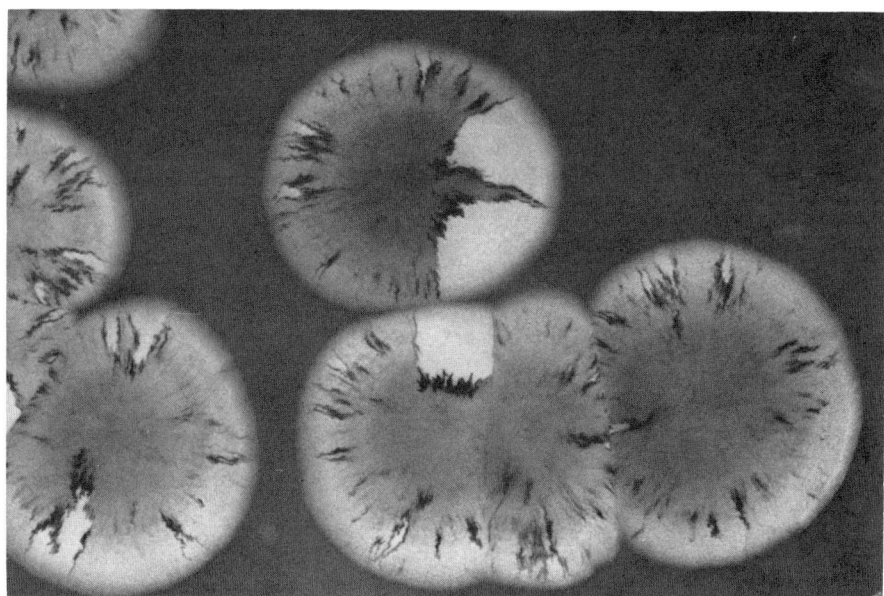

Figure 5
Sectoring due to plasmid loss in *E. coli* colonies. Each colony arose from a single colony-forming unit (probably a single bacterium) of a strain with an R388::Mud*lac* plasmid growing on X-gal indicator medium. The white, darkly-bordered sectors were composed of bacteria that had lost the R388::Mud*lac* plasmid. Note how extrawide sectors frequently formed from coincident plasmid losses. (Reprinted, with permission, from Shapiro and Brinkley 1984.)

DISCUSSION

The role of transposable elements as mutagenic agents in bacteria is well known. As the specific examples discussed above reveal, this mutagenic activity displays considerable sophistication. Elements such as IS*1*, IS*2*, IS*5*, and Tn*1000* can modulate expression of genetic information in a variety of ways. These include the movement of transcriptional termination signals, the creation of new promoters, and the enhancement or silencing of existing promoters. Study of prokaryotic genetic systems outside *E. coli* continue to provide further instances of regulatory changes involving transposable elements (see also, Scordilis et al. 1987), and it is hardly to be doubted that we will continue to be enlightened by the versatility of transposons and IS elements as modulators of genetic function. In this respect, they resemble eukaryotic transposable elements.

Bacterial transposable elements are also involved in the control of genomic stability. When they are active, the genome is unstable. It is highly significant that all transposable elements that have been analyzed in detail are subject to elaborate control circuits (Kleckner 1983; Toussaint and Resibois 1983), and two instances are cited above where the activities of Tn3, IS1, and IS5 are known to be highly sensitive to structural changes in particular components of the genome. The examples of how Mu promotes the generation of hybrid protein coding sequences and blocks plasmid inheritance illustrate how specifically a transposable element can respond to changes in bacterial living conditions. Since their activities are so sensitive and have so often been useful to the cells that harbor them, it seems that prokaryotic transposable elements are teaching us to look at genome reorganization as a meaningful process with positive significance for bacterial survival.

ACKNOWLEDGMENTS

I thank Nancy Cole for technical assistance and for printing Figures 3 and 4. Research in my laboratory is supported by grants from the National Science Foundation (DCB-8416998 and DMB-8401410).

REFERENCES

Amster, O. and A. Zamir. 1986. Sequence rearrangements may alter the in vivo superhelicity of recombinant plasmids. *FEBS Lett.* **197**: 93.

Amster, O., D. Salomon, and A. Zamir. 1982. A cloned immunoglobulin cDNA fragment enhances transposition of IS elements into recombinant plasmids. *Nucl. Acids Res.* **10**: 4525.

Brennan, M. and K. Struhl. 1980. Mechanisms of increasing expression of a yeast gene in *Escherichia coli*. *J. Mol. Biol.* **136**: 333.

Casadaban, M.J. 1976. Transposition and fusion of the *lac* genes to selected promoters in *Escherichia coli* using bacteriophages lambda and Mu. *J. Mol. Biol.* **104**: 541.

DiNardo, S., K.A. Voelkel, S. Sternglanz, A.E. Reynolds, and A. Wright. 1982. Escherichia coli DNA topoisomerase I mutants have compensatory mutations in DNA gyrase genes. *Cell* **31**: 43.

Echols, H. 1986. Multiple DNA-protein interactions governing high-precision DNA transactions. *Science* **233**: 1050.

Glansdorff, N., D. Charlier, and M. Zafarullah. 1981. Activation of gene expression by IS2 and IS3. *Cold Spring Harbor Symp. Quant. Biol.* **45**: 153.

Grindley, N.D.F. and R. Reed. 1985. Transpositional recombination in prokaryotes. *Annu. Rev. Biochem.* **54**: 863.

Heffron, F. 1983. Tn3 and its relatives. In *Mobile genetic elements* (ed. J.A. Shapiro), p. 223. Academic Press, New York.

Iida, S., R. Meyer, and W. Arber. 1983. Prokaryotic IS elements. In *Mobile genetic elements* (ed. J.A. Shapiro), p. 159. Academic Press, New York.

Jaurin, B. and S. Normark. 1983. Insertion of IS2 creates a novel *ampC* promoter in *Escherichia coli. Cell* **32:** 809.
Kleckner, N. 1983. Transposon Tn10. In *Mobile genetic elements* (ed. J.A. Shapiro), p. 261. Academic Press, New York.
Mizuuchi, K. and R. Craigie. 1986. Mechanism of bacteriophage Mu transposition. *Annu. Rev. Genet.* **20:** 385.
Muster, C.J., J.A. Shapiro, and L.A. MacHattie. 1983. Recombination involving transposable elements: The role of target molecule replication in Tn1△Ap-mediated replicon fusion. *Proc. Natl. Acad. Sci.* **80:** 2314.
Reynolds, A.E., J. Felton, and A. Wright. 1981. Insertion of DNA activates the cryptic *bgl* operon in *E. coli* K12. *Nature* **293:** 625.
Reynolds, A.E., S. Mahadevan, J. Felton, and A. Wright. 1985. Activation of the cryptic *bgl* operon: Insertion sequences, point mutations, and changes in superhelicity affect promoter strength. In *Genome rearrangement* (ed. M. Simon and I. Herskowitz), p. 265. Alan R. Liss, New York.
Reynolds, A.E., S. Mahadevan, S.F.J. LeGrice, and A. Wright. 1984. Enhancement of bacterial gene expression by insertion elements or by mutation in a CAP-cAMP binding site. *J. Mol. Biol.* **191:** 85.
Schaefler, S. and M.A. Malamy. 1969. Taxonomic investigations on expressed and cryptic phospho-β-glucosidases in Enterobacteriaceae. *J. Bacteriol.* **99:** 422.
Scordilis, G.E., H. Ree, and T.G. Lessie. 1987. Identification of transposable elements which activate gene expression in *Pseudomonas cepacia. J. Bacteriol.* **169:** 8.
Shapiro, J.A. 1984a. Observations on the formation of clones containing *araB-lacZ* cistron fusions. *Mol. Gen. Genet.* **194:** 79.
―――. 1984b. Transposable elements, genome reorganization and cellular differentiation in Gram-negative bacteria. *Symp. Soc. Gen. Microbiol.* (Part 2) **36:** 169.
―――. 1984c. The use of Mud*lac* transposons as tools for vital staining to visualize clonal and non-clonal patterns of organization in bacterial growth on agar surfaces. *J. Gen. Microbiol.* **130:** 1169.
―――. 1985a. Mechanisms of DNA reorganization in bacteria. *Int. Rev. Cytol.* **93:** 25.
―――. 1985b. Scanning electron microscope study of *Pseudomonas putida* colonies. *J. Bacteriol.* **164:** 1171.
―――. 1985c. Back cover. *Abstracts on Molecular genetics of bacteria and phages.* Cold Spring Harbor Laboratory, Cold Spring Harbor, New York.
―――. 1986. Control of *Pseudomonas putida* growth on agar surfaces. In *The bacteria,* volume X (ed. J. Sokatch), p. 27. Academic Press, New York.
―――. 1987a. Organization of developing *E. coli* colonies viewed by scanning electron microscopy. *J. Bacteriol.* **197:** 142.
―――. 1987b. Some lessons of phage Mu. In *Phage Mu* (ed. N. Symonds et al.), p. 251. Cold Spring Harbor Laboratory, Cold Spring Harbor, New York.
Shapiro, J.A. and P. Brinkley. 1984. Programming of DNA rearrangements involving Mu prophages. *Cold Spring Harbor Symp. Quant. Biol.* **49:** 313.
Stokes, H.W. and B.G. Hall. 1984. Topological repression of gene activity by a transposable element. *Proc. Natl. Acad. Sci.* **81:** 6115.

Surette, M.G., S.J. Buch, and G. Chaconas. 1987. Transposasomes: Stable protein-DNA complexes involved in the in vitro transposition of bacteriophage Mu DNA. *Cell* **49:** 253.

Toussaint, A. and A. Resibois. 1983. Phage Mu: Transposition as a life-style. In *Mobile genetic elements* (ed. J.A. Shapiro), p. 105. Academic Press, New York.

The Role of the Bacterial Host in the Mechanism and Regulation of Tn*10* Transposition

DENISE ROBERTS,* DONALD MORISATO, AND NANCY KLECKNER
Department of Biochemistry and Molecular Biology
Harvard University
Cambridge, Massachusetts 02138

OVERVIEW

Transposon Tn*10* can promote a variety of genetic alterations in *Escherichia coli*, including insertion mutations, chromosomal deletions, and chromosomal rearrangements (Kleckner et al. 1979; Ross et al. 1979). We describe below several ways in which the occurrence of these Tn*10*- and IS*10*-promoted events is under control of the host cell. First, DNA methylation by the host enzyme *dam* methylase reduces transposase expression. Second, methylation prevents transposase action at inner termini. Third, host-DNA-binding proteins' integration host factor (IHF) and/or hemagglutinating unit (HU) are necessary for transposase action at the outer termini. Finally, we present evidence that the host cell responds to transposition by induction of a λ prophage and hence, probably, the cellular SOS system.

INTRODUCTION

Genetic experiments suggest that Tn*10* transposes by a nonreplicative "cut-and-paste" mechanism in which Tn*10* is cut, intact, from a donor site and reinserted unaltered into a target site. First, transposition of a heteroduplex element in which the two parental DNA strands are differentially marked by single-base mismatches produces a mixed colony that contains Tn*10* at a single insertion site but with the genetic information of both parental strands (Bender and Kleckner 1986). Second, the genetic and physical structures of intrachromosomal deletions and deletion-inversions (Kleckner et al. 1979; Ross et al. 1979) can be most simply explained as products of break-join events. Finally, a Tn*10*-promoted structure has been identified in vivo whose formation involves double-stranded excision of the transposon from its donor molecule (Morisato and Kleckner 1984). This nonreplicative mechanism contrasts with mechanisms proposed for other bacterial transposable elements in which replication, with production of a "cointegrate" intermediate,

*Present address: Department of Biology, MIT, Cambridge, Massachusetts 02139.

is an obligatory step in the transposition process (for further reference, see Shapiro 1979).

Wild-type Tn*10* consists of inverted repeats of an insertion sequence, IS*10* (Fig. 1). IS*10*-right is a functional transposable element and can transpose independently of the rest of Tn*10*. For experimental purposes, Tn*10* derivatives (mini-Tn*10* elements) have been constructed that lack transposase, but which can be complemented to transpose by transposase provided in *trans*.

Transposition of Tn*10* and IS*10* requires a protein product, transposase, made from the single, open reading frame of IS*10*-right and expressed from the promoter pIN. Overlapping the -10 region of pIN is a sequence (GATC) that is recognized and methylated by *dam* methylase.

Sites at the ends of Tn*10* and IS*10* are essential for transposition. The inner and outer termini, defined according to their positions in wild-type Tn*10*, are each active in transposase-dependent events, and are similar but not identical in sequence. A 23-bp sequence, presumed to be the site of transposase action, is imperfectly repeated at the inner and outer termini. At the inner terminus, but not at the outer terminus, this sequence contains a *dam* methylation site (GATC). At the outer terminus, between bp 30 and 44, there is a sequence homologous to the binding consensus sequence of IHF (Craig and Nash 1984). DNA sequences in this region are necessary for efficient outer terminus activity; deletion derivatives that retain only the outer 27 bp are severely reduced for transposition (Way and Kleckner 1984). A similar IHF binding site is not found at the inner terminus. These differences between the outer and inner termini result in differential regulation by host proteins.

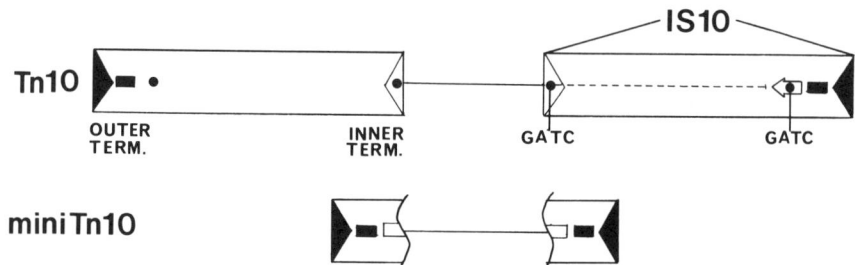

Figure 1

Structures of Tn*10*, IS*10*, and mini-Tn*10*. Transposase (- - - -) is expressed from the promoter, pIN (open arrow). (△) Inner termini; (▲) outer termini; (■) homology to the IHF-binding consensus sequence; and (●) Dam methylation sites (GATC). Mini-Tn*10* consists of inverted repeats of the indicated portion (70 bp) of the outer terminus. The left IS*10* contains several base differences from the right IS*10* (Halling et al. 1982) and is defective for transposition (Foster et al. 1981).

RESULTS AND DISCUSSION

The Role of *dam* Methylase in Tn*10*/IS*10* Transposition

The two *dam* methylation sites in IS*10* put Tn*10* and IS*10* under host control (Roberts et al. 1985). Specifically, the promoter and inner terminus are less active when the sites are methylated, and more active when the sites are unmethylated or hemimethylated. This is most simply illustrated by comparing IS*10* transposition frequencies in Dam$^+$ and Dam$^-$ strains. The transposition frequency of wild-type IS*10* is 200 times higher in Dam$^-$ than in Dam$^+$ strains (Table 1).

Increased IS*10* transposition in Dam$^-$ strains is directly due to loss of methylation at the two GATC sites in IS*10*. When the methylation sites are altered by point mutations that prevent methylation without destroying promoter or terminus activity, IS*10* transposition is identical in Dam$^+$ and Dam$^-$ strains. Transposition frequencies of singly mutant IS*10* elements show intermediate effects in Dam$^+$ relative to Dam$^-$, demonstrating that each methylation site contributes to the effect.

This comparison of Dam$^+$ and Dam$^-$ strains essentially compares fully methylated DNA to unmethylated DNA. Although unmethylated DNA is never found in wild-type cells, hemimethylated DNA does arise, for example, following DNA synthesis. We were interested in the possibility that methylation could be used to modulate IS*10* activity in wild-type (Dam$^+$) strains, and thus tested promoter and terminus activity in hemimethylated DNA.

The activity of hemimethylated pIN was compared to unmethylated and fully methylated pIN in vitro by abortive initiation assays (Roberts et al. 1985). In this assay, unmethylated pIN is 12-fold more active than methylated pIN (Table 2), consistent with in vivo measurements. There are two different possible hemimethylated forms of pIN methylated on opposite strands. Each hemimethylated species is more active than methylated pIN and slightly less active than unmethylated pIN.

Table 1
IS*10* Transposition in Dam$^-$ and Dam$^+$ Strains

Genotype of IS*10*		Transposition frequencies $\times 10^{-5}$		
pIN	inner term	Dam$^+$	Dam$^-$	Dam$^-$/Dam$^+$
wt[a]	wt[a]	1.6	356.0	220.0
wt[a]	GATC$^-$	1.9	51.0	26.0
GATC$^-$	wt[a]	4.9	31.0	10.0
GATC$^-$	GATC$^-$	5.8	4.7	0.8

[a] wt = wild-type.

Table 2
Effect of Methylation on pIN Promoter Activity

Strand methylation (top/bottom)	Relative promoter activity in abortive initiation assays (%)
+/+	8
−/−	100
−/+[a]	61
+/−[b]	38

[a] Orientation I.
[b] Orientation II.

We measured the activity of the hemimethylated inner terminus by an in vivo assay (Roberts et al. 1985). High frequency of recombination (HFR) strains of *E. coli* can transfer their chromosomal DNA to a recipient cell. A single DNA strand is transferred from a unique site on the donor chromosome, and the complementary strand is synthesized in the recipient. If the donor strain is Dam$^+$, the transferred strand will be methylated. Synthesis of the complementary strand will generate hemimethylated DNA, analogous to that produced at a replication fork. If the recipient is Dam$^-$, the DNA will remain hemimethylated. Finally, if IS*10* is inserted close to the origin of transfer, it will be moved efficiently into the recipient cell and can transpose from the transferred DNA segment into the recipient chromosome. By this kind of transfer experiment, we can generate hemimethylated IS*10* and measure its behavior in transposition assays.

We first compared transposition of IS*10* elements inserted at the same site but in opposite orientations (orientation I and orientation II). With IS*10* in orientation I, one strand is transferred as the methylated strand, and with IS*10* in orientation II, the other strand is transferred as the methylated strand. Synthesis of the complementary strands following transfer will generate the two hemimethylated species of IS*10*. For wild-type IS*10*, transposition is 60 times higher in orientation I than in orientation II (Table 3). This difference is due to the methylation sites; the orientation difference disappears with IS*10* elements carrying the two methylation site mutations. Furthermore, we can demonstrate that most of the orientation difference is due to the methylation site at the inner terminus by comparing transposition frequencies of IS*10* elements carrying single methylation site mutations (Roberts et al. 1985).

This comparison shows that one hemimethylated species is more active than the other. However, the biologically relevant comparison is actually between hemimethylated IS*10* and fully methylated IS*10*, since these are the two forms that exist alternately in wild-type cells. We cannot make this comparison directly in the same type of experiment, but we can estimate the

Table 3
Transposition of Hemimethylated IS*10* after HFR Transfer

Genotype of IS*10*			Relative transposition frequencies[a]	
pIN	inner term	Orientation	Dam$^-$ recipient	Dam$^+$ recipient
wt[b]	wt[b]	I	120	6
		II	2	1
GATC$^-$	GATC$^-$	I	20	80
		II	100	100

[a]Normalized to the double GATC$^-$ mutant in orientation II.
[b]wt = wild-type.

effect by comparing transposition after transfer into Dam$^+$ and Dam$^-$ recipient strains. In a Dam$^+$ recipient, the DNA is initially hemimethylated, but becomes fully methylated by the action of *dam* methylase in the recipient. If methylated IS*10* is less active than hemimethylated IS*10*, methylation in the recipient should reduce transposition. Consistent with this proposal, for wild-type IS*10* in the active orientation (orientation I), transposition in a Dam$^+$ recipient is 20-fold lower than in a Dam$^-$ recipient. This is likely to be an underestimate of the difference between hemimethylated and fully methylated IS*10*, because in the Dam$^+$ recipient, IS*10* may transpose before methylation occurs, and thus the transposition observed is from a mixed population of hemimethylated and fully methylated elements.

From these results, we suggest that IS*10* is regulated by methylation. Normally, chromosomal DNA is fully methylated, and IS*10* activity is low. Immediately after a replication fork passes, two hemimethylated IS*10* elements are generated. One element is fully active, with an active promoter and inner terminus. The second element has an inactive inner terminus, and thus will itself be inactive, but has an active promoter and can possibly aid transposition of the active IS*10*. Shortly after replication, action of *dam* methylase will regenerate fully methylated DNA, and IS*10* will return to its inactive state. This type of regulation is not specific to Tn*10*; Tn*5* and Tn*903* have *dam* methylation sites in similar positions, and transposition and transposase expression are increased in Dam$^-$ strains (Roberts et al. 1985; J. Yin et al., pers. comm.).

Several biological implications follow from this model. First, methylation may provide a mechanism by which transposition levels can be kept low, while retaining a potential for more efficient transposition with a simple and rapid mode of activation. Second, methylation may have the effect of limiting transposition to a brief interval of the cell cycle. Third, if most transposition does occur immediately after replication, there will usually be two copies of

IS*10* and two copies of the donor chromosome at the time of transposition. If IS*10* is excised from a donor molecule by double strand breaks at the ends of the element, transposition may generate a gapped donor molecule. Regulation of transposition by methylation may serve to couple transposition to chromosomal replication and, thus, ensure there is always one intact copy of the chromosome. This last point may be an advantage for elements, such as Tn*10*, that transpose by a nonreplicative cut-and-paste mechanism.

A Cellular Response to Transposition

As described above, Tn*10* transposition may leave a gapped donor molecule. Following transposition, this broken donor could be degraded, resealed, or repaired by recombination. Below, we present evidence for induction of a cellular response that is consistent with the occurrence of damaged DNA at some point in the transposition process (D. Roberts and N. Kleckner, in prep.).

In cultures of a strain lysogenic for the bacteriophage λ, spontaneous phage induction results in the release of free phage particles. In the presence of plasmids that promote a high level of transposition, a significant increase in the production of phage particles is observed (Table 4). The plasmids that show this effect contain a mini-Tn*10* element (see Fig. 1) and transposase read from a strong promoter (pTac). We believe that prophage induction occurs as the result of Tn*10* transposition. Both transposase and Tn*10* termini are required; increased phage production is eliminated by point mutations in the terminus that reduce Tn*10* activity in transposition assays and by deletions in the transposase gene. Furthermore, the level of phage release is roughly proportional to the level of transposition (data not shown).

Normal prophage induction occurs by activation of RecA protease, which cleaves λ-repressor, resulting in phage excision and production of phage particles (Roberts and Roberts 1975). In vitro, RecA protease activity is stimulated by single-stranded DNA (Craig and Roberts 1980), which has led to the suggestion that the in vivo signal for RecA protease activation may be single-stranded DNA generated by DNA damage or by the process of repair-

Table 4
Correlation between Transposition and Prophage Induction

Tn*10* sequences on plasmid		Transposition × 10^{-3}		Phage/ml × 10^6	
pTac:t'ase	mini-Tn*10*	−IPTG	+IPTG	−IPTG	+IPTG
+	+	1.2	34.0	4.4	47.0
+	mutant	<0.1	0.6	3.4	6.3
t'ase Δ	+	<0.1	<0.1	4.0	3.5

ing damaged DNA. Induction of λ by Tn*10* plasmids appears to occur through the normal pathway; the plaques produced are turbid, and induction is blocked by RecA⁻ mutations and by cleavage-resistant cI mutations (data not shown).

Activated RecA protease also cleaves LexA repressor, the general negative regulator of a set of genes collectively referred to as SOS functions (Witkin 1976; Witkin and Roegner-Maniscalco, this volume). RecA protease cleaves LexA repressor even more readily than λ-repressor (Bailone et al. 1979; Moreau et al. 1980; Little et al. 1980), thus in cells containing Tn*10* plasmids, there may be induction of some SOS genes. Increased expression of SOS genes has not been detected (O. Huisman, pers. comm.). However, this may be due to the fact that only a subset of cells in the culture are undergoing transposition at any point in time.

We suggest that transposition, or a closely correlated process, produces a signal that activates RecA protease to cleave λ-repressor; one possible source of an inducing signal could be a broken or degraded donor chromosome. Increased levels of SOS functions, which include recombination and repair enzymes, could help the cell to repair the donor chromosome following transposon excision. Even if the SOS genes, per se, are not induced, these data suggest that the cell can detect and respond to transposition, at least by induction of λ and possibly in other ways.

Role of IHF and HU in Tn*10*/IS*10* Transposition

A requirement for host factors in Tn*10* transposition has been revealed by studies in vitro (Morisato and Kleckner 1987). In a partially purified system, we have observed Tn*10* transposition and the related intramolecular reaction, transposon circle formation. In the latter event, first described in vivo, the transposon segment is excised from the donor site, and one of the two strands is joined to create a circular structure (Morisato and Kleckner 1984).

The requirements for transposon circle formation in vitro have been characterized (Morisato and Kleckner 1987). Circle formation with substrates containing two outer termini (mini-Tn*10*, Fig. 1) requires host factors in addition to transposase and intact Tn*10* termini. The *E. coli* proteins IHF and HU have been identified as the host factors capable of participating in the reaction (Fig. 2). Addition of purified IHF or HU results in production of the circular product, and the effects of the two proteins are roughly additive. IHF is more effective in stimulating circle formation, and about five times more circular product is made in the presence of IHF than with HU. IHF and HU are the only two host proteins found to stimulate the reaction, and they fully account for the host stimulation observed with crude cell extracts.

IHF and HU are required for circle formation with two outer termini but

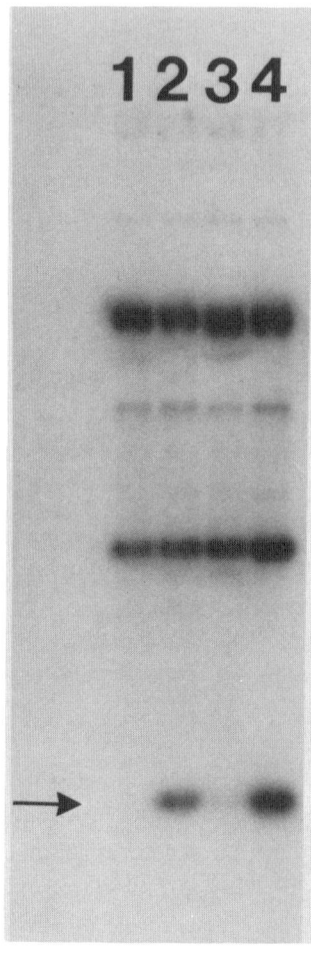

Figure 2
In-vitro host-factor requirements for circle formation by a transposon segment containing two outer termini. The position of the circular product is indicated by the arrow; larger bands are parental plasmid species. Reactions were carried out in the presence of transposase and (*1*) no additional host proteins; (*2*) purified IHF; (*3*) cell extract of an IHF$^-$ strain; (*4*) IHF plus an IHF$^-$ extract.

not with two inner termini. With a substrate containing two unmethylated inner termini, circle formation occurs efficiently in the absence of added host factors (Fig. 3). Also, as predicted from the studies with *dam* methylase, the reaction with two inner termini is sensitive to the state of methylation; methylated ends are inactive, and unmethylated ends are active. A substrate that has one inner terminus and one outer terminus is subject to both types of regulation; circle formation requires an unmethylated substrate and IHF or HU (data not shown).

Measurements of Tn*10* activity in vivo support the in vitro results (Roberts 1986). The frequency of Tn*10*-promoted chromosomal rearrangements is

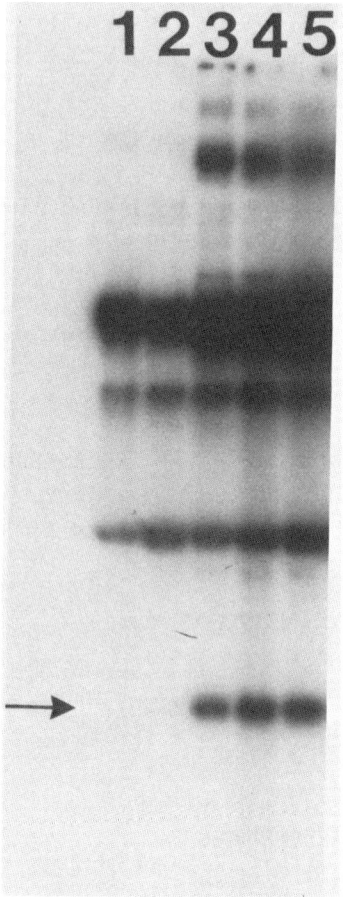

Figure 3
In vitro circle formation by a transposon segment containing two inner termini. Reactions were carried out with methylated inner termini (*1* and *2*) and with unmethylated inner termini (*3–5*). The position of the circular product is indicated by the arrow. For the methylated substrate, reactions contained transposase alone (*1*) and transposase plus complete host extract (*2*). For the unmethylated substrate, reactions contained transposase alone (*3*), transposase plus IHF (*4*), and transposase plus HU (*5*).

reduced fivefold in IHF^- strains relative to IHF^+, consistent with the fivefold greater activity of IHF relative to HU in vitro. Furthermore, a decrease in the frequency of rearrangements is only observed with elements that use at least one outer terminus in the event; no decrease is observed when two inner termini are used. The in vivo importance of HU for Tn*10* transposition has not yet been assessed.

IHF and HU are small, basic, heterodimeric DNA-binding proteins. The intracellular roles of these two proteins are not completely understood, but an involvement of each protein has been demonstrated for several different events. IHF is a site-specific DNA-binding protein (Craig and Nash 1984). IHF is required in vivo and in vitro for efficient λ-integration (Nash and Robertson 1981; Miller and Friedman 1980). IHF mutants are pleiotropic;

effects include altered expression of several genes, for example λcII (Peacock et al. 1984) and the *E. coli ilv* operon (Friden et al. 1984). HU binds nonspecifically to DNA, forming histone-like structures in vitro (Rouviere-Yaniv et al. 1979) and altering the pitch of the DNA helix (Broyles and Pettijohn 1986). HU is required for several in vitro systems, including chromosomal DNA replication (Dixon and Kornberg 1984), Mu transposition (Craigie et al. 1985), and site-specific DNA inversion (Kahmann et al. 1985; Johnson et al. 1986).

The roles of IHF and HU in Tn*10* transposition have not yet been determined. Full activity of two inner termini in the absence of host factors suggests that the activity for DNA strand breaking and rejoining is contained completely within transposase itself, and that IHF and HU act as accessory proteins that are required for activity of the outer termini. This requirement puts Tn*10* transposition under the control of host proteins. It is not known how IHF and HU are regulated, or whether levels of these proteins fluctuate under different conditions. We expect that Tn*10* activity will be affected by conditions that result in changes in IHF and HU levels in the cell.

ACKNOWLEDGMENTS

This work was supported by grants to N.K. from the National Science Foundation (PCM-83-03415 and DMB-85-18930) and the National Institutes of Health (GM-25326).

REFERENCES

Bailone, A., A. Levine, and R. Devoret. 1979. Inactivation of prophage λ repressor *in vivo*. *J. Mol. Biol.* **131**: 553.

Bender, J. and N. Kleckner. 1986. Genetic evidence that Tn10 transposes by a non-replicative mechanism. *Cell* **45**: 801.

Broyles, S. and D. Pettijohn. 1986. Interaction of the *Escherichia coli* HU protein with DNA. Evidence for formation of nucleosome-like structures with altered DNA helical pitch. *J. Mol. Biol.* **187**: 47.

Craig, N.L. and H.A. Nash. 1984. *E. coli* integration host factor binds to specific sites in DNA. *Cell* **39**: 707.

Craig, N.L. and J.W. Roberts. 1980. *E. coli* recA protein-directed cleavage of phage λ repressor requires polynucleotide. *Nature* **283**: 26.

Craigie, R., D.J. Arndt-Jovin, and K. Mizuuchi. 1985. A defined system for the DNA strand-transfer reaction at the initiation of bacteriophage Mu transposition: Protein and DNA substrate requirements. *Proc. Natl. Acad. Sci.* **82**: 7570.

Dixon, N.E. and A. Kornberg. 1984. Protein MHU in the enzymatic replication of the chromosomal origin of *Escherichia coli*. *Proc. Natl. Acad. Sci.* **81**: 424.

Foster, T.J., M.A. Davis, D.E. Roberts, K. Takeshita, and N. Kleckner. 1981. Genetic organization of transposon Tn10. *Cell* **23:** 201.

Friden, P., V. Voelkel, R. Sternglanz, and M. Freudlich. 1984. Reduced expression of the isoleucine and valine enzymes in integration host factor mutants of *Escherichia coli*. *J. Mol. Biol.* **172:** 573.

Halling, S.M., R.W. Simons, J.C. Way, R.B. Walsh, and N. Kleckner. 1982. DNA sequence organization of IS10-right of Tn10 and comparison with IS10-left. *Proc. Natl. Acad. Sci.* **79:** 2608.

Johnson, R., M. Bruist, and M. Simon. 1986. Host protein requirements for in vitro site specific DNA inversion. *Cell* **46:** 531.

Kahmann, R., F. Rudt, C. Koch, and G. Mertens. 1985. G inversion in bacteriophage Mu DNA is stimulated by a site within the invertase gene and a host factor. *Cell* **41:** 771.

Kleckner, N., K. Reichardt, and D. Botstein. 1979. Inversions and deletions of the *Salmonella* chromosome generated by the translocatable tetracycline resistance element Tn10. *J. Mol. Biol.* **127:** 89.

Little, J.W., S.H. Edmiston, L.Z. Pacelli, and D.W. Mount. 1983. Cleavage of the *Escherichia coli* LexA protein by the RecA protease. *Proc. Natl. Acad. Sci.* **77:** 3225.

Miller, H.I. and D.I. Friedman. 1980. An *E. coli* gene product required for λ site specific recombination. *Cell* **20:** 711.

Moreau, P.L., M. Fanica, and R. Devoret. 1980. Induction of prophage λ does not require full induction of RecA protein synthesis. *Biochimie* **62:** 687.

Morisato, D. and N. Kleckner. 1984. Transposase promotes double strand breaks and single strand joins at Tn10 termini in vivo. *Cell* **39:** 181.

———. 1987. Tn10 transposition and circle formation *in vitro*. *Cell* **51:** 101.

Nash, H.A. and C.A. Robertson. 1981. Purification and properties of the *Escherichia coli* protein factor required for λ integrative recombination. *J. Biol. Chem.* **256:** 9246.

Peacock, S., H. Weissbach, and H.A. Nash. 1984. *In vitro* regulation of phage λ *c*II gene expression by *Escherichia coli* integration host factor. *Proc. Natl. Acad. Sci.* **81:** 6009.

Roberts, D. 1986. "Isolation and characterization of mutants of *E. coli* that are affected for transposition of Tn10." Ph.D. thesis, Harvard University, Cambridge, Massachusetts.

Roberts, D., B.C. Hoopes, W.R. McClure, and N. Kleckner. 1985. IS10 transposition is regulated by DNA adenine methylation. *Cell* **43:** 117.

Roberts, J.W. and C.W. Roberts. 1975. Proteolytic cleavage of bacteriophage λ repressor in induction. *Proc. Natl. Acad. Sci.* **72:** 147.

Ross, D., J. Swan, and N. Kleckner. 1979. Physical structures of Tn10-promoted deletions and inversions: Role of 1400 bp inverted repetitions. *Cell* **16:** 721.

Rouviere-Yaniv, J., M. Yaniv, and J. Germond. 1979. *E. coli* DNA binding protein HU forms nucleosome-like structure with circular double-stranded DNA. *Cell* **17:** 265.

Shapiro, J.A. 1979. Molecular model for the transposition and replication of bacteriophage Mu and other transposable elements. *Proc. Natl. Acad. Sci.* **76:** 1933.

Way, J.C. and N. Kleckner. 1984. Essential sites at transposon Tn10 termini. *Proc. Natl. Acad. Sci.* **81:** 3452.

Witkin, E. 1976. Ultraviolet mutagenesis and inducible DNA repair in *Escherichia coli*. *Bacteriol. Rev.* **40:** 869.

SOS Mutagenesis and Replisome Reactivation in Bacteria: Options Acquired by Transposition?

EVELYN M. WITKIN AND VIVIEN ROEGNER-MANISCALCO
Waksman Institute of Microbiology
Rutgers State University of New Jersey
Piscataway, New Jersey 08854

OVERVIEW

In *Escherichia coli*, mutagenesis by ultraviolet light (UV) and many carcinogenic chemicals requires the RecA and UmuDC proteins, which are induced by DNA damage as components of the survival-enhancing SOS response. We show that RecA and UmuDC can interact, at least functionally, not only in UV mutagenesis but also in induced replisome reactivation (IRR), a newly recognized SOS function required for recovery from inhibition of DNA synthesis caused by UV damage. We propose that error-prone translesion DNA replication is possible at every site of blocked replication, although it actually occurs only at sites that are refractory to accurate modes of repair or tolerance of UV damage. We discuss evidence that *umuDC* and at least one of its plasmid-borne analogs, *mucAB* on plasmid pKM101, are (or once were) carried on transposons, and that SOS mutability (and possibly enhanced replisome reactivation activity) may be gained or lost by transposition.

INTRODUCTION

DNA damage by many radiations and carcinogenic chemicals induces the pleiotropic SOS response in *E. coli*, a set of phenotypes, many of which enhance DNA repair and cell survival, resulting from expression of at least 17 genes (for reviews, see Witkin 1976, 1984; Little and Mount 1982; Walker 1984). RecA protein, activated by interaction with single-stranded DNA and a nucleotide cofactor (Craig and Roberts 1980) promotes proteolytic cleavage of LexA protein, the common repressor of bacterial SOS genes. Figure 1 illustrates regulation of typical genes in the LexA regulon. SOS mutagenesis (the ability to be mutagenized by SOS-inducing agents) requires two of the induced proteins, RecA and UmuDC. Neither $recA^-$ (Witkin 1969) nor umu^- mutants (Kato and Shinoura 1977) are mutable by UV. RecA* (the activated form shown in Figure 1) is required not only for its antirepressor activity but also has a more direct role in the mutagenic process (Witkin and

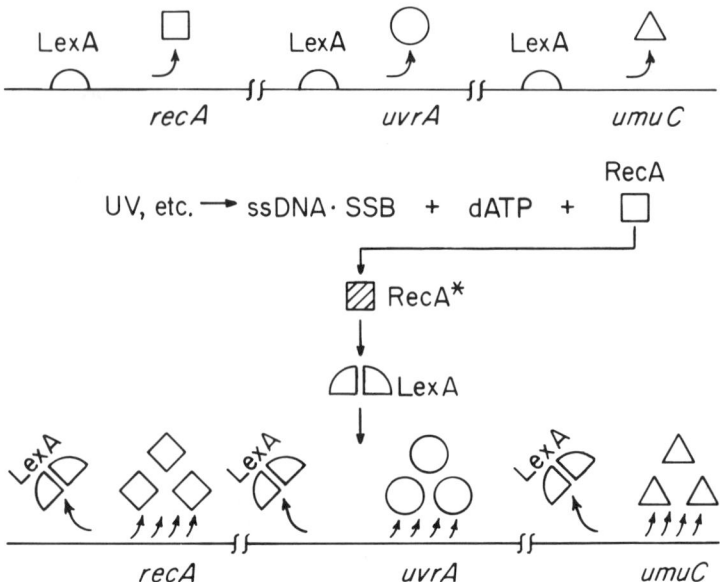

Figure 1
Regulation of the SOS response in *E. coli*. Top line shows repression of DNA damage-inducible genes by LexA protein in undamaged wild-type cell; DNA damage activates RecA protein to RecA*; RecA* causes cleavage of LexA, and at least 17 LexA-controlled genes scattered around the chromosome are derepressed. The SOS response enhances several kinds of DNA repair, increases recombination activity, elevates spontaneous and induced mutation rates, delays cell division, and relaxes the requirements for initiation of DNA synthesis. (The *uvrA* gene product is required for excision repair.) In lysogenic strains, RecA* also cleaves prophage repressors, resulting in cell lysis and release of mature bacteriophage particles.

Kogoma 1984; Ennis et al. 1985). The specific functions of RecA* and UmuDC are not yet known, but together they are thought to mediate translesion DNA replication, the error-prone incorporation of nucleotides opposite noncoding template lesions (Witkin 1976; Bridges and Woodgate 1985; Lu et al. 1986), a mechanism that could account for the targeting of mutations, induced by UV and several carcinogens, opposite bases specifically damaged by the agents in question (Miller 1982).

RecA protein binds UV-irradiated double-stranded DNA increasingly as the UV dose increases and inhibits proofreading by pol III holoenzyme or by its epsilon subunit (*dnaQ* protein) in vitro (Lu et al. 1986). These observations led Lu and his colleagues to propose that RecA*, bound to the single-stranded DNA flanking a bulky lesion, facilitates the stable misincorporation of a nucleotide by pol III opposite the lesion. UmuDC, according to the two-step model of UV mutagenesis (Bridges and Woodgate 1985), is necessary for resumptiom of replication after the misincorporation step mediated by RecA.

It has long been known that DNA synthesis on templates containing UV damage is discontinuous, producing segments of daughter-strand DNA roughly corresponding in length to the distance between pyrimidine dimers (Rupp and Howard-Flanders 1968). Postreplication repair, mainly via RecA-dependent recombination between sister molecules, then joins the daughter strands to generate intact high molecular weight DNA (Rupp et al. 1971). Replication is blocked at or just before pyrimidine dimers and other bulky lesions, and must be reinitiated at a point downstream after each block. Recently, the ability to recover from UV-induced blockage of DNA replication has been shown to be an SOS-inducible phenotype, IRR, requiring RecA plus post-UV synthesis of at least one other protein (IRR factor) (Khidhir et al. 1985). Because Khidhir and his colleagues found that *umuC* mutants are IRR-proficient, they concluded that UmuDC protein is not necessary for recovery from UV-induced inhibition of DNA synthesis.

In our laboratory, we have combined mutant alleles of *recA* and *umuC* or *umuD* as a means of seeking genetic evidence for functional or physical interaction between their products. We found that double mutants combining *recA718* (an allele that does not reduce recombination, SOS inducibility, or UV mutagenesis) (Witkin et al. 1982; McCall et al. 1987) with any of three *umu*⁻ alleles are synergistically UV-sensitive. In this paper, we show that the extreme UV sensitivity of the double mutant is associated with a synergistic defect in ability to recover from UV-induced inhibition of DNA synthesis. We discuss the implications of these results for SOS mutagenesis, as well as the growing evidence that *umuDC* and its analogs on plasmids, such as *mucAB*, are (or were) carried on transposons.

RESULTS

The Synergistic UV Sensitivity of Double Mutants *recA718 umuC36*

Figure 2 shows the extreme UV sensitivity of double mutants combining *recA718* with each of three *umu*⁻ alleles: *umuC36*, *umuD44*, and *umuC122*::Tn5, compared to the relatively moderate sensitivities of the single mutant strains. The double mutant is far more sensitive than would be expected if the sensitivities of the two single mutants acted additively.

IRR in the Double Mutant

After ruling out excessive DNA degradation after UV irradiation as the cause of the extreme UV sensitivity of the double mutant (data not shown), we

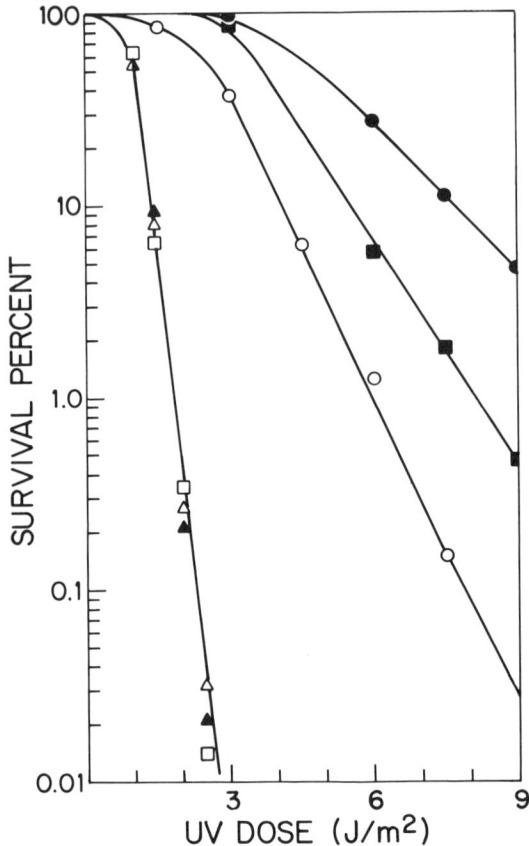

Figure 2
Synergistic UV sensitivity of double mutants combining *recA718* and *umuC36*, *umuC122*::Tn5, or *umuD44*. (●)Strain SC18-RP (*recA⁺ umuDC⁺*); (■)SC18-RP-UM36 (*recA⁺ umuC36*); (○) SC18 (*recA718 umuDC⁺*); (△) SC18-UM36 (*recA718 umuC36*); (▲) SC18-UM122 (*recA718 umuC122*::Tn5); (□)SC18-D44 (*recA718 umuD44*).

considered the possibility that it might be defective in ability to recover from the inhibition of DNA synthesis caused by UV. We examined the rate of DNA synthesis by the method of Khidhir et al. (1985), pulse-labeling exponentially growing bacteria for 2 minutes with [^{3}H]thymidine at various times before and after exposing the bacteria to UV (3 J/m^2). Figure 3 shows that, whereas neither mutant allele alone prevents IRR, no recovery is detected in the double mutant *recA718 umuC36*. The double mutant is synergistically deficient in IRR activity, a defect that accounts for its extreme UV sensitivity.

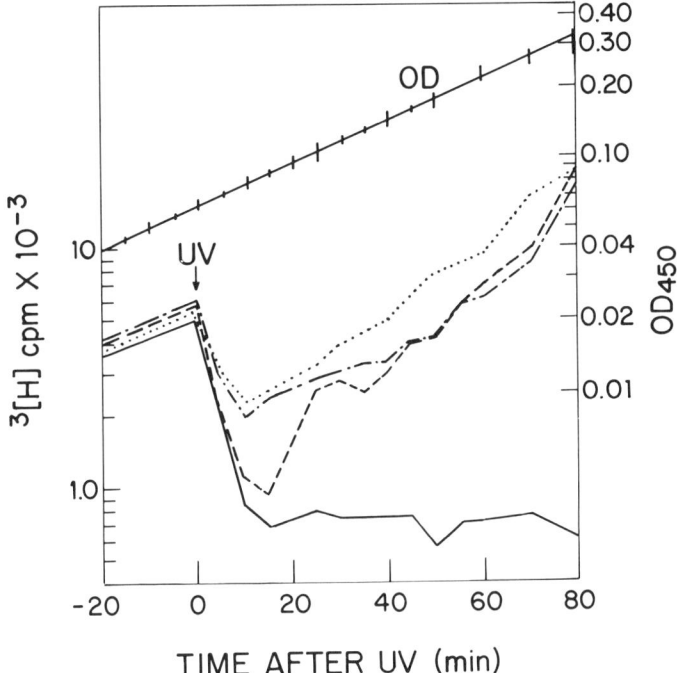

Figure 3
Synergistic loss of ability to recover form inhibition of DNA synthesis after UV irradiation in double mutants combining *recA718* and *umuC36*. (- - - -) SC18-RP1 (*recA$^+$ umuDC$^+$*); (- - -) SC18-RP-UM36 (*recA$^+$ umuC36*); (- - -) SC18 (*recA718 umuDC$^+$*); (—) SC18-UM36 (*recA718 umuC36*). Vertical lines on optical density (OD) curve define the range of values obtained for all four strains.

DISCUSSION

Correction or Complementation of RecA718 by UmuDC

There are two possible interpretations of the basis of IRR deficiency in the double mutant *recA718 umuC36*, based on the known requirement of the recovery process for RecA protein. UmuDC$^+$ protein either corrects a conditional defect in the ability of RecA718 to effect IRR or complements the unconditional defect of RecA718 by a RecA-independent activity of its own. Direct interaction between RecA and UmuD is indicated by recent evidence for RecA*-dependent proteolytic processing of UmuD in vitro (H. Shinagawa; H. Echols; both pers. comm.). Our current evidence, however, does not distinguish between the two possibilities.

IRR and SOS Mutagenesis

Whether RecA718 and UmuDC interact directly to promote IRR or not, our results have an important implication for SOS mutagenesis. Heretofore, the role of UmuDC in UV mutagenesis has been assumed to be limited to sites where a UV-induced mutation actually occurs, probably by a process of error-prone translesion replication. Target sites for UV mutagenesis are a small subset of sites where replication is transiently blocked by a UV photoproduct, most of which are repaired or tolerated by error-free mechanisms (excision repair or recombinational postreplication repair) (Witkin 1976). Our results indicate that UmuDC and RecA must both be present, in close proximity, not only at target sites for UV mutagenesis, but at every site of transiently blocked replication. Thus, mutagenic translesion replication, which requires these two proteins, is potentially available, although rarely utilized, wherever a UV photoproduct blocks the progress of a replication fork. We propose that translesion replication is relatively slow and inefficient, compared to error-free modes of repair or lesion tolerance, except at special "trouble-spots," that for one reason or another, are refractory to excision repair or recombinational repair and where translesion replication has the advantage or is the only possible mode of survival. For example, a region where two photoproducts are close together on opposite strands might well require SOS processing (Witkin 1976). A prediction, based on this proposal, is that mutations or postirradiation treatments that specifically decrease the rate of excision repair should result in increased UV mutagenesis. Two mutations have just this effect: In *mfd* mutants (George and Witkin 1975) and in a *dna*B mutant (Bridges et al. 1976), excision of pyrimidine dimers is very slow, although neither strain is UV-sensitive, and both mutants are markedly UV hypermutable. Posttreatment with low concentrations of caffeine slows pyrimidine dimer excision without decreasing survival and also causes UV hypermutability (George 1974). These observations support the idea that the level of UV mutagenesis is determined by competition between error-free and error-prone modes of handling UV damage.

Analogs of UmuDC on Plasmids

Many naturally occurring plasmids enhance the UV resistance and UV mutability of *E. coli* and other enterobacteria and confer UV mutability upon strains that are naturally UV nonmutable (for review, see Upton and Pinney 1983; Walker 1984). For example, pKM101 carries an analog of the *umuDC* operon, *mucAB*, which complements $umuC^-$ or $umuD^-$ mutants of *E. coli*, restoring their UV mutability and UV resistance. The *mucAB* operon is regulated, like *umuDC*, by bacterial RecA and LexA proteins (Elledge and Walker 1983), and shares about 35% sequence homology with its bacterial analog (Perry et al. 1985). When compared at equal gene dosages, MucAB is more efficient and less dependent upon RecA* activity than UmuDC (Blanco et al. 1986).

Possible Transposon Origin of UmuDC and MucAB

DNA damage-inducible SOS-like responses, including induction of a RecA-like protein, have been observed in numerous species of enterobacteria (Sedgwick and Goodwin 1985). However, Sedgwick and Goodwin have shown that in many of these species (including some species of *Salmonella, Shigella, Proteus,* and even *Escherichia*) the DNA damage-inducible response does not include UV mutability, as if UmuDC proteins are absent or inactive (see also Walker 1984). Sedgwick (1985) has summarized evidence suggesting that the UmuDC operon of *E. coli* may have originally been acquired on a transposon: (1) Sequences that hybridize with *umuDC* sequences of *E. coli* are absent from the genomes of most of the species of UV-nonmutable enterobacteria surveyed, which could mean that *E. coli* may have acquired these genes relatively recently; (2) the positions of five different restriction enzyme sites around the *umuDC* genes of 39 wild isolates of *E. coli* were found to be highly variable, a common feature of transposon insertions, whether indicative of different chromosomal locations or of deletions flanking the insertion; and (3) the region 5' to the *umuDC* operon, where the sequence variations begin, contains a concensus sequence found in several transposon termini, including those of Tn*3*, Tn*1000*, Mu, and Tn*951*.

The *mucAB* operon on pKM101 is surrounded by inverted repeats (Langer and Walker 1981), suggesting that these genes, too, are (or once were) carried on a transposon. The immediately selectable feature of mucAB or UmuDC activity is more likely to be its ability to increase survival of bacteria sustaining DNA damage than its enhancement of mutability, although the transient increase in genetic variability could contribute to population fitness.

Does MucAB Enhance IRR Activity?

Because *umu*⁻ mutants of *E. coli* are only moderately UV-sensitive, it is reasonable to attribute their sensitivity to loss of translesion replication activity. However, many of the species of enterobacteria that lack UV mutability are far more UV-sensitive than *umu*⁻ mutants of *E. coli*, yet are restored to the same degree of UV resistance when plasmid pKM101 is introduced (Sedgwick and Goodwin 1985). The discrepancy between their extreme UV sensitivity without the plasmid and their greatly increased UV resistance with the plasmid strongly suggests that the *mucAB* genes are contributing to UV resistance by more than one mechanism. If this resistance were due only to acquisition of translesion replication activity, one would expect UV-induced mutations to have an improbably high frequency, far higher than is actually observed. We propose that MucAB may be able to enhance IRR activity, as well as translesion replication activity, in the UV-nonmutable species lacking *umuDC* genes. We have shown that UmuDC can enhance IRR activity in the presence of RecA718, a protein that is functional in most other RecA-dependent activities but is specifically deficient in recovery of DNA synthesis after UV irradiation. UV-nonmutable species, lacking both chromosomal *umuDC*-like genes and plasmids carrying *umuDC* analogs, may not have been significantly challenged, in their evolutionary histories, by UV or other environmental agents that cause replication-blocking DNA lesions. The absence of exposure to this kind of stress could also account for their RecA-like proteins not having acquired great proficiency in overcoming such replication blocks. The wild-type RecA proteins of such strains may be more like RecA718 than like RecA$^+$ of *E. coli* in this regard. Analysis of the IRR activity of UV-sensitive, UV-nonmutable species, with and without pKM101, could clarify the contribution of MucAB to the recovery process.

Transposons contribute powerfully to mutagenesis by insertion. In bacteria, they may also do so by carrying genes required for mutagenic translesion DNA replication, which, like other plasmid-borne genes that confer resistance to environmental stress, can then be rapidly acquired or discarded.

ACKNOWLEDGMENT

This work was supported by grant AI-10778 (to E.M.W.) from the National Institutes of Health.

REFERENCES

Blanco, M., G. Herrera, and V. Aleixandre. 1986. Different efficiency of UmuDC and MucAB proteins in UV light induced mutagenesis in *Escherichia coli*. *Mol. Gen. Genet.* **205:** 234.

Bridges, B.A. and R. Woodgate. 1985. Mutagenic repair in *Escherichia coli:* Products of the *recA* gene and of the *umuD* and *umuC* genes act at different steps in UV-induced mutagenesis. *Proc. Natl. Acad. Sci.* **82:** 4193.

Bridges, B.A., R.P. Mottershead, and A.R. Lehman. 1976. Error-prone DNA repair in *Escherichia coli* IV. Excision repair and radiation-induced mutation in a *dnaB* strain. *Biol. Zentralbl.* **95:** 393.

Craig, N.L. and J.W. Roberts. 1980. *E. coli recA* protein-directed cleavage of phage lambda repressor requires polynucleotide. *Nature* **283:** 26.

Elledge, S.J. and G.C. Walker. 1983. The *muc* genes of pKM101 are induced by DNA damage. *J. Bacteriol.* **155:** 1306.

Ennis, D.G., B. Fisher, B. Edmiston, and D.W. Mount. 1985. Dual role for *Escherichia coli* RecA protein in SOS mutagenesis. *Proc. Natl. Acad. Sci.* **82:** 3325.

George, D.L. 1974. "Characterization of a mutant *Escherichia coli* B/r defective in mutation frequency decline." Ph.D. thesis, Rutgers, State University of New Jersey, New Brunswick, New Jersey.

George, D.L. and E.M. Witkin. 1975. Ultraviolet light-induced responses of an *mfd* mutant of *Escherichia coli* B/r having a slow rate of dimer excision. *Mutat. Res.* **28:** 347.

Kato, T. and Y. Shinoura. 1977. Isolation and characterization of mutants of *Escherichia coli* deficient in induction of mutations by ultraviolet light. *Mol. Gen. Genet.* **156:** 121.

Khidhir, M.A., S. Casaregola, and I.B. Holland. 1985. Mechanism of transient inhibition of DNA synthesis in ultraviolet-irradiated *E. coli:* Inhibition is independent of *recA* whilst recovery requires RecA itself and an additional, inducible SOS function. *Mol. Gen. Genet.* **199:** 133.

Langer, P.J. and G.C. Walker. 1981. Restriction endonuclease cleavage map of pKM101: Relationship to parental plasmid R46. *Mol. Gen. Genet.* **182:** 268.

Little, J.W. and D.W. Mount. 1982. The SOS regulatory system of *Escherichia coli*. *Cell* **29:** 11.

Lu, C., R.H. Scheuermann, and H. Echols. 1986. Capacity of RecA protein to bind preferentially to UV lesions and inhibit the editing subunit (ϵ) of DNA polymerase III: A possible mechanism for SOS-induced targeted mutagenesis. *Proc. Natl. Acad. Sci.* **83:** 619.

McCall, J.O., E.M. Witkin, T. Kogoma, and V. Roegner-Maniscalco. 1987. Constitutive expression of the SOS response in *recA718* mutants of *Escherichia coli* requires amplification of RecA718 protein. *J. Bacteriol.* **169:** 728.

Miller, J.H. 1982. Carcinogens induce targeted mutations in *Escherichia coli*. *Cell* **31:** 5.

Perry, K.L., S.J. Elledge, B.B. Mitchell, L. Marsh, and G.C. Walker. 1985. *umuDC* and *mucAB* operons whose products are required for UV light- and chemical-induced mutagenesis: UmuD, MucA, and LexA proteins share homology. *Proc. Natl. Acad. Sci.* **82:** 4331.

Rupp, W.D. and P. Howard-Flanders. 1968. Discontinuities in the DNA synthesized in an excision-defective strain of *Escherichia coli* following ultraviolet irradiation. *J. Mol. Biol.* **31:** 291.

Rupp, W.D., C.E. Wilde, D.L. Reno, and P. Howard-Flanders. 1971. Exchanges between DNA strands in ultraviolet-irradiated *Escherichia coli*. *J. Mol. Biol.* **61:** 25.

Sedgwick, S.G. 1985. A transposon origin for the *E. coli umuDC* operon. *Br. J. Cancer* **51:** 607 (Abstr.).

Sedgwick, S.G. and P.A. Goodwin. 1985. Differences in mutagenic and recombinational DNA repair in enterobacteria. *Proc. Natl. Acad. Sci.* **82:** 4172.

Upton, C. and R.J. Pinney. 1983. Expression of eight unrelated Muc$^+$ plasmids in eleven repair-deficient *E. coli* strains. *Mutat. Res.* **112:** 261.

Walker, G.C. 1984. Mutagenesis and inducible responses to deoxyribonucleic acid damage in *Escherichia coli*. *Microbiol. Rev.* **48:** 60.

Witkin, E.M. 1969. The mutability toward ultraviolet light of recombination-deficient strains of *Escherichia coli*. *Mutat. Res.* **8:** 9.

———. 1976. Ultraviolet mutagenesis and inducible DNA repair in *Escherichia coli*. *Bacteriol Rev.* **40**: 869.

———. 1984. The SOS response: Implications for cancer. In *Genes and cancer* (ed. J.M. Bishop et al.), p. 99. Alan R. Liss, New York.

Witkin, E.M. and T. Kogoma. 1984. Involvement of the activated form of RecA protein in SOS mutagenesis and stable DNA replication in *Escherichia coli*. *Proc. Natl. Acad. Sci.* **81**: 7539.

Witkin, E.M., J.O. McCall, M.R. Volkert, and I.E. Wermundsen. 1982. Constitutive expression of SOS functions and modulation of mutagenesis resulting from resolution of genetic instability at or near the *recA* locus of *Escherichia coli*. *Mol. Gen. Genet.* **185**: 43.

Mutational Effects of Transposable Element Insertions

Mobile DNA Elements and Spontaneous Gene Mutation

MELVIN M. GREEN
Department of Genetics
University of California
Davis, California 95616

OVERVIEW

The molecular genetic analysis of presumptive spontaneous mutations in *Drosophila melanogaster* demonstrates an inordinate fraction to be causally associated with insertions of mobile DNA elements. The operational definition of spontaneous mutation as an autonomous event is examined in the context of those genetical-environmental conditions that "drive" mobile DNA elements.

INTRODUCTION

"The study of spontaneous mutation is laborious, at best, but it seems an indispensable preliminary to the interpretation of mutations induced experimentally, and it may be the only approach now open for the study of gene evolution." (Stadler 1946).

Natural selection is central to the process of Darwinian evolution. In principle, selection acts on genetic variation preexisting in the evolving species. Such genetic variation, it has long been presumed, arises through the process of spontaneous gene mutation. Thus, the nature and mode of origin of spontaneous gene mutations are fundamental to the selection event and basic to evolution. In the broad sense, spontaneous gene mutation is a key biological phenomenon for as Dobzhansky repeatedly emphasized, "...nothing in biology makes sense except in the light of evolution."

Before discussing spontaneous mutation, it is necessary to make clear what is meant by this phenomenon. A useful empirical definition states that spontaneous gene mutations are those that occur de novo independent of external agencies (i.e., are intrinsic to the gene and are independent of known extrinsic factors). The limitations of this definition will be considered subsequently.

Twenty-five years ago, it was possible to define the nature and causes of spontaneous gene mutation with reasonable certainty. Information from the genetic code, studies of genetic fine structure, and experiments in chemical mutagenesis, when taken together, established that gene mutations are associated with bp changes in the DNA. It is appropriate here to cite the classical deduction of Watson and Crick (1953), "It seems plausible to us that a spontaneous mutation, which as implied earlier, we imagine to be a change in the sequence of bases, is due to a base occurring very occasionally in one of the less likely tautomeric forms, at the moment when the complementary chain is being formed." Thus, transitions and transversions, bp losses and additions that arose either through errors in replication or repair of DNA represented the nature and causes of gene mutation. By extrapolation, spontaneous gene mutations were no different from mutagen-induced mutations. They were distinguished only by their comparatively rare occurrence. Coincidentally, McClintock (1950) established yet another mechanism of gene mutation, mutation via transposable "foreign" DNA that inserts into specific genes and is recognized as a gene mutation by virtue of an accompanying phenotypic change. This "foreign" DNA is now known by a number of synonyms: controlling element, insertion sequence, mobile DNA element, nomadic gene, and so forth. The evidence now at hand demonstrates these elements have a major role in generating spontaneous mutations, thereby vindicating the deductions made by McClintock from the results of genetic experiments that were too long ignored or misunderstood.

In the discussion that follows, mutation data drawn primarily from *D. melanogaster* will be emphasized and, where appropriate, parallelisms with comparable data from other organisms will be included.

DISCUSSION

Intergenic and Intragenic Spontaneous Mutation

Two general classes of spontaneous mutational events associated with a change in phenotype have long been recognized in *D. melanogaster*. They are, broadly speaking, intergenic and intragenic events. Intergenic events comprise those phenotypic changes associated with the comparatively large loss or large duplication of specific chromosome regions, often cytologically identifiable. Noteworthy examples are the X chromosome Notch wing (N) mutation where 13 of 20 spontaneous mutations are associated with a definable cytological loss (Lindsley and Grell 1968) and the X-chromosome Beadex-recessive wing (Bx) mutation, in which each of three spontaneous mutations is associated with the tandem duplication of the Bx^+ gene (Green 1953 and unpubl.). There is good reason to believe that these intergenic changes, albeit relatively rare and associated with a restricted set of genes, arise via unequal crossing-over mediated in part by mobile DNA elements. In the main intergenic mutational events are of minor relevance. An exception is spontaneous mutation at the aleurone and plant color receptor (R) locus in maize, where a spontaneous mutation rate of $49/10^5$ gametes can be causally associated with unequal crossing-over (Stadler and Emmerling 1956).

The molecular cloning of a number of *D. melanogaster* genes reveals that contrary to expectation an inordinate number of spontaneous mutations are causally associated with the insertion of mobile DNA elements. Prefatory to considering the molecular genetic information, a brief résumé of the current knowledge of *D. melanogaster* mobile DNA elements seems advisable. A detailed, in depth description of *D. melanogaster* mobile DNA elements has been compiled by Finnegan and Fawcett (1986). For purposes of consideration here, answers are needed to two questions: how many mobile DNA elements occur in the *D. melanogaster* genome, and what is the molecular organization of the elements?

Arithmetically, the number of different families of mobile elements can be estimated from the following assumptions. Assume mobile elements make up all of the middle, repetitive DNA in the fly genome. Assume further their average length is 5 kb and their average copy number is 25. Calculating from these numbers, the estimate number of mobile element families is 50 + . It is of interest to note that in situ hybridization to *D. melanogaster* polytene chromosomes with 12 different mobile elements established hybridization to 300 different bands in a total of approximately 2000 (Ananiev et al. 1984). These results, extrapolated to 50 families of elements, mean that about 50% of the euchromatin sites (bands) contain mobile elements. To date, 33 different elements have been described.

Overall, *D. melanogaster* mobile elements vary in length from 2 to 10 kb. Based on the structure of their termini, they can be classed into several classes: (1) copia-like elements, with long direct terminal repeats; (2) foldback elements, with long inverted terminal repeats; and (3) F elements, without terminal repeats. Variation in the length of repeats have been described and Finnegan and Fawcett (1986) should be consulted for details.

The cloning and molecular characterization of spontaneous mutations at ten different *D. melanogaster* loci provides instructive information on the nature of spontaneous mutation. The ten loci were selected because a minimum of four spontaneous mutants had been analyzed. The mutants arose as single events in a variety of laboratory stocks. The following loci are included: X chromosome–*su(s)* (suppressor-of-sable), *w* (white eye color), *N* (Notch wing), *ct* (cut wing), *v* (vermilion eye color), *f* (forked bristles), *Bx*; autosomal loci–*bx* (bithorax halteres), *bxd* (bithoraxoid halteres), and *ry* (rosy eye color).

The compilation which follows is drawn primarily from Inoue and Yamamoto (1987) and Mattox and Davidson (1984). Among a total of 72 spontaneous mutations, 57 or almost 80% are insertion mutations. Among 41 insertion mutations, in which the mobile element has been identified, 39 are one or other of four copia-like elements: copia, 412, gypsy/mdg4, or B104/roo. An apparent disproportionate number of mutations are associated with the gypsy/mdg4 element: 25 of 41 mutations are gypsy insertion, 9 are roo/B104, 4 are 412, and 2 are copia. Some loci seem to be hot spots for gypsy insertion. Among the ten spontaneous *bx* mutations, all of which are insertions, nine are gypsy. Among six spontaneous *bxd* mutations, five are gypsy, and similarly five of seven *su(s)* mutations are gypsy inserts. These results are surprising because they bear no direct relationship to copy number in the genome. (Recorded copy number/genome listed by Rubin [1983] is: gypsy approximately 10, roo/B104 approximately 100, 412 approximately 40, and copia approximately 20–60). As a first approximation these findings suggest that only a limited number of mobile elements are mobile. This proposition will be considered below in some detail.

Are the *D. melanogaster* spontaneous mutation data unique? Two examples of spontaneous mutations drawn from bacteriophages λ and P1 demonstrate they are not. Thus, Lieb (1981) reported 15 of 25 spontaneous mutations in the λ-repressor gene (*cI*) are insertions, 7 of $IS1$, 3 or $IS2$, and 5 of $IS5$. In bacteriophage P1, Sengstad and Arber (1982) found the majority of spontaneous mutations are $IS2$ insertions.

Do mutations, isolated from flies in the wild, mirror the *D. melanogaster* data reported above? At present, there is insufficient information to answer this question because no systematic isolation and characterization has been carried out of mutations borne by wild flies. The one such detailed analysis is that of the second chromosome lethal giant larva mutation, *l(2)gl*, of *D. melanogaster* described by Mechler et al. (1985). Among 13 *l(2)gl* mutations isolated from wild flies collected in the USSR and California, 11 are deletions, 1 is an insertion of the copia-like B104/roo, and 1 is a complex deletion/insertion. At issue in this case is the question of whether these *l(2)gl* mutations are really spontaneous. There is good reason to believe that they may be induced by mitotic recombination (MR) mutator elements (Green and Shepherd 1979) that are widespread and frequently occurring in wild *D. melanogaster* flies and which cause chromosome deletions and other chromosome rearrangements.

Although the case for the association of spontaneous mutations with mobile-element insertion is substantial, it should be noted that there are exceptions. Noteworthy is the finding of Farabaugh et al. (1978) that among 140 spontaneous mutations of the *lacI* gene of *Escherichia coli*, only 2 were insertions.

These data suffice to illustrate one useful generalization: some genes are hot spots for mobile-element insertion, others are cold spots. Much depends upon the molecular structure of the gene and/or of the particular mobile element.

RESULTS

Mutation and Mobile Element Usage

Based upon the data presented above, it was suggested that a limited number of mobile elements are associated with the production of spontaneous mutations. This suggestion assumes, of course, that the mutation data are representative. The comparison of the amounts of middle repetitive DNA in *D. melanogaster* and its sibling species, *Drosophila simulans*, permits an independent test of this notion. Dowsett and Young (1982) found *D. simulans* to contain about one-seventh the middle repetitive DNA of *D. melanogaster*. This is indeed an unexpected finding because the two species are chromosomally homosequential and produce viable hybrids. If, as Dowsett and Young speculated, ". . .rearrangements of nomadic DNA prove to be a primary agent of mutagenesis, spontaneous mutation may be found to vary several fold between species in a fashion that reflects quantitative differences in the nomadic DNA contents of their genomes." Put simply, the spontaneous mutation rate of *D. simulans* should be significantly lower than that of *D. melanogaster* provided the mutational propensity of all mobile elements is, on the average, equal. A test of this proposition was undertaken by estimating the spontaneous mutation rate of X-chromosome-recessive-lethal mutations in *D. simulans* males (Eeken et al. 1987). To detect X-linked recessive lethal mutations the classical *D. melanogaster C1B* and *Basc* methods, both of which employ inverted X chromosomes, were, in principle, duplicated. Two appropriate X-chromosome inversions, one with a recessive lethal á la *C1B* and one homozygous viable á la *Basc*, were induced with X rays, and each was used to capture X chromosomes of males to determine the X-chromosome recessive lethal mutation rate. Two independent experiments in which together a total of 5000 X chromosomes were scored for lethal mutations gave a spontaneous mutation rate in *D. simulans* males of 0.3% or precisely that of *D. melanogaster*. In the face of significant differences in amounts of middle repetitive DNA and presumed numbers of mobile elements, how can the mutation rate data be rationalized? Here, two points are relevant. First, recessive lethal mutations in *D. melanogaster* are prone to insertion mutagenesis, just as are visible mutations. Second, spontaneous insertion mutations are known in *D. simulans* for at least two genes: *w*, where the molecular analysis of seven *w* mutants established five to be insertions (Inoue and Yamamoto 1987), and facet eye (*fa*), where the single spontaneous mutant is associated with a mobile element insertion (Kidd and Young 1986). Among the possible explanations for the equal spontaneous mutation rates of *D. melanogaster* and *D. simulans*, two seem probable. The first alleges that the average number of mobile DNA elements in the two species is identical, and a large part of the middle repetitive DNA of *D. melanogaster* is something other than mobile elements. The second alleges that there exist highly mobile and nearly immobile DNA elements in *D. melanogaster*, whereas *D. simulans* contains only the former and in number equal to that in *D. melanogaster*. For the present there are no compelling reasons that militate for one or the other explanation. There is a bias toward the second explanation because, among the identified *D. melanogaster* mobile elements, only a minority have been associated with spontaneous mutations. The molecular analysis of additional spontaneous mutations could change the situation significantly.

The Mobility of Mobile DNA Elements

The association of spontaneous mutations with mobile DNA elements perforce poses the question, "What makes mobile elements move?" McClintock provided a model for mobile element mobility by deducing that in the activator-dissociation (*Ac-Ds*) system governing mutation in maize, *Ac* is autonomous in movement and *Ds* is driven by *Ac*. In the absence of *Ac*, *Ds* is stationary. In other words, some mobile DNA elements are self-propelled, others are driven.

In *D. melanogaster* subsequent to the isolation of some mobile DNA elements, the mobility of these elements was deduced from the results of in situ hybridization experiments with polytene chromosomes. The fact that the polytene chromosome sites of in situ hybridization of a particular mobile element were different when different stocks were compared, this time the process of transposition and integration of *D. melanogaster* mobile elements is incompletely understood. There are a few facts that are instructive. The movement and integration of some mobile elements is under genetic control. Thus, P element mutagenesis is controlled by MR elements, mappable genetic elements that are also capable of chromosome breakage. (The question of P element mutagenesis will be considered in detail elsewhere in this volume.) What is pertinent here is the observation that in addition to P sequences, at least one other mobile element, copia, has moved to generate mutations at two gene loci, *w* (Rubin et al. 1982) and *ry* color (Coté et al. 1986), under the influence of the MR element. The precise nature of the MR element remains to be elucidated.

The movement of inducer (I) elements is also under genetic control, since mobility occurs when a cross is made between an I male and R female. Specific genotypes determine the I and R states. Details of the I-R mutagnesis system are discussed elsewhere in this volume.

Finally, there is evidence that mobility of the mobile element, hobo, occurs among the progeny of a cross between flies carrying the complete hobo sequence and those lacking the complete sequence (Streck et al. 1986).

The foregoing information on the P, *I*, copia, and hobo mobile elements argues for the exercise of caution when concluding that a mobile element associated mutation arose spontaneously, *sensu strictu*. Implicit in spontaneous mutation is autonomy of origin; spontaneous mutations associated with P, *I*, copia, and hobo could be driven.

Mass Mutation

The recent reports of "bursts" or "explosions" of mutations in *D. melanogaster* are on two points of interest here. First, they provide a possible explanation for the independently occurring mutational bursts, described by Goldschmidt (1939), Tiniakov (1939), and Neel (1942). Mutational bursts, described by Lim (1979) and Gerasimova (1981), are associated with mobile element transposition and insertion, especially the copia-like element, gypsy. By inference the observations of Goldschmidt, Tiniakov, and Neel are comparable situations. Along the same lines, Golubovsky (1980) has argued that the frequent extraction of particular allelic mutations from wild flies over several consecutive seasons of collecting are the outcome of "mutation fashions" synonymous with mutational bursts. However designated—"burst of mutation," "transposition explosion," or "mutation fashion"—all these phenomena share one feature in common, a sudden transient mode of origin. An unpredictable event takes place: quiescent mobile elements without obvious explanation become highly mobile. Often, as suddenly as the mutational bursts arise, they as suddenly subside. Precisely what brings about the change in mobile element mobility is unclear. It is possible that the phenomenon of genomic stress (McClintock 1978) underlies the activation of mobile elements. If so, what remains unclear is the nature of the stress element. Golubovsky has suggested that some kind of infectious agent may be involved, a point that will be discussed below. What is apparent is that the occurrence of a burst of mutation with the associated mobile-element-insertion mutation means that such mutations isolated subsequent to the burst are not autonomous, although they are interpreted as spontaneous mutations as already defined. In this connection, another word of caution is in order. The isolation from wild flies collected in a specific habitat of a number of allelic mutations has been interpreted as a mutational burst. Because wild *D. melanogaster* populations undergo annual crashes, genetic drift rather than a mutational burst might favor the occurrence of an inordinate number of allelic mutations in the subsequent fly populations.

"Foreign" DNA and Mutation

In a particularly intriguing series of experiments, Gershenson (1986) reported that a variety of RNA and DNA viruses were highly mutagenic when injected into adult *D. melanogaster* males. Mutagenicity was measured by assaying for

X and 2nd chromosome recessive lethal mutations. A number of interesting results emerge from this study: (1) Increased induced mutation rates as much as tenfold greater than controls were observed; (2) DNA viruses, insect viruses, were mutagenically more potent than RNA viruses; (3) as a mutagen, viral DNA was more effective than intact DNA virus; and (4) induced lethal mutations were site-specific rather than randomly distributed. In a parallel experiment Gazaryan et al. (1984) found that Rous sarcoma virus DNA was mutagenic when injected into the poleplasm of *D. melanogaster* embryos. (The poleplasm cells differentiate into the adult gonad.) Mutagenesis was site-specific for an array of mutations with a visible phenotype. Furthermore, virus-specific DNA was detected in flies bearing a Rous-sarcoma-DNA-induced mutation.

A possible explanation for the mutagenic effects of viruses and viral DNA described comes from comparable observations made in maize. For some time, it has been known that an increased incidence of mutations occurs among the progeny of maize plants infected with the barley stripe mosaic virus, an RNA virus (Sprague et al. 1963). In one series of experiments designed to elucidate the nature of the viral-induced mutations, Freeling and associates selected for alcohol dehydrogenase negative (ADH^-) mutations among virus-infected maize plants (Mottinger et al. 1984). Cloning of two induced ADH^- mutations revealed both to be insertion mutations. However, the insertion proved to be unrelated to the virus but involved a copia-like mobile element called BS-1, normally present in 1–5 copies in maize plants (Johns et al. 1985). These results are interpreted within the context of McClintock's genomic stress concept. The viral infection generates stress to which the plant responds, in part, by increased mobility of resident mobile DNA elements.

By extrapolating from the maize experiments a tentative explanation can be proffered for the findings of Gazaryan et al. (1984) and Gershenson (1986) as follows. The injected virus or virus DNA acts as an inducer of genomic stress. The response by the fly is increased mobile element movement and integration, thereby increasing the occurrence of mutations.

SUMMARY

Taking the spontaneous mutation data of *D. melanogaster* at face value, an inordinate fraction are associated with the insertion of mobile DNA elements. However, the stipulation that a particular mobile-element-associated mutation arose spontaneously is confounded by the existence of genetic-environmental situations in which mobile element mobility and integration increases. There is the distinct possibility that the specification of a mobile-DNA-element-associated mutation as spontaneous may often be spurious, if spontaneous is delimited to autonomous movement and integration.

REFERENCES

Ananiev, E.V., V.E. Barski, Y.V. Ilyin, and M.V. Rysic. 1984. The arrangement of transposable elements in the polytene chromosomes of *Drosophila melanogaster*. *Chromosoma* **90:** 366.

Coté, B., W. Bender, D. Curtis, and A. Chovnick. 1986. Molecular mapping of the *rosy* locus in *Drosophila melanogaster*. *Genetics* **112:** 769.

Dowsett, A.P. and M.W. Young. 1982. Differing levels of dispersed repetitive DNA among closely related species of *Drosophila*. *Proc. Natl. Acad. Sci.* **79:** 4570.

Eeken, J.C.J., A.W.M. de Jong, and M.M. Green. 1987. The spontaneous mutation rate in *Drosophila simulans*. *Mutat. Res.* (in press).

Farabaugh, P.J., U. Schmeissner, M. Hofer, and J.H. Miller. 1978. Genetic studies of the *lac* repressor VII on the molecular nature of spontaneous hot spots in the *lac I* gene of *Escherichia coli*. *J. Mol. Biol.* **126:** 847.

Finnegan, D.J. and D.H. Fawcett. 1986. Transposable elements in *Drosophila melanogaster*. *Oxf. Surv. Eukaryotic Genes* **3:** 1.

Gazaryan, K.G., D.S. Nabirochkin, A.G. Tatosyan, A.K. Shakhbazyan, and E.N. Shibanova. 1984. Genetic effects of injection of Rous sarcoma virus DNA into poleplasm of early *Drosophila melanogaster* embryos. *Nature* **311:** 392.

Gerasimova, T.I. 1981. Genetic instability at the *cut* locus of *Drosophila melanogaster* induced by the *MRh12* chromosome. *Mol. Gen. Genet.* **184:** 544.

Gershenson, S.M. 1986. Viruses as environmental mutagenic factors. *Mutat. Res.* **167:** 203.

Goldschmidt, R.B. 1939. Mass mutation in the Florida stock of *Drosophila melanogaster*. *Am. Nat.* **73:** 547.

Golubovsky, M.D. 1980. Mutational process and microevolution. *Genetika* **52/53:** 139.

Green, M.M. 1953. The Beadex locus in *Drosophila melanogaster*: Genetic analysis of the mutant Bx^{r49k}. *Z. Vererbungsl.* **85:** 435.

Green, M.M. and S.H.Y. Sheperd. 1979. Genetic instability in *Drosophila melanogaster*: The induction of specific chromosome 2 deletions by *MR* elements. *Genetics* **92:** 823.

Inoue, Y.H. and M. Yamamoto. 1987. Insertional DNA and spontaneous mutation at the *white* locus in *Drosophila simulans*. *Mol. Gen. Genet.* **209:** 94.

Johns, M.A., J.P. Mottinger, and M. Freeling. 1985. A low copy number, copia-like transposon in maize. *EMBO J.* **4:** 1093.

Kidd, S. and M.W. Young. 1986. Transposon-dependent mutant phenotypes at the *Notch* locus of *Drosophila*. *Nature* **323:** 89.

Lieb, M. 1981. A fine structure map of spontaneous and induced mutations in the lambda repressor gene, including insertions of IS elements. *Mol. Gen. Genet.* **184:** 364.

Lim, J.K. 1979. Site-specific instability in *Drosophila melanogaster:* The origin of the mutation and cytogenetic evidence for cite specificity. *Genetics* **93:** 681.

Lindsley, D.L. and E.H. Grell. 1968. Genetic variations of *Drosophila melanogaster.*. *Carnegie Inst. Wash. Publ.* **627:** 1.

Mattox, M.W. and N. Davidson. 1984. Isolation and characterization of the *Beadex* locus of *Drosophila melanogaster:* A putative *cis*-acting negative regulatory element for the *heldup-a* gene. *Mol. Cell. Biol.* **4:** 1343.

McClintock, B. 1950. The origin and behavior of mutable loci in maize. *Proc. Natl. Acad. Sci.* **36:** 344.

———. 1978. Mechanisms that rapidly reorganize the genome. *Stadler Genet. Symp.* **10:** 25.

Mechler, B.M., W. McGinnis, and W.J. Gehring. 1985. Molecular cloning of *l(2)giant larva*, a recessive oncogene of *Drosophila melanogaster*. *EMBO J.* **4:** 1551.

Mottinger, J.P., M.A. Johns, and M. Freeling. 1984. Mutations of the *Adh-1* gene in maize following infection with barley stripe mosaic virus. *Mol. Gen. Genet.* **195:** 367.

Neel, J.V. 1942. A study of a case of high mutation rate in *Drosophila melanogaster*. *Genetics* **27:** 519.

Rubin, G.M. 1983. Dispersed repetitive DNAs in *Drosophila*. In *Mobile genetic elements* (ed. J. Shapiro), p. 329. Academic Press, New York.

Rubin, G.M., M.G. Kidwell, and P.M. Bingham. 1982. The molecular basis of P-M hybrid dysgenesis: The nature of induced mutations. *Cell* **29:** 987.

Sengstad, C. and W. Arber. 1982. Is-2 insertion is a major cause of spontaneous mutagenesis of bacteriophage P1: Non-random distribution of target sites. *EMBO J.* **2:** 67.

Sprague, G.F., H.H. McKinney, and L. Greeley. 1963. Virus a mutagenic agent in maize. *Science* **141:** 1052.

Stadler, L.J. 1946. Spontaneous mutation at the *R* locus in maize I, the aleurone and plant color effects. *Genetics* **31:** 377.

Stadler, L.J. and M.H. Emmerling. 1956. The relation of unequal crossing over to the interdependence of R^r elements (P) and (S). *Genetics* **41:** 124.

Streck, R.D., J.E. MacGaffey, and S.K. Beckendorf. 1986. The structure of hobo transposable elements and their insertion sites. *EMBO J.* **5:** 3615.

Tiniakov, G.G. 1939. Highly mutable stock from a wild population of *Drosophila melanogaster*. *C.R. Acad. Sci. USSR* **22:** 609.

Watson, J.D. and F.R.C. Crick. 1953. The structure of DNA. *Cold Spring Harbor Symp. Quant. Biol.* **18:** 123.

Molecular Genetic Studies of the Mouse Dilute Locus: Analysis of Two Dilute Alleles and Dilute-suppressor

MARJORIE C. STROBEL, PETER K. SEPERACK, KAREN J. MOORE,
NEAL G. COPELAND, AND NANCY A. JENKINS
Mammalian Genetics Laboratory
BRI-Basic Research Program
NCI-Frederick Cancer Research Facility
Frederick, Maryland 21701

OVERVIEW

The original mouse dilute mutation (d) was identified solely by its variant coat color. However, most other d alleles manifest the additional phenotypes of neurological disorders and juvenile lethality. The over 200 alleles at the dilute locus provide a strong genetic basis for the molecular analysis, which will allow a correlation between these phenotypes and expression patterns of the genomic d locus.

Initial studies have focused on two d mutations, the original d and a dilute-lethal allele, $d^{l\;20J}$. The original d mutation, now termed dilute-viral or d^v, is causally associated with an ecotropic murine leukemia provirus, $Emv\text{-}3$, whereas the $d^{l\;20J}$ mutation is correlated with a small deletion. Analysis of germline and somatic d^+ revertants of d^v has elucidated the probable mechanism by which proviral excision causes phenotypic reversion. Together, d^v and $d^{l\;20J}$ have provided molecular entree to the genomic d locus, allowing the initial identification of dilute transcripts. Finally, the effect of dilute-suppressor (dsu), an unlinked, semidominant mutation that suppresses the dilute coat color defect of d^v, on the $d^{l\;20J}$ allele has been examined. These studies have allowed a hypothesis to be drawn concerning the mechanism of suppression by dsu.

INTRODUCTION

Numerous mouse phenotypic variants have been recognized and propagated by mouse fanciers and mouse geneticists. Over the years, genetic analysis of these variants has correlated these phenotypes with the mutation of specific genetic loci. In some cases, the identification of numerous alleles of a locus has allowed fine-structure genetic mapping of the relevant chromosome region. However, subsequent molecular elucidation of these genetically defined loci has been impeded by the difficulties in gaining molecular access to a discrete region in the mouse genome. The mouse d locus is among those rare loci for which both extensive genetic analysis and molecular access are available.

The murine d mutation was identified as a coat color variant of the mouse fancy, characterized by a "diluted" coat pigmentation. This coat color phenotype has been correlated with a recessive mutation (d) that maps to mouse chromosome 9 (Silvers 1979). While melanin synthesis in d/d homozygotes is normal, the morphology of the dilute melanocytes is abnormal (Russell 1948). Wild-type, neural crest-derived melanocytes are highly dendritic cells with widely dispersed melanin granules; however, d/d melanocytes are virtually adendritic with pigment granules clumped near the cell nucleus. Since the dendrites are the conduits for pigment granule movement from the melanocytes to the hair bulb and hair shaft, the adendritic dilute melanocytes provide only an uneven release of melanosomes. Therefore, the predominant pigment deposition occurs only at the base of the hair shaft, and the hair tips are virtually devoid of melanin, giving an overall dilution of the mouse's coat color (Markert and Silvers 1956).

Experiments in which developing melanoblasts are transplanted to the neutral environment of the anterior eye chamber show the d mutation is cell-autonomous. While transplanted, wild-type melanoblasts elaborate dendrites, the d/d melanoblasts, in general, remain adendritic upon transplantation (Markert and Silvers 1959). The inherent inability of dilute melanoblasts to appropriately differentiate suggests the d mutation may be affecting a component of the cytoskeleton or cell surface critical for dendrite formation and/or granule transport.

Since the elucidation of the original d mutation, over 200 additional d alleles, both induced and spontaneous, have been identified. Animals homozygous for the original d allele exhibit only the dilute coat color phenotype; in contrast, homozygotes for the radiation-induced dilute-opisthotonic (d^{op}) or spontaneous d^l mutations manifest additional phenotypes: opisthotonus, a neurological disorder typified by convulsive arching of the head and neck, and juvenile lethality at about three weeks of age (Russell 1952; Russell 1971). In an extensive complementation analysis of the radiation-induced alleles, the coat color, neurological and lethality phenotypes were genetically inseparable, implying they are ascribable to mutation of a single functional unit (Russell 1971). Furthermore, the genetic analysis of several spontaneous d alleles in our laboratory has emphasized that d gene products are essential for normal melanocyte morphology, neurological development, and juvenile survival.

Another feature of the mouse d locus is the existence of a suppressor of the dilute coat color defect (Sweet 1983). This suppressor was initially recognized by the appearance of intensely colored animals in a stock homozygous for the original d allele. By genetic and molecular criteria, the nearly wild-type coat color and dendritic melanocytes of these animals are due not to a reversion of the d mutation, rather to the mutation of an unlinked locus. The *dsu* locus maps to mouse chromosome 1. Originally reported to be a recessive mutation (Sweet 1983), recent genetic analysis has shown the suppression phenotype is associated with a semidominant mutation at the *dsu* locus (K.J. Moore et al., in prep.).

Our laboratory is utilizing the rich collection of d alleles as the basis for understanding, at the molecular level, the gene products of and genomic organization at the dilute locus. Ultimately, this approach will define the regions and/or expression patterns of the d functional unit critical for correct cell type- and stage-specific expression of the d gene products. Initial studies have focused on three spontaneous mutants: the original d allele, the $d^{1\ 20J}$ mutation, and *dsu*. Elucidation of the molecular defect in d and $d^{1\ 20J}$ mutations has provided molecular access to and precise definition of a discrete genomic region critical for d gene function. The genetic analysis of the interaction between *dsu* and these two d alleles has suggested a potential mechanism of suppression by the *dsu* gene product.

RESULTS

The Original *d* Allele

Initial insight into the nature of the original d mutation was afforded by the cosegregation of d and *Emv-3*. *Emv-3* is the sole ecotropic provirus in the dilute brown albino (DBA)/2J inbred mouse strain, which is homozygous for d. Furthermore, *Emv-3* is present in six additional inbred strains that carry d, whereas it is absent from hundreds of mouse strains that are wild-type at the d locus (Jenkins et al. 1981).

A causal relationship between the *Emv-3* proviral integration and the d mutation was implied by the germline reversion of this d allele: Revertant animals, wild-type in coat color, arose in a homozygous d stock at high-frequency (3.9×10^{-6} reversions/gamete, Schlager and Dickie 1971). In contrast, no revertants of other spontaneous or induced d mutations have been observed. Examination, with an ecotropic virus-specific probe, of DNA derived from germline d^+ revertant animals showed *Emv-3* sequences were absent, implying phenotypic reversion is caused by proviral excision (Jenkins et al. 1981). Because of the association between the original d allele and *Emv-3*, this mutation has been termed d^v.

The presence of the provirus provided a molecular entree to the d region, allowing the isolation of genomic clones that span the integration site and the identification of unique sequence probes from this region (Copeland et al. 1983). Using one of these probes (p0.3, Rinchik et al. 1986) in Southern blot analysis of wild-type, d^v and d^+ germline revertant DNAs, a polymorphic EcoRI restriction pattern is seen: p0.3 recognized a 9-kb fragment in $+/+$, an 18-kb fragment in d^v/d^v, and a 9.5-kb fragment in d^+/d^+ mouse DNAs (Fig. 1). Further analysis showed the increased size of the d^+ revertant fragment was due to the retention of a single, 0.5-kb proviral long terminal repeat (LTR) at the former integration site. The presence of the single LTR, in a phenotypically wild-type animal, implies the d^v mutation must be caused by the insertion of Emv-3 in a noncoding region at or near d. DNA sequence

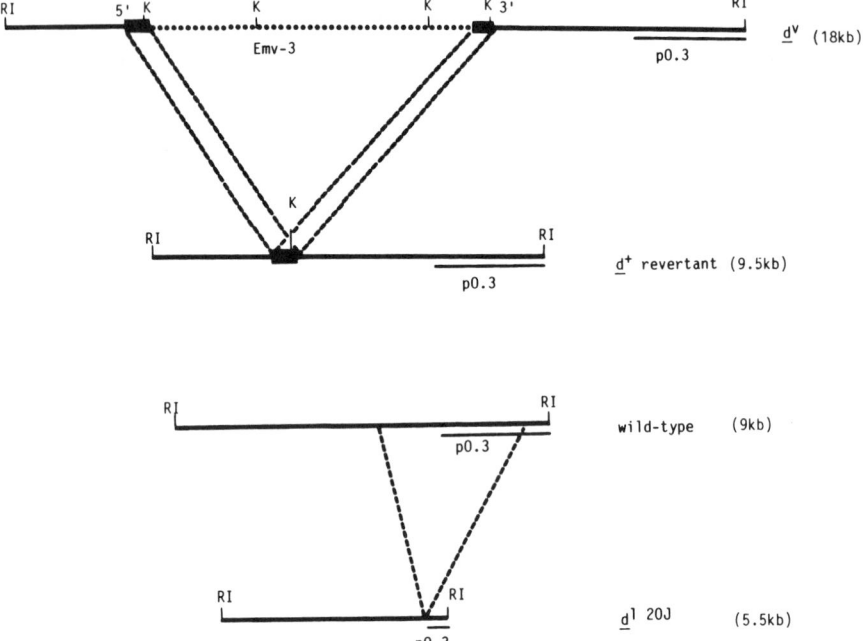

Figure 1
Physical maps of the region flanking the proviral integration site in different dilute alleles. Size of these EcoRI (RI) restriction fragments was determined by genomic Southern blot analysis using the 2.6-kb unique sequence probe, p0.3. The precise organization of these fragments was determined by restriction mapping of the cloned fragment. KpnI (K) restriction enzyme sites are provided for reference. Dashed lines delineate sequences removed from the d^v and wild-type alleles to generate, respectively, the d^+ revertant and $d^{1\ 20J}$ chromosomes. Thin line, sequence homologous to p0.3; dotted line, Emv-3 provirus; shaded boxes, proviral LTRs.

analysis of the region immediately flanking the integration site confirmed this hypothesis. No significant open reading frames were observed within 600 bases of the proviral integration site (Hutchison et al. 1984).

Retention of a single LTR in the d^+ revertant chromosome suggests the proviral excision event proceeded by a homologous recombination mediated by the proviral LTRs. Conceivably, the d^+ revertant allele could be generated by two qualitatively different recombination events: an exchange between LTRs on a single chromosome, leading to the excision of a proviral circle, or an uneven crossing over between LTRs on sister chromatids or chromosomes. Each proposed recombination event would generate the LTR-containing 9.5-kb fragment in the d^+ revertant allele; however, the predicted structure of the homologous chromosome would differ. In an intrachromosomal recombination, the unaffected chromosome would retain the 18-kb d^v allele; whereas an uneven exchange between homologs could generate one chromosome with a 27-kb EcoRI restriction fragment containing a tandem duplication of the 9-kb provirus.

The recent identification of the first somatic d^+ revertant has allowed elucidation of the probable reversion mechanism (Seperack et al. 1988). Originally, the somatic revertant mouse was recognized by its mosaic coat, which contained patches of both wild-type and dilute pigmented hairs. The animal, also, was mosaic at the DNA level. All adult tissues, representing each embryonic lineage, contained both the 9.5-kb fragment (d^+) and 18-kb fragment (d^v) alleles (Fig. 1). Furthermore, the d^+ revertant allele was transmitted in the germline: 25% of this animal's progeny (33 revertant offspring/126 total progeny) carried the d^+ allele, implying 50% of the premeiotic germ cells in the somatic revertant animal were heterozygous d^+/d^v.

The detection of only the d^v and d^+ revertant alleles provides strong evidence for proviral excision proceeding by an intrachromosomal homologous recombination event. Since the revertant allele was present in all adult tissues and the germline, the reversion event must have occurred very early in this animal's development, possibly at the first cleavage or two-cell stage. Therefore, it is formally possible that the reciprocal product of an uneven exchange (i.e., the presumptive 27-kb allele) could have segregated only into extraembryonic lineages and, therefore, is not represented in the adult tissues.

With the identification of both germline and somatic d^+ revertants, an in vivo reversion frequency for this mammalian retrotransposon-induced mutation can be calculated (Table 1). In over 1,100,000 d^v/d^v animals examined at The Jackson Laboratory over the last five years, 11 germline and 1 somatic d^+ revertants have been identified, giving a germline frequency of 4.5×10^{-6} reversions/gamete and a somatic frequency of 9×10^{-7} reversions/animal.

Table 1
Germline and Somatic d^+ Reversion Frequencies in DBA Mice (d^v/d^v)

Total no.		No. of d^+ revertants		Reversion	
animals	gametes	somatic	germline	somatic (per animal)	germline (per gamete)
1,115,818	2,231,636	1	11	9×10^{-7}	4.5×10^{-6}

All d^+ revertable animals arose over the last five years in the DBA breeding colony of The Jackson Laboratory, Bar Harbor, Maine.

Dilute-lethal, d^{120J}

The association of *Emv-3* with the d^v allele has provided molecular access to the d gene region. Since the d^v proviral integration site was within noncoding sequence, this allele could not define a precise genomic region as dilute-specific. However, such a region has been defined by the clarification of the genomic defect in the spontaneous d^{120J} allele. Using DNA derived from animals carrying this d^1 allele, a unique *Eco*RI restriction fragment of 5.5 kb was recognized by the p0.3 molecular probe (Fig. 1). Subsequent restriction mapping showed this genomic fragment represented a deletion breakpoint fusion fragment. The deletion began 2 kb 3' of the proviral integration site of the d^v allele and extended 3' for 3.5 kb, removing most of the sequence present in the p0.3 probe. DNA sequence analysis of the cloned deletion breakpoint fusion fragment indicates it is a simple deletion (M. Strobel et al., in prep.).

Animals homozygous for the d^{120J} allele exhibit the full range of dilute-associated phenotypes: diluted coat color, opisthotonus, and lethality. The association of these phenotypes with a deletion suggested the deleted sequences, possibly the p0.3 probe, were a critical part of the d gene. When the p0.3 molecular probe was used in Northern blot analysis of RNA derived from the B16 mouse melanoma cell line (wild-type at d), transcripts were detected. Therefore, B16 RNA was used to construct a cDNA library. Using the p0.3 probe, a 2.5-kb cDNA clone was isolated.

When this cDNA probe was used in Northern analysis of B16 RNA, three transcripts, approximately 11, 9, and 7 kb in size, were recognized (Fig. 2). A similar transcription pattern was seen in brain RNA derived from the C57BL/6J mouse ($+/+$ at d). As shown in Figure 2, brain RNA was a rich source of dilute-specific transcripts. At longer exposure, these three RNAs were weakly detected in kidney, spleen, and thymus, whereas no transcripts were found in liver.

The cDNA probe has been used to examine the transcripts in tissue RNAs from DBA/2J mice homozygous for either the d^v or the d^+ revertant allele.

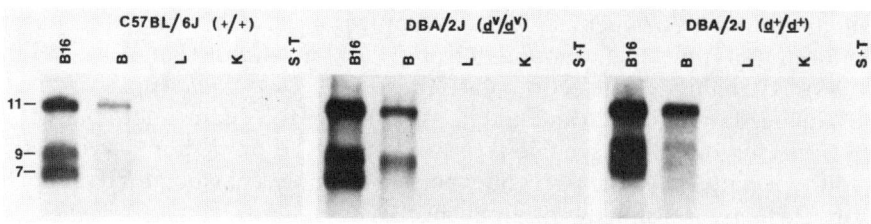

Figure 2
Northern blot analysis of RNA derived from the B16 mouse melanoma cell line and tissues of C57BL/6J and DBA/2J mice. Total RNA was prepared from the B16 mouse melanoma cell line (wild-type at d). Also, total RNA was made from brain (B), liver (L), kidney (K), and spleen and thymus ($S + T$) of three mouse strains: C57BL/6J, wild-type at d; DBA/2J, homozygous for the d^v allele; and DBA/2J, homozygous for a germline d^+ revertant allele. These RNAs were probed with the 2.5-kb cDNA described in the text. Transcript size was estimated relative to an RNA marker ladder.

As shown in Figure 2, d^v/d^v brain RNA exhibited a unique pattern of transcripts: Only two predominant RNAs were detected, one of 11 kb and a single smaller transcript of about 8 kb. In contrast, d^+/d^+ animals exhibited a virtually wild-type pattern. While the precise structure of these RNAs is presently unknown, these results imply the proviral insertion in the d^v allele is affecting only production of the 9- and 7-kb RNAs. Since the *only* phenotype associated with the d^v allele is the coat color defect, the apparent abolition of these transcripts in the d^v allele suggests that they may be critical for normal d gene function in melanocytes. In contrast, the 11-kb transcript appears to be qualitatively unaffected, suggesting that this gene product may be critical for normal neurological function. It is interesting that an 11-kb transcript was the predominant RNA detected in the undifferentiated rat PC-12 cell line (data not shown), a neural crest-derived cell that assumes a neuronal phenotype upon treatment with nerve growth factor (Greene and Tischler 1976).

Interaction of *dsu* with the d^v and $d^{l\ 20J}$ Alleles

Initially, *dsu* was recognized by the examination of virtually wild-type colored animals that arose in a homozygous d^v stock. Melanocytes from these exceptional animals exhibited a highly dendritic, wild-type morphology; however, genetic and molecular analyses showed these animals were not d^+ revertants. Rather, the suppressor phenotype segregates independently of d and is ascribable to a mutation that maps to chromosome 1 (Sweet 1983). Initially reported to be recessive, recent genetic analysis has shown the *dsu* mutation is semidominant (K.J. Moore, in prep.).

Since *dsu* was identified by its effect on the retrotransposon-induced d^v allele, the *dsu* product could be mediating its suppressor effect in a manner similar to that documented for the suppressors of other retrotransposon-induced mutations in *Drosophila* (Kubli 1986) and yeast (Roeder and Fink 1983). By analogy to the action of suppressors of *Drosophila* copia- and yeast TY-induced mutations, *dsu* could be suppressing the dilute color coat defect by restoring the dilute-specific transcription pattern in d^v homozygotes to a wild-type pattern. Rather, recent experiments imply *dsu* mediates its suppressor effects by a distinctly different mechanism.

Genetic crosses were established to examine the effect of *dsu* on animals homozygous for the $d^{l\ 20J}$ deletion mutation (K.J. Moore, in prep.). As previously described, these d^l homozygotes display the full range of dilute-associated phenotypes causally associated with a defined deletion. Juvenile $d^{l\ 20J}$ homozygotes are dilute in coat color, exhibit severe opisthotonic convulsions, and die by three weeks of age. In contrast, animals homozygous for both $d^{l\ 20J}$ and *dsu* are nearly wild-type in coat color, but manifest the dilute-associated neurological disorder and lethality. Therefore, the mutant *dsu* product rescues *only* the coat color defect of both the d^v and d^l alleles, implying *dsu* can suppress dilute mutations *only* in melanocytes. Furthermore, *dsu* can suppress this mutant dilute phenotype, whether it is causally associated with a retrotransposon insertion or a deletion. Since the deletion allele probably represents a functionally null mutation, the mutant *dsu* gene product may be compensating for the mutant *d* gene product in melanocytes.

DISCUSSION

The rich collection of *d* alleles provides a superb background for a detailed molecular analysis of the *d* locus. Ultimately, this analysis will allow elucidation of the cell type- and stage-specific transcription pattern and the genomic organization of the *d* locus.

The association between *Emv-3* and the original *d* allele (d^v) has provided initial molecular access to the *d* region of mouse chromosome 9. Furthermore, the examination of both germline and the first somatic d^+ revertants has clarified the mechanism of reversion of the d^v allele. All available data indicate proviral excision occurs by an intrachromosomal homologous recombination through the proviral LTRs. The only detectable recombination products are a single LTR at the former integration site on the affected chromosome (d^+ allele) and an intact provirus on the homolog (d^v allele). An intrachromosomal homologous recombination mechanism of proviral excision is in accord with other observations. Solo LTRs, presumably the remains of ancestral excision events, are present in the mouse genome (Wirth et al. 1983, 1984); whereas tandemly arrayed proviruses have not been observed. Also, an intrachromosomal recombination event has been proposed for the reversion of retrotransposon-induced mutations in *Drosophila* (Zachar et al. 1985) and yeast (Roeder and Fink 1983).

The detection of both germline and somatic revertants has allowed a determination of an in vivo reversion frequency for both meiotic and mitotic lineages, 4.5×10^{-6} reversion/gamete and 9×10^{-7} reversion/animal, respectively. Since germline reversions appear to occur during meiosis, rather than premeiotic growth of the germ cells, the difference between these reversion frequencies may reflect the relative probabilities of meiotic and mitotic recombination. In comparison, a reversion frequency of 1×10^{-5}–2×10^{-5} has been determined for the germline excision of a *Drosophila* copia element (at the white-apricot [w^a] locus, Carbonare and Gehring 1985; Zachar et al. 1985) and mitotic excision of a yeast TY element (TY*912* at HIS*4*, Winston et al. 1984). Furthermore, the in vitro excision frequency in cells containing an integrated murine leukemia provirus (MLV) has been estimated at 10^{-7} to 10^{-8} excisions/cell/generation (Varmus et al. 1981), a rate similar to that calculated for somatic reversion of d^v.

Together, the d^v and d^{120J} mutations have defined a genomic region critical for *d* gene function and, as such, have allowed an initial examination of the *d* gene products. Three transcripts of 11, 9, and 7 kb are recognized in the B16 mouse melanoma cell line and the C57BL/6J mouse (both wild-type at *d*). In contrast, this transcription pattern appears to be altered in the retrotransposon-induced d^v allele; an 11-kb and an approximately 8-kb RNA are detected. However, a d^+ germline revertant exhibits an apparent wild-type pattern. These data suggest the proviral insertion is affecting the transcription pattern at dilute, as has been observed for other retrotransposon-induced mutations (Kubli 1986).

In contrast to the vast majority of *d* alleles, d^v homozygotes display only the dilute coat color defect. This observation suggests the proviral insertion may be critically affecting only the part of the genomic *d* locus necessary for maintenance of normal melanocyte morphology. Ongoing studies are directed toward defining the structure of and relationship between the multiple dilute transcripts. Potentially, examination of the transcription patterns exhibited by the various *d* alleles at different stages in development will allow precise correlation of the dilute-associated phenotypes with the alteration or abolition of specific *d* gene transcripts.

Finally, genetic analysis of the interaction between *dsu* and the d^v and d^{120J} alleles implies the *dsu* gene product can suppress *only* the coat pigmentation phenotype of *d* mutations. Because *dsu* can suppress the diluted coat color of a d^{120J} homozygote, an animal carrying a deletion of dilute-specific information, the mechanism of *dsu* suppression may be a compensatory one: The *dsu* gene product may be compensating for the absence of or a defect in a *d* product in melanocytes. Further studies will clarify the mechanism by which the mutant *dsu* gene product can compensate for *d* mutations to restore normal melanocyte morphology.

ACKNOWLEDGMENTS

Our special thanks to the numerous individuals at The Jackson Laboratory, Bar Harbor, Maine, who continue to recognize and to provide to this laboratory new spontaneous dilute mutations. This research is supported by the National Cancer Institute, Department of Health and Human Services, under contract number CO-23909 with Bionetics Research, Inc. M.C.S. is the recipient of a postdoctoral fellowship from the National Institutes of Health (HD-07014).

REFERENCES

Carbonare, B.D. and W.J. Gehring. 1985. Excision of *copia* element in a revertant of the *white-apricot* mutation of *Drosophila melanogaster* leaves behind one long terminal repeat. *Mol. Gen. Genet.* **199:** 1.

Copeland, N.G., K.W. Hutchison, and N.A. Jenkins. 1983. Excision of the DBA ecotropic provirus in dilute coat-color revertants of mice occurs by homologous recombination involving the viral LTRs. *Cell* **33:** 379.

Greene, L.A. and A.S. Tischler. 1976. Establishment of a nonadrenergic clonal line of rat adrenal pheochromocytoma cells which responds to nerve growth factor. *Proc. Natl. Acad. Sci.* **73:** 2424.

Hutchison, K.W., N.G. Copeland, and N.A. Jenkins. 1984. Dilute-coat color locus of mice: Nucleotide sequence analysis of the d^{+2J} and d^{+Ha} revertant alleles. *Mol. Cell. Biol.* **4:** 2899.

Jenkins, N.A., N.G. Copeland, B.A. Taylor, and B.K. Lee. 1981. Dilute (*d*) coat color mutation of DBA/2J mice is associated with the site of integration of an ecotropic MuLV genome. *Nature* **293:** 370.

Kubli, E. 1986. Molecular mechanisms of suppression in *Drosophila*. *Trends Genet.* **2:** 204.

Markert, C.L. and W.K. Silvers. 1956. The effects of genotype and cell environment on melanoblast differentiation in the house mouse. *Genetics* **41:** 429.

―――. 1959. Effects of genotype and cellular environment on melanocyte morphology. In *Pigment cell biology* (ed. M. Gordon), p. 241. Academic Press, New York.

Rinchik, E.M., L.B. Russell, N.G. Copeland, and N.A. Jenkins. 1986. Molecular genetic analysis of the *dilute-short ear* (D-SE) region of the mouse. *Genetics* **112:** 321.

Roeder, G.S. and G.R. Fink. 1983. Transposable elements in yeast. In *Mobile genetic elements* (ed. J.A. Shapiro), p. 299. Academic Press, New York.

Russell, E.S. 1948. A quantitative histological study of the pigment found in the coat-color mutants of the house mouse. II. Estimates of the total value of pigment. *Genetics* **33:** 228.

Russell, L.B. 1971. Definition of functional units in a small chromosomal segment of the mouse and its use in interpreting the nature of radiation-induced mutations. *Mutat. Res.* **11:** 107.

Russell, W.K. 1952. X-ray induced mutations in mice. *Cold Spring Harbor Symp. Quant. Biol.* **16:** 89.

Schlager, G. and M.M. Dickie. 1971. Natural mutation rates in the house mouse. Estimates for five specific loci and dominant mutations. *Mutat. Res.* **11:** 89.

Seperack, P.K., M.C. Strobel, D.J. Corrow, N.A. Jenkins, and N.G. Copeland. 1988. Somatic and germline reverse mutation rates of the retrovirus-induced dilute coat-color mutation of DBA mice. *Proc. Natl. Acad. Sci.* **85:** 189.

Silvers, W.K. 1979. *The coat colors of mice.* Springer-Verlag, New York.

Sweet, H.O. 1983. Dilute-suppressor, a new suppressor gene in the house mouse. *J. Hered.* **74:** 305.

Varmus, H.E., N. Quintrell, and S. Ortiz. 1981. Retroviruses as mutagens: Insertion and excision of a nontransforming provirus alter expression of a resident transforming provirus. *Cell* **25:** 23.

Winston, F., D.T. Chaleff, B. Valent, and G.R. Fink. 1984. Mutations affecting TY-mediated expression of the HIS4 gene of *Saccharomyces cerevisiae*. *Genetics* **107:** 179.

Wirth, T., M. Schmidt, T. Baumruker, and I. Horak. 1984. Evidence for mobility of a new family of mouse middle repetitive DNA elements (LTR-IS). *Nucleic Acids Res.* **12:** 3603.

Wirth, T., K. Gloggler, T. Baumruker, M. Schmidt, and I. Horak. 1983. Family of middle repetitive DNA sequences in the mouse genome with structural features of solitary retroviral long terminal repeats. *Proc. Natl. Acad. Sci.* **80:** 3327.

Zachar, A., D. Davison, D. Garza, and P.M. Bingham. 1985. A detailed developmental and structural study of the transcription effects of insertion of the *copia* transposon into the white locus of *Drosophila melanogaster*. *Genetics* **111:** 495.

Regulation of the Maize Suppressor-Mutator Element

NINA FEDOROFF, PATRICK MASSON, AND JODY BANKS
Carnegie Institution of Washington
Department of Embryology
Baltimore, Maryland 21210

OVERVIEW

The nature of the regulatory mechanisms governing expression of the transposable maize suppressor-mutator (*Spm*) element has been deduced from genetic and molecular studies. Small, extensively deleted, transposition-defective *Spm* (*dSpm*) elements insert into and near genes in such a way that gene expression is under either the negative or positive control of a separate nondefective element, implying that the element encodes a *trans*-acting gene product that interacts with element sequences. The properties of the *a-m2 Spm* insertion alleles of the maize *a* locus suggest that the element-encoded regulatory mechanism is a positive one. *Spm* elements also undergo reversible genetic inactivation, possibly by methylation of the element, and there is evidence that the element can exist in two different states of inactivity. One of the inactive states is relatively unstable and readily reversed by the element's positive regulatory function, whereas the other inactive state, termed cryptic, is very stable and can be reversed by the element's positive regulatory function only at a low frequency.

INTRODUCTION

The maize *Spm* element family comprises a group of transposable elements that interact genetically (for review, see Fedoroff 1983). The *Spm* element and the cognate enhancer (*En*) element are virtually identical, 8.3-kb transposons (Pereira et al. 1986; Masson et al. 1987). *dSpm* elements are common, and most of those analyzed are deletion derivatives of the intact element (Schiefelbein et al. 1985; Schwarz-Sommer et al. 1985; Masson et al. 1987). *dSpm* elements transpose in the presence of a fully functional *Spm* element, indicating that the element encodes a *trans*-acting transposition function.

 Spm and *dSpm* insertion mutations have been studied extensively, revealing that there are different types of interactions between a mutant locus and an *Spm* element located elsewhere in the genome (for review, see McClintock 1965; Fedoroff 1983). *dSpm* insertion mutations have been useful in studying the genetic interactions among *Spm* elements, as well as the nature of the genetic mechanisms that regulate the element's activity. In the following discussion, we outline the evidence that the *Spm* element has a positive autoregulatory mechanism and that the element can be genetically inactivated by a negative mechanism. We also discuss the interactions between the positive and negative regulatory mechanisms.

RESULTS

The Genetic Properties of *dSpm* Insertion Mutations

There are two different types of *dSpm* insertion mutations, which we have designated *Spm*-suppressible and *Spm*-dependent (Masson et al. 1987). In both types of mutant alleles, a *dSpm* element inserted in or near a gene mediates a change in gene expression when a fully functional *Spm* element is present. In an *Spm*-suppressible allele, the gene with the *dSpm* insertion mutation continues to be expressed. Recent studies on one such mutation at the maize *bronze* locus reveal that the inserted element is transcribed, but that the corresponding sequence is almost completely spliced from the transcript (Kim et al. 1987). When a fully functional *Spm* element is present in the same genome with an *Spm*-suppressible allele, however, the affected locus is not expressed. In an *Spm*-dependent insertion mutation, such as those discussed in detail below, the mutant gene is expressed only in the presence of a fully functional *Spm* element. Thus, the properties of both *Spm*-suppressible and *Spm*-dependent insertion mutations indicate that an element-encoded gene product can interact with the *dSpm* element at the affected locus.

The *a-m2* Alleles of the *a* Locus

The *a-m2* alleles comprise a group of derivatives isolated from a mutant allele in which a fully functional *Spm* element inserted 0.1 kb upstream of the *a* gene's transcription start site (Masson et al. 1987; Schwarz-Sommer et al. 1987). The gene and element are transcribed divergently, and the gene is expressed at a reduced level relative to the wild type (Fig. 1a). McClintock selected a number of derivatives of the original allele with heritable changes affecting either the element or expression of the locus, and several of these have been subjected to molecular analysis (Masson et al. 1987). Some of the derivatives originated by internal deletions within the element (Fig. 1b). Expression of the *a* gene and excision of the element in such derivatives are both dependent on the presence of a fully functional *Spm* element elsewhere in the genome (Fig. 1b). Thus, a gene product encoded by the element interacts with element sequences not only to promote excision of the elements but also to activate expression of the adjacent gene.

Spm-dependent *a* gene expression is observed with all of the *dSpm* deriva-

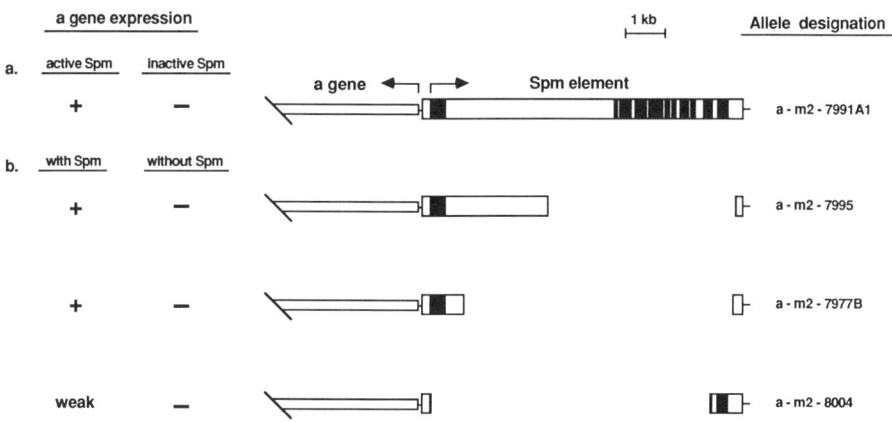

Figure 1
Structure of the *a* locus in several *a-m2* alleles. The complete 8.3-kb *Spm* element is inserted 0.1 kb upstream of the *a* gene. As indicated by the arrows, the gene and the element are transcribed divergently. The *Spm* element's intron/exon structure is assumed to be identical to that of the similar *En* element (Pereira et al. 1986) and is represented by the unshaded (intron) and shaded blocks (exons) within the element. The structure of the element in several derivatives of the original *Spm* insertion shown in part a is represented in part b of the figure.

tives depicted in Figure 1b. Hence, the element sequences that mediate *a* gene expression are near element ends. Since the insertion probably disrupts the gene's own regulatory region and because the gene is expressed in derivatives in which the length of the inserted element varies from 1.1 kb to 8.3 kb, it appears likely that sequences at the element's left end are serving as promoter or enhancer sequences (or both) for the *a* gene (Masson et al. 1987). This inference is supported by the observation that the *a-m2-8004* allele, in which the intraelement deletion removes most of the first exon and part of the first intron present in the *a-m2-7977B* allele (Fig. 1b), shows a lower level of *a* gene expression than does the latter. We have suggested that the ability of an *Spm*-encoded gene product to activate *a* gene expression is dependent on the operation of an element-encoded positive regulator that normally functions to activate expression of the element (Masson et al. 1987).

An Inactive *Spm* Element

We have isolated a derivative of the *a-m2-7991A1* allele (Fig. 1a) in which the inserted element is inactive but not defective. The inactive element does not promote its own excision, nor does it support expression of the adjacent *a* gene. The inactive element is genetically distinguishable from a *dSpm* element because it can be reactivated. The percentage of progeny kernels showing spontaneous reactivation of the element, as well as renewed *a* gene expression, varies among plants from 0% to more than 70% (Table 1). The percentage of progeny kernels showing reactivation of the element is higher when the inactive element is present in the male parent than when it is present in the female parent (Table 1). Regardless of the spontaneous frequency of activation, the inactive element can be reactivated in all progeny kernels receiving a weakly active *Spm* element, where the activity can be distinguished from that of the standard *Spm* element of the *a-m2-7991 A1* allele (Table 1). In kernels receiving both the inactive element and the weakly active element, the inactive element becomes fully active. Hence, the weakly active element can *trans*-activate expression of the inactive element in much the same way that an *Spm* element can *trans*-activate expression of the *a* gene of the *a-m2* alleles (Fig. 1b).

Cryptic *Spm* Elements

Spm elements exist in the maize genome in a stably inactive or "cryptic" form. Such elements become spontaneously active at a very low frequency. However, the frequency of activation can be enhanced in several ways. These include the introduction of a broken chromosome (McClintock 1945, 1946,

Table 1
Spontaneous and *Spm*-promoted Reactivation of an Inactive *Spm* Element

Plant number	Parent carrying inactive *Spm*	Active *Spm* elements[a]	Active *Spm* in presence of weak *Spm* element[b]
1	♀	0	—[c]
2	♂	51.7	100
3	♀	0	—
4	♀	0	—
	♂	0	—
5	♀	0.4	—
	♂	73.5	—
6	♀	0	—
	♂	30.1	100
7	♀	0.6	—
8	♀	8.2	—
	♂	—	100
9	♀	19.4	—
	♂	42.9	100

[a] Expressed as the percentage of kernels receiving the inactive *Spm* element that exhibit activity of the element, as judged from expression of the *a* gene of the *a-m2-7991 A1* allele and somatic excision of the *Spm* element.
[b] Expressed as the percentage of kernels receiving the inactive *Spm* element that exhibit full *Spm* activity.
[c] Not determined.

1947), X-ray or UV-irradiation (Neuffer 1966; Bianchi et al. 1969), or the introduction of another *Spm* element (Fedoroff 1986). The frequency at which a cryptic element is converted to a heritably active form is approximately 0.4% in the presence of a weakly active *Spm* element (Fedoroff 1986). In similar material, the spontaneous frequency of activation was estimated to be about 1 in 100,000 progeny kernels. Thus, a weakly active *Spm* element can promote the conversion of a cryptic element to a stably active element.

Although our studies on the molecular mechanism of element inactivation have just begun, we have preliminary evidence that a cryptic element is extensively methylated, whereas an inactive element is partially methylated (J. Banks and N. Fedoroff, unpubl.). Similar observations have been made for other maize transposable elements (Chandler and Walbot 1986; Schwartz and Dennis 1986). There is a strikingly GC-rich region near the element's left end, corresponding to most of the element's first untranslated exon (Pereira et al. 1986). Figure 2 shows the distribution at the element's left end of the CG and CNG residues that are commonly methylated in plant DNA (Gruenbaum et al. 1981). That the GC-rich region is important in element expression is suggested by the earlier observation that *a* gene expression is less intense in the *a-m2-8004* allele, from which most of the GC-rich region is missing (Fig. 2), than it is in the *a-m2-7977B* allele, which contains all of the GC-rich region (Fig. 1). Moreover, we observe that the full-length *a-m2-8167B* element (not shown), which also exhibits a low level of *a* gene expression, is methylated in the GC-rich region, as judged by its insensitivity to cleavage by restriction enzymes whose ability to cleave DNA is sensitive to methylation (J. Banks and N. Fedoroff, unpubl.). These observations suggest that methylation may be involved in maintaining the cryptic and inactive states.

DISCUSSION

There is both direct and indirect evidence that the *Spm* element encodes a positive regulatory gene product. Indirect evidence is provided by the ability

Figure 2
Distribution of the CGs and CNGs at the left end of the *Spm* element. The horizontal line represents the left end of the *Spm* element. The vertical lines represent the positions of CG dinucleotides and CNG trinucleotides (Masson et al. 1987). The transcription start site identified by Pereira et al. (1986) is indicated by the arrow below the diagram. The positions of the left deletion end point in two *a-m2* alleles (*a-m2-8004* and *a-m2-7977B*) are indicated by the arrows above the diagram.

of *Spm* to *trans*-activate expression of the *a* gene in mutants with a *dSpm* element inserted just upstream of the transcription initiation site (Masson et al. 1987; Schwarz-Sommer et al. 1987). In mutants of this type, the inserted *dSpm* element functions as an *Spm*-dependent promoter or enhancer for the adjacent *a* gene, suggesting that the element-encoded positive regulatory protein interacts with element sequences to promote expression of the gene. Although element sequences at and upstream of the element's transcription initiation site at nucleotide 209 (Pereira et al. 1986) suffice to mediate some *Spm*-dependent *a* gene expression, maximal expression is observed in mutants that have at least an additional 1 kb of sequence from the element's left end, including the element's GC-rich first exon and part of the first intron. The level of *Spm*-dependent *a* gene expression decreases when this region is either deleted or methylated. These observations suggest that the GC-rich region is involved in regulation of gene expression by the element's positive regulatory gene product and that methylation of the GC-rich region can have an inhibitory effect on expression.

Direct evidence that the *Spm* element encodes a positive regulatory function comes from the observation that an active element can activate an inactive element (McClintock 1957, 1958, 1959; Table 1). We have made a distinction between two types of inactive elements, one of which is unstably inactive and the other of which is stably inactive (cryptic). We have described an inactive derivative of an *Spm* element that spontaneously returns to an active state at a relatively high frequency. Such an inactive element is fully active in the presence of a weakly active *Spm* element, implying that it is activated in *trans* by an element-encoded gene product (Table 1). By contrast, a cryptic element is not invariably activated in the presence of a weakly active *Spm*. Nonetheless, a cryptic element becomes active in a small fraction of progeny containing a weakly active *Spm* element. Of these, about half remain heritably active, indicating that the interaction between the cryptic element and a gene product encoded by the active element promotes the conversion of the cryptic element to an active element (Fedoroff 1986).

We conclude that the *Spm* element encodes a positive autoregulatory gene product, in addition to its *trans*-acting transposition function. The properties of *Spm*-dependent alleles suggest that the positive autoregulatory gene product interacts with sequences in the vicinity of the element's transcription start site to promote element expression, probably at the transcriptional level. We conclude from the analysis of inactive and cryptic *Spm* elements that there is a negative control mechanism that prevents expression of the element and that the element's positive autoregulatory gene product can directly activate an inactive element but not a cryptic element. The element's regulatory gene product also appears capable of promoting the conversion of a cryptic element to a heritably active element.

ACKNOWLEDGMENTS

This work was supported by National Institutes of Health (NIH) grant 5-RO1-GM34296-02 and fellowships from the NIH (J.B.), Fonds National de la Recherche Scientifique of Belgium, and Pioneer Hi-Bred International (P.M.).

REFERENCES

Bianchi, A., F. Salamini, and F. Restaino. 1969. Concomitant occurrence of different controlling elements. *Maize Genet. Coop. Newsl.* **43:** 91.

Chandler, V.L. and V. Walbot. 1986. DNA modification of a maize transposable element correlates with loss of activity. *Proc. Natl. Acad. Sci.* **83:** 1767.

Fedoroff, N. 1983. Controlling elements in maize. In *Mobile genetic elements* (ed. J. Shapiro), p. 1. Academic Press, New York.

―――. 1986. Activation of *Spm* and modifier elements. *Maize Genet. Coop. Newsl.* **60:** 18.

Gruenbaum, Y., T. Naveh-Many, H. Cedar, and A. Razin. 1981. Sequence specificity of methylation in higher plant DNA. *Nature* **292:** 860.

Kim, H.-Y., J.W. Schiefelbein, V. Raboy, D.B. Furtek, and O.E. Nelson. 1987. RNA splicing permits expression of a maize gene with a *defective suppressor-mutator* transposable element insertion in an exon. *Proc. Natl. Acad. Sci.* **84:** 5863.

Masson, P., R. Surosky, J.A. Kingsbury, and N.V. Fedoroff. 1987. Genetic and molecular analysis of the *Spm-dependent a-m2* alleles of the maize *a* locus. *Genetics* **117:** 117.

McClintock, B. 1945. Cytogenetic studies of maize and Neurospora: Induction of mutations in the short arm of chromosome 9 in maize. *Carnegie Inst. Wash. Year Book* **44:** 108.

―――. 1946. Maize genetics. *Carnegie Inst. Wash. Year Book* **45:** 176.

―――. 1947. Cytogenetic studies of maize and Neurospora: The mutable *Ds* locus in maize. *Carnegie Inst. Wash. Year Book* **46:** 146.

―――. 1957. Genetic and cytological studies of maize. *Carnegie Inst. Wash. Year Book* **56:** 393.

―――. 1958. The suppressor-mutator system of control of gene action in maize. *Carnegie Inst. Wash. Year Book* **57:** 415.

―――. 1959. Genetic and cytological studies of maize: Further studies of the Spm system. *Carnegie Inst. Wash. Year Book* **58:** 452.

―――. 1965. The control of gene action in maize. *Brookhaven Symp. Biol.* **18:** 162.

Neuffer, M.G. 1966. Stability of the suppressor element in two mutator systems at the *A1* locus in maize. *Genetics* **52:** 521.

Pereira, A., H. Cuypers, A. Gierl, Z. Schwarz-Sommer, and H. Saedler. 1986. Molecular analysis of the *En/Spm* transposable element system of *Zea mays*. *EMBO J.* **5:** 835.

Schiefelbein, J.W., V. Raboy, N.V. Fedoroff, and O.E. Nelson, Jr. 1985. Deletions within a defective suppressor-mutator element in maize affect the frequency and developmental timing of its excision from the *bronze* locus. *Proc. Natl. Acad. Sci.* **82:** 4783.

Schwartz, D. and E. Dennis. 1986. Transposase activity of the *Ac* controlling element in maize is regulated by its degree of methylation. *Mol. Gen. Genet.* **205:** 476.

Schwarz-Sommer, Z., A. Gierl, R. Berndtgen, and H. Saedler. 1985. Sequence comparison of "states" of *a1-m1* suggests a model of *Spm* (*En*) action. *EMBO J.* **4:** 2439.

Schwarz-Sommer, Z., N. Shepherd, E. Tacke, A. Gierl, W. Rohde, L. Leclercq, M. Mattes, R. Berndtgen, P.A. Peterson, and H. Saedler. 1987. Influence of transposable elements on the structure and function of the *A1* gene of *Zea mays*. *EMBO J.* **6:** 287.

The Functional Potential of the Human LINE-1 Family of Interspersed Repeats

MAXINE F. SINGER, JACEK SKOWRONSKI,* THOMAS G. FANNING, AND
SKORN MONGKOLSUK
Laboratory of Biochemistry
National Cancer Institute
Bethesda, Maryland 20892

OVERVIEW

LINE-1s (L1s) are likely to be mammalian movable elements that transpose through a mechanism involving an RNA intermediate and an element-encoded reverse transcriptase. Such a mechanism implies the occurrence of L1 transcription and translation. Earlier, we identified unit length L1 transcripts in the cytoplasmic, poly(A)$^+$ RNA of a human teratocarcinoma cell line, NTera2D1. The RNAs represent only one L1 strand, the one with long open reading frames (ORFs), and initiate (5' end) with the first residue of the genomic L1 family members. They have now been characterized further by sequence analysis of cDNAs. Although each of 19 cDNAs is slightly different in sequence and, thus, is probably transcribed from a different family member, their overall structures are similar and consistent with the RNAs being intermediates in reverse-transcription-mediated transposition. In the cDNA consensus sequence, an approximately 5-kb reading frame is broken in two (to form a 5' ORF 1 and a 3' ORF 2) by a short region containing two in-frame stop codons. ORF 2 (3852 bp) includes regions that could encode a polypeptide with homology to both reverse transcriptase and a nucleic-acid-binding domain. RNA transcribed from at least one of the cDNAs can be translated in vitro, suggesting that the cDNA represents a functional mRNA.

INTRODUCTION

All mammalian genomes that have been analyzed contain a large family of interspersed, repeated sequences now termed L1 (e.g., L1Hs and L1Md for the *Homo sapiens* and *Mus domesticus* families, respectively) (Burton et al. 1986). Typically, the longest family members are from 6 to 7 kbp, depending on species, and appear to represent the unit-length elements. Other family members are truncated.

---

*Present address: Cold Spring Harbor Laboratory, Cold Spring Harbor, New York 11724

Banbury Report 30: Eukaryotic Transposable Elements as Mutagenic Agents
© Cold Spring Harbor Laboratory. 0-87969-230-8/88. $1.00 + .00

It is likely that both unit-length and truncated members were inserted in dispersed genomic locations by transposition because (1) many are surrounded by target-site duplications (of varying lengths) and (2) alleles of several mammalian genes differ by the presence or absence of an L1 (for review, see Skowronski and Singer 1986). Recently, several elements known to be transposable in *Drosophila melanogaster* were shown to have structures that are very similar to L1 units, thereby lending support to the notion that L1s are movable elements (Fawcett et al. 1986; DiNocera and Casari 1987; P.P. DiNocera, pers. comm.). Unlike other known transposable elements, none of the members of this phylogenetically widespread group have terminal repeats, either direct or inverted. However, they all contain at least two ORFs, of which one could, conceptually, encode a reverse transcriptase-like polypeptide (Fawcett et al. 1986; Hattori et al. 1986; Loeb et al. 1986; DiNocera and Casari 1987; Fanning and Singer 1987). If, as has been suggested, transposition of these elements depends on the encoded reverse transcriptase, then at least one of the elements in each genome should be competent for transcription and translation.

We previously detected a discrete, 6.5-kb, cytoplasmic poly(A)$^+$ L1 RNA in NTera2D1 cells (Skowronski and Singer 1985). The RNA anneals with probes covering virtually an entire genomic L1 unit and contains only the strand with the long ORFs. Primer extension experiments demonstrated that the 5' most residue in the RNA corresponds to the first (5') base in the human genomic unit-length L1s (J. Skowronski et al., in prep.). We have now sequenced one 5975-bp-long cloned cDNA (cD11) derived from the NTera2D1 transcripts and have obtained partial sequence data for 18 other cDNAs. The data indicate that a subset of genomic L1s are transcribed in these cells and that at least one of them produces translatable RNA.

RESULTS

The overall structure of cD11 is summarized in Figure 1. Partial sequence data on 18 other cDNAs indicate that they vary from the cD11 structure in an occasional base (less than 5% divergence on the average). This indicates that many different unit-length genomic L1s are transcribed in the NTera2D1 cells.

cD11 has the overall structure of a genomic unit-length L1 with the exception of 32 bp at the 5' end and about 50 bp at the 3' end that are missing, presumably because of incomplete copying during cDNA synthesis. Thus, if the RNA is transcribed from a typical genomic unit, it is not spliced. About 50% of human L1 units are 132 bp longer than the others because of an extra segment inserted at about residue 760 (Hattori et al. 1985). cD11 lacks this 132 bp, as does a random sample of the other cDNAs, indicating that the

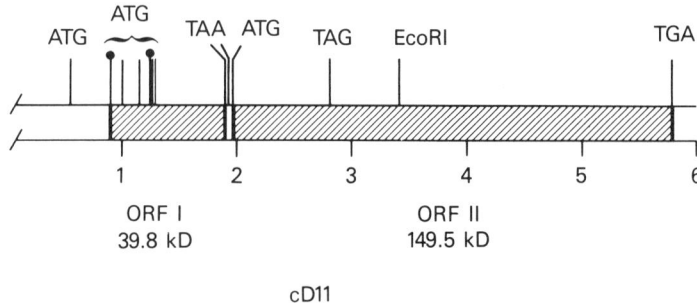

Figure 1
Landmarks on the L1 cD11. The slashed regions indicate the long ORFs. All stop codons within the reading-frame region are indicated. ATGs that are positioned near the start of the ORFs are shown; those with ● are in a context associated with efficient translation initiation. In addition, the ATG in the 5′ leader is indicated; it is not in frame with the others (see text). One conserved EcoRI site is marked. Below the diagram is given the size of the polypeptides that could be encoded by the 2 ORFs (assuming ORF 2 was not aborted by the TAG stop codon).

NTera2D1 RNAs were transcribed from family members lacking the extra base pairs.

Following an 800-residue-long 5′ segment with no significant ORF, cD11 has an 1122-residue-long ORF (ORF 1). The 37th codon is an ATG that is within the sequence ANNATGG and could thus initiate translation efficiently (Kozak 1986a). Considering all three possible frames, only one ATG codon occurs in the 800 residues preceding ORF 1, and it is not in a favored context for initiation. Moreover, the frame it initiates ends with a stop codon after 20 codons. To test whether the ORF 1 in cD11 is translatable, the segment of cD11 from the 5′ end to the EcoRI site (Fig. 1) was inserted in an appropriate vector, and capped RNA was synthesized with T3 RNA polymerase. When the RNA is added to a reticulocyte lysate, a polypeptide of about 42 kD is synthesized. If segments of ORF 1 are deleted from the constructions, the resulting RNA yields an appropriately shortened polypeptide, indicating that the observed polypeptide was the expected product. These experiments suggest that cD11 represents a functional mRNA for synthesis of the ORF 1 polypeptide. The predicted protein is markedly hydrophilic (36% R, K, H, E, plus D) and has no significant homology to proteins recorded in the March 1987 version of GenBank. An 80-amino-acid-long region near the carboxyl terminus is about 50% homologous to the protein predicted by the analogous region of mouse genomic L1, L1Md (Loeb et al. 1986; Mottez et al. 1986).

After the TAA stop codon that ends ORF 1, there are 33 residues followed by another TAA and then, still in-frame, ORF 2 begins. This arrangement recurs in the other cDNAs that were analyzed and in the consensus sequence of human genomic L1s (Y. Sakaki and A.F. Scott, pers. comm.). Inspection of the translation products produced by the capped RNA described in the preceding paragraph shows that ORF 2 products are undetectable. Thus, neither a polypeptide corresponding to reinitiation of translation after the two TAA stop codons nor a fusion protein representing readthrough of the pair of stops was seen. It is possible that with the greater sensitivity provided by antibodies to the predicted proteins such products will be detectable. At present, we can only conclude that if ORF 2 is translated, it is at a much reduced efficiency compared to ORF 1. When ORF 1 is deleted from the constructions used for RNA synthesis, a polypeptide corresponding in size to that predicted for ORF 2 appeared, suggesting that ORF 2 of cD11 is inherently translatable.

The first ATG codon in ORF 2 occurs after 10 codons. Its context is not optimal (ATAATGA). From this ATG, there are 1275 codons to the first stop codon (TGA) that is present in all the cDNAs and thus appears to end ORF 2. One TAG stop codon interrupts the cD11 ORF 2 after 287 codons. Thus, ORF 2 of cD11 is not competent to produce the 1275-amino-acid-long polypeptide. Our partial sequence data on the other cDNAs indicate that at least 11 of them also have at least one stop codon or frameshift in ORF 2. These are scattered in random positions. The codon corresponding to the TAG in cD11 ORF 2, for example, is a TGG in several other cDNAs and is thus not a consensus sequence. We do not presently know whether any of the cytoplasmic poly(A)$^+$ RNAs in the NTera2D1 cells can be translated to give the ORF 2 product. This is an interesting question because the protein predicted by ORF 2 has several provocative features.

Overall, there is about 60% similarity between the proteins predicted by the ORF 2s of genomic L1Mds and cD11 (no murine cytoplasmic poly(A)$^+$ RNAs have been analyzed). However, the extent of similarity varies markedly from one region to another within ORF 2. In some regions, similarity is as much as 90%, counting only identical amino acids. Moreover, this high degree of regional conservation holds even when L1Md, L1Hs (a consensus sequence very similar to cD11; Y. Sakaki et al., pers. comm.), and L1 sequences from rabbits (L1Oc; Demers et al. 1986) and cats (L1Fc; Fanning and Singer 1987) are compared (Fanning and Singer 1987). Several of these regions (labeled A–G in Fig. 2) together include the peptide segments that were previously identified as similar to regions conserved in known or suspected reverse transcriptases in retroviruses and retrotransposons (Hattori et al. 1986; Loeb et al. 1986; Fanning and Singer 1987). Two other well-conserved regions (labeled H and I in Fig. 2) are markedly similar to regions in transferrin (Hattori et al. 1986) and nucleic-acid-binding proteins (Fanning and Singer 1987), respectively.

cD11 has 192 residues following the consensus TGA stop codon at the end of ORF 2. Most of the other cDNAs extend somewhat further and have 204

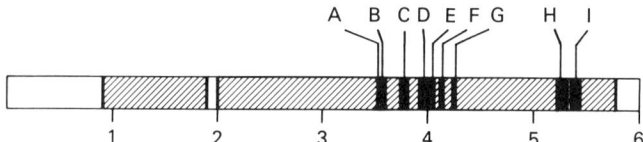

Figure 2
Regions in human L1s that predict polypeptides with homology to retrovirus reverse transcriptases (A–G), transferrin (H), and nucleic acid-binding proteins (I). The *Eco*RI site (Fig. 1) is at about residue 3400.

bp in this 3'-trailer region, followed by a variable length, A-rich region that typically includes at least one polyadenylation signal, AATAAA. The consensus sequence for the 204-bp segment is different from that compiled from random human genomic L1s, suggesting that a subset of genomic L1s, termed subset T, is transcribed to give the poly(A)$^+$ cytoplasmic NTera2D1 RNA. Seven of the 19 cDNAs extend from 20 to 50 residues into the A-rich region. Each of them has a somewhat different sequence in this region, and even the position of the AATAAA is variable; several of the cDNAs have more than one AATAAA. In two of the cDNAs, a tandem repetition of the structure T(A)$_n$ occurs as it does in various genomic L1s.

DISCUSSION

L1s, together with the related *Drosophila* elements I (Fawcett et al. 1986), F (DiNocera and Casari 1987), and G (DiNocera et al. 1986), the ingi element of *Trypanosoma brucei* (Kimmel et al. 1987) and R2 of *Bombyx mori* (Burke et al. 1987), appear to define a new class of movable elements (for review, see Finnegan and Fawcett 1986). The I and F *Drosophila* elements have been demonstrated to be transposable experimentally. The evidence that the mammalian L1s are movable depends on indirect evidence from alleles that differ by the presence and absence of L1s (for review, see Skowronski and Singer 1986). The common features of this class of elements include the absence of terminal repeats, the formation of variable-size target-site duplications, and the presence of at least one ORF that can encode a protein with homology to known reverse transcriptases. Reverse transcription, mediated by an element-encoded reverse transcriptase, is widely considered to be a step in transposition. We suggest calling this family of elements class II retrotransposons, to distinguish them from the related, but clearly different, retrotransposons such as Ty, copia, and IAP, all of which appear to transpose by means of an element-encoded reverse transcriptase but have long terminal repeats (LTRs) (Baltimore 1984).

One intermediate in the proposed transposition mechanism is a full-length transcript of the element. The data summarized here demonstrate that such transcripts occur in NTera2D1 human teratocarcinoma cells. Transcripts recently detected in mouse cells may also be full length (Dudley 1987). These transcripts could serve as templates for copying by reverse transcription. However, the data obtained thus far leave open the question of whether any of the RNAs can be translated to give active reverse transcriptase.

Several aspects of the conservation of L1s are notable. First is the conservation of structure and high copy number among mammals. Second, within a species, the maintenance of long ORF regions suggests that one or more family members may be competent, in principle, to produce polypeptides. In mice, ORF 1 and ORF 2 are open even in randomly selected genomic family members (Loeb et al. 1986). In humans, such translatable elements appear to be rarer, and even most of the cytoplasmic poly(A)$^+$ RNAs represented in our cDNA clones appear to contain single-base-pair changes that close ORF 1 or ORF 2, or both, prematurely. As pointed out, cD11 could, in principle, be an mRNA for ORF 1 but not for ORF 2. We do not know whether ORF 1 or ORF 2 is open in any of the other cDNAs. Overall, however, the conservation of the reading frames throughout mammals is notable and raises the possibility that selective pressure(s) has operated to maintain at least one functional L1 in each mammalian genome. If so, L1s, and perhaps certain other class II retrotransposons, are not disposable transposables, but rather encode essential genetic elements. Moreover, the possibility that reverse transcriptase is ubiquitous in mammals considerably expands the likely impact of the enzyme in shaping mammalian genomes (Baltimore 1984).

L1 elements are very abundant in mammalian genomes constituting on the order of 3% or more of the DNA. However, in humans most of them are incapable of independent expression for one of several reasons. First, most are truncated; only about 4×10^3 (Adams et al. 1980; Grimaldi et al. 1984) of the almost 10^5 human elements (Hwu et al. 1986) are full length. Second, of the full-length elements, only a subset are transcribed and processed to form poly(A)$^+$ cytoplasmic RNA. Third, most of the full-length cytoplasmic poly(A)$^+$ RNAs in NTera2D1 cells probably cannot be functional mRNAs because of stop codons in the ORFs, although one or more (e.g., cD11) could serve as an mRNA for ORF 1. If ORF 2 is translatable in one or more of these RNAs, its translation is likely to be at a very low level. This is because of the need either to suppress the translational stop codons that separate ORFs 1 and 2 (Jacks and Varmus 1985; Yoshinaka et al. 1985) or to reinitiate translation at one of the downstream AUGs (Kozak 1986b). Thus, although L1s succeed in reaching high copy numbers, at most only a very few are likely to be transcriptionally or translationally active.

ACKNOWLEDGMENTS

We thank Ron Thayer for help and Gail Gray for preparing the manuscript. P.P. DiNocera, A.F. Scott, and Y. Sakaki were generous with unpublished information.

REFERENCES

Adams, J.W., R.E. Kaufman, P.J. Kretschmer, M. Harrison, and A.W. Nienhuis. 1980. A family of long reiterated DNA sequences, one copy of which is next to the human beta globin gene. *Nucleic Acids Res.* **8:** 6113.

Baltimore, D. 1984. Retroviruses and retrotransposons: The role of reverse transcription in shaping the eukaryotic genome. *Cell* **40:** 481.

Burke, W.D., C.C. Calalang, and T.H. Eickbush. 1987. The site-specific ribosomal insertion element type II of *Bombyx mori* (R2Bm) contains the coding sequence for a reverse transcriptase-like enzyme. *Mol. Cell. Biol.* **7:** 2221.

Burton, F.H., D.D. Loeb, C.F. Voliva, S.L. Martin, M.H. Edgell, and C.A. Hutchison III. 1986. Conservation throughout Mammalia and extensive protein-encoding capacity of the highly repeated DNA long interspersed sequence one. *J. Mol. Biol.* **187:** 291.

Demers, G.W., K. Brech, and R.C. Hardison. 1986. Long interspersed L1 repeats in rabbit DNA are homologous to L1 repeats of rodents and primates in an open-reading-frame region. *Mol. Biol. Evol.* **3:** 179.

DiNocera, P.P. and G. Casari. 1987. Related polypeptides are encoded by *Drosophila* F elements, I factors and mammalian L1 sequences. *Proc. Natl. Acad. Sci.* **84:** 5843.

DiNocera, P.P., F. Graziani, and G. Lavorgna. 1986. Genomic and structural organization of *Drosophila melanogaster* G elements. *Nucleic Acids Res.* **14:** 675.

Dudley, J.P. 1987. Discrete high molecular weight RNA transcribed from the long interspersed repetitive element L1Md. *Nucleic Acids Res.* **15:** 2581.

Fanning, T. and M.F. Singer. 1987. The LINE-1 DNA sequences in four mammalian orders predict proteins that conserve homologies to retroviral proteins. *Nucleic Acids Res.* **15:** 2251.

Fawcett, D.H., C.K. Lister, E. Kellett, and D.J. Finnegan. 1986. Transposable elements controlling I-R hybrid dysgenesis in *D. melanogaster* are similar to mammalian LINES. *Cell* **47:** 1007.

Finnegan, D.J. and D.H. Fawcett. 1986. Transposable elements in *Drosophila melanogaster*. *Oxf. Surv. Euk. Genes* **3:** 1.

Grimaldi, G., J. Skowronski, and M.F. Singer. 1984. Defining the beginning and end of KpnI family segments. *EMBO J.* **3:** 1753.

Hattori, M., S. Hidaka, and Y. Sakaki. 1985. Sequence analysis of a *KpnI* family member near the 3'-end of the human β-globin gene. *Nucleic Acids Res.* **13:** 1753.

Hattori, M., S. Kuhara, O. Takenaka, and Y. Sakaki. 1986. L1 family of repetitive DNA sequences in primates may be derived from a sequence encoding a reverse transcriptase-related protein. *Nature* **321:** 625.

Hwu, H.R., J.W. Roberts, E.H. Davidson, and R.J. Britten. 1986. Insertion and/or deletion of many repeated DNA sequences in human and higher ape evolution. *Proc. Natl. Acad. Sci.* **83:** 3875.

Jacks, T. and H.E. Varmus. 1985. Expression of the Rous sarcoma virus *pol* gene by ribosomal frameshifting. *Science* **230:** 1237.

Kimmel, B., O.K. Ole-Moiyoi, and J.R. Young. 1987. Ingi, a 5.2 kbp dispersed sequence element from *Trypanosoma brucei* that carries half of a smaller mobile element at either end and has homology with mammalian LINES. *Mol. Cell. Biol.* **7:** 1465.

Kozak, M. 1986a. Point mutations define a sequence flanking the AUG initiator codon that modulates translation by eukaryotic ribosomes. *Cell* **44:** 283.

———. 1986b. Bifunctional messenger RNAs in eukaryotes. *Cell* **47:** 481.

Loeb, D.D., R.W. Padgett, S.C. Hardies, W.R. Shehee, M.B. Comer, M.H. Edgell, and C.A. Hutchison III. 1986. The sequence of large L1Md element reveals a tandemly repeated 5′-end and several features found in retrotransposons. *Mol. Cell. Biol.* **6:** 168.

Mottez, E., P.K. Rogan, and L. Manuelidis. 1986. Conservation in the 5′-region of the long interspersed mouse L1 repeat: Implications of comparative sequence analysis. *Nucleic Acids Res.* **14:** 3119.

Skowronski, J. and M.F. Singer. 1985. Expression of a cytoplasmic LINE-1 transcript is regulated in a human teratocarcinoma cell line. *Proc. Natl. Acad. Sci.* **82:** 6050.

———. 1986. The abundant LINE-1 family of repeated DNA sequences in mammals: Genes and pseudogenes. *Cold Spring Harbor Symp. Quant. Biol.* **51:** 457.

Yoshinaka, Y., I. Katoh, T.D. Copeland, and S. Oroszlan. 1985. Murine leukemia virus protease is encoded by the *gag-pol* gene and is synthesized through suppression of an amber termination codon. *Proc. Natl. Acad. Sci.* **82:** 1618.

Factors Affecting Retrotransposition of Intracisternal A-particle Proviral Elements

EDWARD L. KUFF
Laboratory of Biochemistry
National Cancer Institute
National Institutes of Health
Bethesda, Maryland 20892

OVERVIEW

Although intracisternal A-particle (IAP) proviral elements are abundant in the mouse genome, a number of factors restrict their potential for retrotransposition. The most obvious restrictions are imposed by genetic defects in the elements themselves and by the inhibitory effects of genomic methylation. Sequestration of elements in genetically inert heterochromatin may also be a major limiting factor. Even when transcription of competent provirus(es) has been initiated, the location and other properties intrinsic to the particles themselves may hamper reverse transcription and/or integration of provirus. Most of the IAP elements found to be transposed have contained major deletions involving both the *gag* and *pol* coding regions; retrotransposition of such elements must depend upon coexpression of competent IAP genes. A number of factors that limit retrotransposition are common to both IAPs and the Ty elements of yeast and may be general mechanisms for maintaining equilibrium between reiterated endogenous transposable elements and their host genome.

INTRODUCTION

Mus musculus contains approximately 1000 IAP-related proviral elements per haploid genome (Lueders and Kuff 1977). IAP elements are actively transcribed in many mouse tumor cells (Kuff and Lueders 1988), and particles may accumulate to levels of several thousand per cell in some tumor types. IAPs are also found in oocytes and preimplantation mouse embryos (Yotsuyanagi and Szollosi 1984). Transcripts and occasional particles are seen in many normal tissues, particularly the thymus (Kuff and Fewell 1985). The number and chromosomal location of the active IAP elements have not been defined in any cell type.

Both somatic and germ line transpositions of IAP elements have been reported (Table 1). Most of these insertions were found because they affected the function of cellular genes at the target sites. Several were detected as

Table 1
Transpositions of IAP Elements

Cellular gene at target site	Size (kb) and type[a] of inserted element	Effect	Cell type	Reference
κ-light chain	7.0; I	Abnormal splice; abolish protein	Hybridoma Sp6	Hawley et al. (1984); Kuff et al. (1983)
κ-light chain	4.9; IΔ1	Reduce mRNA; reduce protein	Hybridoma Sp6	Hawley et al. (1984); Kuff et al. (1983)
c-*mos*	4.2; IΔ3	Activate gene	Myeloma XRPC-24	Canaani et al. (1983)
c-*mos*	4.2; IΔ3	Slight activation or none	Myeloma NSI	Cohen et al. (1983); Gattoni-Celli et al. (1983)
c-*myc*	Unknown	None	Myeloma J558	Greenberg et al. (1985)
About 60 sites; unknown	3.8; IIB	Unknown	Myeloma MOPC-315	Shen-Ong and Cole (1984)
IL-3	4.8; IΔ1	Activate gene	Leukemia WEH1-3B	Ymer et al. (1985, 1986)
Renin-2	3.0; IΔ4	Unknown	DBA/2 germ line	Burt et al. (1984)
Pseudo-αglobin	3.4; IΔ3	None	BALB/c germ line	Lueders et al. (1982)

[a]See Fig. 1.

silent rearrangements. Shen-Ong and Cole (1984) found multiple new insertions of a particular IAP element in the genome of the MOPC-315 myeloma. This observation is the strongest present evidence that IAP transpositions involve the generation of new proviral copies, presumably through reverse transcription, rather than rearrangements of preexisting genomic elements. Free proviral copies could not be detected in MOPC-315 (Shen-Ong and Cole 1984) or in an IAP-rich neuroblastoma cell line (A. Feenstra and E. Kuff, unpubl.); however, Grigoryan et al. (1985) reported cloning of a closed circular IAP provirus from the Erhlich ascites carcinoma. In the following, we will assume as the most likely eventuality that IAP elements transpose through the conventional retroviral mechanisms, although this remains to be formally demonstrated.

Since IAP proviral elements are so numerous in the mouse, they might, if viral functions were intact and unregulated, constitute a major source of genomic instability at both the germ line and somatic levels. A number of factors, however, tend to minimize retrotransposition of IAP elements. These factors, considered below, may be generally representative of mechanisms by which endogenous retrotransposons are regulated in eukaryotic systems.

RESULTS

Table 2 lists some established properties of IAPs and their genetic elements that could restrict the likelihood of successful retrotransposition, and they will be discussed briefly in turn.

Structural Properties of IAP Elements

High Proportion of Deleted Elements

The genetic organization of a full-sized 7-kb genomic IAP element is shown in Figure 1 (Mietz et al. 1987) and beneath it the regions found to be deleted in

Table 2
Factors Tending to Minimize Retrotransposition of IAP Genetic Elements

Structural properties of the genetic elements
    High proportion of deleted forms
    Single-base mutations in open reading frames and primer binding site
Chromosomal effects
    Localization of elements in constitutive heterochromatin
    DNA methylation
Properties of the IAPs themselves
    Unprocessed *gag* and *gag-pol* proteins
    Intracisternal location
Cellular suppressing mechanisms

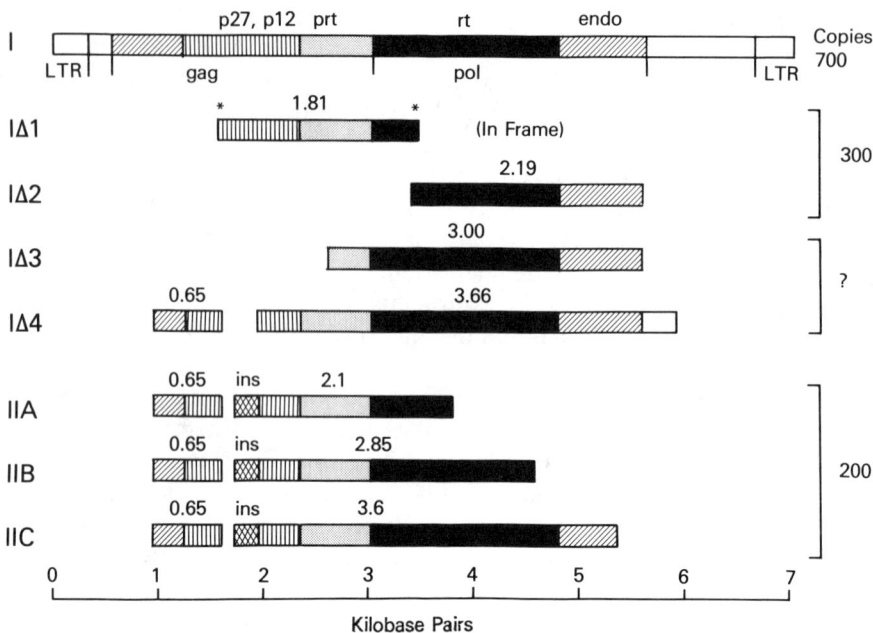

Figure 1

Deletions found in IAP elements. Genetic organization of a full-sized IAP genomic proviral element (top) and regions found to be deleted in a variety of cloned genomic and cDNA forms. The intact IAP provirus is classified as type I (left-hand column) and the varieties of deleted forms as type IΔ 1–4 depending on position and size of major deletions. Type II A–C elements all contain a characteristic small insertion and major deletions of increasing size. The functional defectiveness of the various deleted forms is apparent.

various cloned genomic elements and cDNAs. On the right are very rough estimates of the copy numbers of some variant groups. Type II IAP elements were defined by Shen-Ong and Cole (1982) on the basis of a characteristic small insertion, denoted as AIIins by Lueders and Mietz (1986) who studied these elements in detail. Type II elements all have deletions of various sizes and are defective in *gag* and reverse transcriptase. The type I category of IAP element (lacking AIIins) also includes a large number of deleted forms. Perhaps 40% of all IAP elements are defective in functions required for retrotransposition by virtue of major deletions (Fig. 1).

Single-base Mutations

The complete genomic element pictured in Figure 1, designated MIA14, contains an open *gag* reading frame but has two single-base mutations that interrupt the *pol* reading frame (Mietz et al. 1987). A Syrian hamster full-sized IAP-related genomic element also contains multiple stop codons in both major gene regions (Ono et al. 1985). The proportion of grossly intact elements that are actually pseudogenes is not known. In the absence of any obvious selective pressure for maintaining IAP elements in functional form, inactivation by random small mutations may be widespread.

Mutations in the primer binding site can cause drastic reductions in affinity for the first strand primer, tRNA^{Phe}; e.g., in the cases of IAP-interleukin 3 (IL-3) and $\alpha 25$ (IAP-c-*myc*, Table 1), the calculated binding energies were reduced by 60–65%. Apparently, this did not prevent reverse transcription, since both of these elements were found to have transposed. Multiple IAP RNA species are present in most tumor tissues, and it is possible that IAP-IL-3 and IAP-c-*myc* reverse transcripts were initiated on copackaged RNAs with competent primer binding sites.

Chromosomal Effects

Localization in Genetically Inactive Regions (Constitutive Heterochromatin)

Syrian hamsters contain about 900 genomic elements homologous to the mouse IAP genes. A cloned hamster element was used to study the chromosomal distribution of IAP-related sequences by in situ hybridization (Kuff et al. 1986). More than 50% of the radioactive grains were localized over well-defined regions of constitutive, late-replicating, noncentromeric heterochromatin at a concentration per unit chromosome length more than 5 times greater than the concentration of grains in euchromatic regions. Grains were distributed in the euchromatin of all chromosomes, with some striking local concentrations. In the mouse, IAP sequences were also distributed over the entire chromosome complement (Lueders and Kuff 1977; Kuff et al. 1986). Mouse chromosomes do not have prominent segments of constitutive, noncentromeric heterochromatin favorable for in situ localization. However, Lueders (1987) found that 11 of 12 cloned type II genomic IAP genes had LINE-1 repetitive sequence elements (Singer and Skowronski 1985) in their nearby flanking regions, and that a number of the clones also contained other types of repetitive sequence elements. Thus, type II components in mice may be preferentially located in clusters of repetitive sequence elements, i.e., in chromosome regions with a genetic composition resembling constitutive heterochromatin. In *Drosophila* also, many multicopy transposable elements are concentrated in heterochromatin as judged by in situ hybridization (for references, see Kuff et al. 1986). Germ line fixation of new proviral copies could be favored in chromosome regions that are comparatively lacking in protein-coding genes. Constitutive heterochromatin is considered to be relatively genetically inert, and IAP elements embedded in this milieu may not be subject to activation.

DNA Methylation

There is a strong positive correlation between IAP expression in a variety of cell types and hypomethylation of IAP genomic sequences (Lasneret et al. 1983; Hojman-Montes de Oca et al. 1984; Morgan and Huang 1984; Feenstra et al. 1986). In liver and thymus, the IAP 5' long terminal repeats (LTRs) were nearly all modified at several methylation-sensitive restriction sites, whereas in two IAP-rich tumor cell lines, these same sites were demethylated in 100 or more of the elements (Feenstra et al. 1986). This was not an IAP-specific effect, but rather reflected an extensive genome-wide demethylation in these tumors (A. Feenstra and E. Kuff, unpubl.). In vitro methylation of three clustered *Hha*I sites 5' of the RNA start site of a cloned IAP LTR abolished the promoter activity of this LTR towards a linked chloramphenicol acetyltransferase gene (Feenstra et al. 1986). Thus, demethylation of the 5' LTR is a prerequisite for IAP gene activity. Reduction in the general level of genomic methylation by treatment with 5'-azacytidine has had strong enhancing effects on IAP expression in several cell lines (Lasneret et al. 1983; Hojman-Montes de Oca et al. 1984). Whereas DNA methylation is probably a major mechanism for restriction of IAP expression, there is no present evidence for specific control of particular genomic IAP elements by this means. It is more likely that the IAP elements respond fortuitously to regional or general changes in methylation state programmed to meet other cellular needs.

Properties of IAPs Themselves

Failure to Process the Major Protein Components

gag Gene Product. The major 73-kD IAP structural protein, P73, is equivalent to the unprocessed *gag* precursor polypeptide of conventional retroviruses. The inner particle shell, consisting largely of this one protein (Marciani and Kuff 1973; Wivel et al. 1973), is resistant to most dissociating agents including high salt concentrations and detergents other than sodium dodecyl sulfate. IAPs from some sources are further stabilized by disulfide crosslinking of P73. In contrast, the *gag*-precursor proteins of extracellular retroviruses are processed while or shortly after the particle separates from the cell surface and the reorganized viral core has a much more easily dissociated structure. Integration of newly formed provirus apparently requires interaction between the target chromosomal DNA and an organized nucleoprotein complex derived from the viral core and containing the proviral DNA and requisite enzymatic activity(ies) (Brown et al. 1987). The hyperstable structure formed by the unprocessed IAP *gag* protein may well be incompatible with this process.

Reverse transcriptase. The Mg-dependent IAP reverse transcriptase is active against added homopolymer template primers such as poly(rA): oligo(dT), but heteropolymeric poly(A)-containing RNAs, including IAP-specific and globin mRNAs, are poor exogenous substrates (Wilson and Kuff 1972). The endogenous reaction is weak. In contrast to the reverse transcriptase of extracellular viruses, the IAP activity cannot be solubilized in active form with high salt concentrations or nonionic detergents (Wilson et al. 1974). During isolation of IAPs from a myeloma, the activity cofractionated rigorously with the main particle antigenic component (P73), and no soluble activity could be detected (Wilson et al. 1974). The tight binding of the enzyme to the particles suggests that the *gag-pol* precursor polypeptide, like the *gag* product itself, remains largely unprocessed. This defect might be related to the atypical properties of the enzyme activity.

Intracisternal Location. The *gag* coding region of the IAP specifies a unique amino-terminal hydrophobic segment that resembles a secretory signal sequence and is the likely means for association of nascent IAPs with the ER membrane (Mietz et al. 1987). Particles that have budded into the cisternal spaces are physically separated from the cytosolic precursors required for DNA synthesis, suggesting that reverse transcription is minimal in the intracisternal locale. Activity might be facilitated during turnover of the particles if an organized remnant of the inner core gained access to the cytosolic compartment. This possibility has not been investigated.

Lack of Envelope Protein. The IAP-related genomic elements in both mouse and hamster have multiple stop codons in all reading frames in the DNA region that typically encodes the retroviral envelope proteins (Ono et al. 1985; Mietz et al. 1987). In the case of the mouse genes, where multiple examples are available, the stop codons are seen to be highly conserved in several cloned elements. The lack of a competent envelope gene is consistent with biochemical observations on isolated IAPs from several sources (Marciani and Kuff 1973). However, a proper coat protein, while essential for horizontal transmission, is probably not important for intracellular retrotransposition of IAP elements.

Cellular Suppressing Mechanisms. NIH3T3 cells show extensive demethylation of genomic IAP sequences, yet expression seems to be largely suppressed at the transcriptional level (Wujcik et al. 1984). In normal mouse liver, IAP transcripts are present at very low levels but can be significantly increased by cycloheximide treatment (Dragani et al. 1987). Cycloheximide is generally thought to exert this effect by blocking synthesis of a labile repressor protein. These observations suggest that IAP expression can be down-regulated at several points following demethylation of the regulatory elements in the LTRs, but the mechanisms of control are not defined.

DISCUSSION

Because of the large number of preexisting elements, it has been difficult to estimate the overall frequency of new proviral insertions. Shen-Ong and Cole (1982) showed that a particular structural variant present in about 20 copies in normal BALB/c DNA was increased to 40 and 80 copies, respectively, in the DNAs of two different myelomas. In the more amplified situation, the MOPC-315 tumor line, the new elements were shown to be very similar to one another, as if reverse transcribed from a single IAP RNA species, and integrated at various new locations in the genome (Shen-Ong and Cole 1984). Since the great majority of the IAP elements found to be transposed, including those in MOPC-315, contain major deletions involving both the *gag* and *pol* genes (Table 1), the retrotranspositions must have been mediated by products of competent elements coexpressed in the tumor cells. Depending on the particular combination of active IAP elements, then, different cell types or tumor lines may show characteristic variations in the frequency of transpositions.

In yeast, overexpression of an inducible Ty element led to general activation of the endogenous elements, multiple transpositions within virtually every cell, and a greatly reduced growth rate (Boeke et al. 1985). Accumulation of harmful insertional mutations or deleterious effects of reverse transcription of the cellular mRNAs were suggested as possible reasons for the impaired growth. Transposition was found to be strongly mutagenic for the newly formed Ty elements. Boeke and colleagues noted that transpositions of the 35 endogenous Ty copies were ordinarily rare and suggested that these elements may be largely defective. Subsequently, virus-like particles were found in cells overexpressing the Ty element, and a particle-associated reverse transcriptase activity was observed and identified as the *tyb* gene product (Garfinkel et al. 1985). Particles and enzyme activity were absent in uninduced cells even though appreciable levels of Ty transcripts were observed (5–10% of total poly[A] RNA), again suggesting that the endogenous Ty elements are largely defective. Resemblance of the Ty particles to the mouse IAPs and *Drosophila copia* particles was noted.

Similarities between the Ty and IAP elements are indeed striking. Among them might be mentioned in the present context: a high proportion of functionally defective structural variants (Fig. 1); lack of an envelope (*env*) gene in Ty (Clare and Farabaugh 1985) and a translationally closed *env* region in IAP (Mietz et al. 1987); association of reverse transcriptase activity with virally encoded particles and absence of detectable soluble enzyme (Wilson et al. 1974; Garfinkel et al. 1985); and retention of particles within the cells and lack of horizontal transmission in cocultivation experiments (Minna et al. 1974; Lueders and Kuff 1975; Garfinkel et al. 1985).

IAPs may be more thoroughly sequestered from the general cellular activities by virtue of their intracisternal position than the Ty particles, which are located in the cytoplasmic matrix (Garfinkel et al. 1985). The inducible Ty element used in the experiments cited above also appeared to contain a fully competent reverse transcriptase, in contrast to the atypical, possibly incompletely processed IAP enzyme. These may be reasons why Ty overexpression is inhibitory for cell growth, whereas high levels of IAP expression are apparently not. However, it should be noted that there has been no systematic search for deleterious mutations caused by IAP element transpositions in particle-rich tumors and cell lines. Cells carrying mutations that place them at a selective growth disadvantage with respect to the whole cell population may only be detected by clonal analysis or immunocytochemical techniques.

REFERENCES

Boeke, J.D., D.J. Garfinkel, C.A. Styles, and G.R. Fink. 1985. Ty elements transpose through an RNA intermediate. *Cell* **40:** 491.

Brown, P.O., B. Bowerman, H.E. Varmus, and J.M. Bishop. 1987. Correct integration of retroviral DNA *in vitro*. *Cell* **49:** 347.

Burt, D.W., A.D. Reith, and W.J. Brammer. 1984. A retroviral provirus closely associated with the *Ren-2* gene of DBA/2 mice. *Nucleic Acids Res.* **12:** 8579.

Canaani, E., O. Dreazen, A. Klar, G. Rechavi, D. Ram, J.B. Cohen, and D. Givol. 1983. Activation of the c-*mos* oncogene in a mouse plasmacytoma by insertion of an endogenous intracisternal A-particle genome. *Proc. Natl. Acad. Sci.* **80:** 7118.

Clare, J. and P. Farabaugh. 1985. Nucleotide sequence of a yeast Ty element: Evidence for an unusual mechanism of gene expression. *Proc. Natl. Acad. Sci.* **82:** 2829.

Cohen, J.B., T. Unger, G. Rechavi, E. Canaani, and D. Givol. 1983. Rearrangement of the oncogene c-*mos* in mouse myeloma NSI and hybridomas. *Nature* **306:** 797.

Dragani, T.A., G. Manenti, G. Della Porta, and I.B. Weinstein. 1987. Factors influencing the expression of endogenous retrovirus-related sequences in the liver of B6C3 mice. *Cancer Res.* **47:** 795.

Feenstra, A., J. Fewell, E.L. Kuff, and K. Lueders. 1986. *In vitro* methylation inhibits the promoter activity of a cloned intracisternal A-particle LTR. *Nucleic Acids Res.* **14:** 4343.

Garfinkel, D.J., J.D. Boeke, and G.R. Fink. 1985. Ty element transposition: Reverse transcriptase and virus-like particles. *Cell* **42:** 507.

Gattoni-Celli, S., W.-L.W. Hsiao, and I.B. Weinstein. 1983. Rearranged c-*mos* locus in a MOPC-21 murine myeloma cell line and its persistence in hybridomas. *Nature* **306:** 795.

Greenberg, R., R. Hawley, and K.B. Marcu. 1985. Acquisition of an intracisternal A-particle element by a translocated c-*myc* gene in a plasma cell tumor. *Mol. Cell. Biol.* **5:** 3625.

Grigoryan, M.S., D.A. Kramerov, E.M. Tulchinsky, E.S. Revasova, and E.M. Lukanidin. 1985. Activation of putative transposition intermediate formation in tumor cells. *EMBO J.* **4:** 2209.

Hawley, R.G., M.J. Shulman, and N. Hozumi. 1984. Transposition of two different intracisternal A particle elements into an immunoglobulin kappa-chain gene. *Mol. Cell. Biol.* **4:** 2565.

Hojman-Montes de Oca, F., J. Lasneret, L. Dianoux, M. Canivet, R. Ravicovitch, and J. Peries. 1984. Regulation of intracisternal A particles in mouse teratocarcinoma cells: Involvement of DNA methylation in transcriptional control. *Biol. Cell* **52:** 199.

Kuff, E.L. and J.W. Fewell. 1985. Intracisternal A-particle expression in normal mouse thymus tissue: Gene products and strain-related variability. *Mol. Cell. Biol.* **5:** 474.

Kuff, E.L. and K.K. Lueders. 1988. The intracisternal A-particle gene family: Structure and functional aspects. *Adv. Cancer Res.* (in press).

Kuff, E.L., J.W. Fewell, K.K. Lueders, J.A. DiPaolo, S.C. Amsbaugh, and N.C. Popescu. 1986. Chromosome distribution of intracisternal A-particle sequences in the Syrian hamster and mouse. *Chromosoma* **93:** 213.

Kuff, E.L., A. Feenstra, K. Lueders, L. Smith, R. Hawley, N. Hozumi, and M. Shulman. 1983. Intracisternal A-particle genes as movable elements in the mouse genome. *Proc. Natl. Acad. Sci.* **80:** 1992.

Lasneret, J., M. Canivet, F. Hojman-Montes de Oca, J. Tobaly, R. Emanoil-Ravicovitch, and J. Peries. 1983. Activation of intracisternal A-particles by 5-azacytidine in mouse Ki-Balb cell line. *Virology* **128:** 485.

Lueders, K.K. 1987. Specific association between type-II intracisternal A-particle elements and other repetitive sequences in the mouse genome. *Gene* **52:** 139.

Lueders, K.K. and E.L. Kuff. 1975. Synthesis and turnover of intracisternal A-particle structural protein in cultured neuroblastoma cells. *J. Biol. Chem.* **250:** 5192.

———. 1977. Sequences associated with intracisternal A-particles are reiterated in the mouse genome. *Cell* **12:** 963.

Lueders, K.K. and J.A. Mietz. 1986. Structural analysis of type II variants within the mouse intracisternal A-particle sequence family. *Nucleic Acids Res.* **14:** 1495.

Lueders, K.K., A. Leder, P. Leder, and E. Kuff. 1982. Association between a transposed α-globin pseudogene and retrovirus-like elements in the BALB/c genome. *Nature* **295:** 426.

Marciani, D.J. and E.L. Kuff. 1973. Isolation and partial characterization of the internal structural proteins from murine intracisternal A-particles. *Biochemistry* **12:** 5075.

Mietz, J.A., Z. Grossman, K.K. Lueders, and E.L. Kuff. 1987. Nucleotide sequence of a complete mouse intracisternal A-particle genome: Relationship to known aspects of particle assembly and function. *J. Virol.* **61:** 3020.

Minna, J.D., K.K. Lueders, and E.L. Kuff. 1974. Expression of genes for intracisternal A-particle antigen in somatic cell hybrids. *J. Natl. Cancer Inst.* **52:** 1211.

Morgan, R.A. and R.C. Huang. 1984. Correlation of undermethylation of intracisternal A-particle genes with expression in murine plasmacytomas but not in NIH/3T3 fibroblasts. *Cancer Res.* **44:** 5234.

Ono, M., H. Toh, T. Miyata, and T. Awaya. 1985. Nucleotide sequence of the Syrian hamster intracisternal A-particle gene: Close evolutionary relationship to type B and D oncovirus genes. *J. Virol.* **55:** 387.

Shen-Ong, G.L. and M.D. Cole. 1982. Differing populations of intracisternal A-particle genes in myeloma tumors and mouse subspecies. *J. Virol.* **42:** 411.

―――. 1984. Amplification of a specific set of intracisternal A-particle genes in a mouse plasmacytoma. *J. Virol.* **49:** 171.

Singer, M.F. and J. Skowronski. 1985. Making sense out of LINES: Long interspersed repeat sequences in mammalian genomes. *Trends Biochem. Sci.* **10:** 119.

Wilson, S.H. and E.L. Kuff. 1972. A novel DNA polymerase activity found in association with intracisternal A-type particles. *Proc. Natl. Acad. Sci.* **69:** 1531.

Wilson, S.H., E.W. Bond, A. Matsukage, K.K. Lueders, and E.L. Kuff. 1974. Studies on the relationship between deoxyribonucleic acid polymerase activity and intracisternal A-type particles. *Biochemistry* **13:** 1087.

Wivel, N.A., K.K. Lueders, and E.L. Kuff. 1973. Structural organization of murine intracisternal A-particles. *J. Virol.* **11:** 329.

Wujcik, K.M., R.A. Morgan, and R.C.C. Huang. 1984. Transcription of intracisternal A-particle genes in mouse myeloma and Ltk$^-$ cells. *J. Virol.* **52:** 29.

Ymer, S., W.Q.J. Tucker, H.D. Campbell, and I.G. Young. 1986. Nucleotide sequence of the intracisternal A-particle genome inserted 5' to the interleukin-3 gene in the leukemia cell line WEHI-3B. *Nucleic Acids Res.* **14:** 5901.

Ymer, S., W.Q.J. Tucker, C.J. Sanderson, A.J. Hapel, H.D. Campbell, and I.G. Young. 1985. Constitutive synthesis of interleukin-3 by leukemia cell line WEHI-3B is due to retroviral insertion near the gene. *Nature* **317:** 255.

Yotsuyanagi, Y. and D. Szollosi. 1984. Virus-like particles and related expression in mammalian oocytes and preimplantation embryos. In *Ultrastructure of reproduction* (ed. J. van Blerkom and P.M. Motta), p. 218. Martinus Nijhoff, The Hague, Netherlands.

Two Families of Human Endogenous Retroviral Genomes

ROBERT CALLAHAN
Laboratory of Tumor Immunology and Biology
National Cancer Institute
National Institutes of Health
Bethesda, Maryland 20892

OVERVIEW

Retroviruses are the causative agents of naturally occurring tumors in diverse vertebrate species. In addition to their spread as infectious agents, they can also enter the germ line and be transmitted to subsequent generations of the host as stable Mendelian genetic elements (endogenous retroviruses, Erv). DNA sequences related to infectious retroviruses from several mammalian and avian species represent approximately 0.1% of the human genome. They are found on many, if not all, of the human chromosomes. These retroviral-related sequences can be divided into two diverse families (designated class I and II) based on their relative homology to either mammalian type C or to types A, B, D, and avian type C retroviral genomes. Some of the human endogenous retroviral (Herv) genomes are organized in a manner expected for an integrated proviral genome; others are present in one of two aberrant structures. Expression of Herv genomes has been detected in normal and tumor tissues, as well as in tissue culture cell lines. Although there is no evidence at the present time that they encoded infectious virus, it seems possible that they could be associated with certain inherited diseases or neoplasias.

INTRODUCTION

Four genera of retroviruses have been established on the basis of morphological and biochemical criteria (reviewed in Teich 1982). Defective intracisternal type-A viral particles (IAP) have been observed in early mouse embryos and certain murine tumors (Kuff, this volume). Recently, an infectious murine retrovirus containing extensive homology with the IAP genome has been reported (Callahan et al. 1981). Type B retroviruses are represented by several strains of mouse mammary tumor virus (MMTV). Type C viruses, which are widespread among avian and mammalian species, have been shown to cause leukemia and other tumors. The most recently described type D viruses appear limited to primate species. Although the oncogenicity of type D viruses remains to be established, some members are associated with an immunodeficiency syndrome (Daniel et al. 1984; Marx et al. 1984; Stromberg et al. 1984).

The origins of the four types of retroviruses are obscure. However, nucleotide sequence analysis of recombinant DNA clones containing infectious retroviral genomes demonstrated a major region of homology between the *pol* genes (encoding reverse-transcriptase and an endonuclease) of types A, B, D, and avian type C viruses (Chiu et al. 1984). These studies provide strong support for the concept of two major *pol* gene progenitors, one for mammalian type C viruses and the other for types A, B, D, and avian type C viruses. Other studies have shown that the type C and D retroviruses share antigenic determinants, as do the major internal proteins of type B and D retroviruses (Barbacid et al. 1980). Moreover, certain mammalian type C and type D retroviruses share common determinants in the *env*-encoded gp70 protein (Stephenson et al. 1976; Devare et al. 1978), as well as sequence homology in their p15E coding region (Cohen et al. 1982; Chiu et al. 1984). These patterns of homology indicate that the evolution of recent retrovirus groups involved genetic interactions among their progenitors.

Although many of these retroviruses spread among host populations as infectious agents, they can also integrate into the germline of a host (Teich 1982). This action has at least two consequences for the host. First, the viral genome will be passed on to subsequent generations as a stable Mendelian element (Erv). In fact, all mammalian and avian species studies have been found to contain multiple copies of Erv genomes (Coffin 1982). Although many of the endogenous viral genomes are either not expressed or are biologically inactive, some do encode an infectious retrovirus. Second, viral integration into the germ line is a potentially mutagenic event, which can affect subsequent generations of the host (Jenkins et al. 1981; Copeland et al. 1983).

The potential involvement of members of the Retroviridae family of viruses in humans has represented a major focus of investigation. Early studies using antisera prepared against mammalian type C (Weiss 1982) and MMTV (Sarkar 1980) retroviral proteins suggested that related proteins were expressed in a variety of human neoplasias and normal tissues. In recent years, the use of recombinant DNA technology and more sensitive methodologies for detecting nucleotide sequence homology has led to the identification of two families of Herv genes that are related to the infectious retroviruses of other mammalian and avian species. This chapter describes their structure and our current understanding of their biological activity.

RESULTS

The use of low-stringency blot hybridization conditions and recombinant DNA clones containing retroviral genomes as probes led to the detection and subsequent isolation of recombinant clones of human cellular DNA containing mammalian types A, B, C, and D retroviral-related sequences (Martin et al. 1981b; Bonner et al. 1982; Callahan et al. 1982; Ono 1986). Based on the concept of two *pol* gene families in Retroviridae (Chiu et al. 1984), the organization of mammalian type C-related sequences (class I) will be described separately from those containing types A, B, and D viral-related sequences (class II).

Class I Herv Genomes

Two independent strategies have been used to obtain recombinant clones of human cellular DNA containing type C retroviral-related sequences. In the first approach, a recombinant clone of African green monkey cellular DNA containing murine leukemia virus (MLV)-related sequences (Martin et al. 1981a) was used as a probe to screen a recombinant DNA library of human cellular DNA (Martin et al. 1981b). The MLV-related sequences were found to be organized into three types of structures. In the first type, the MLV-related structural genes were within an 8.8-kb proviral structure bounded by long terminal repeat (LTR)-like elements (Repaske et al. 1983). Restriction enzyme mapping and heteroduplex analysis demonstrated that the "full-length" proviral genomes contain small deletions and/or insertions, suggesting a family of partially related proviral genomes. The LTRs of the full-length proviral genome (512 bp) are 90–95% related and are bounded by direct repeats in the flanking cellular sequences. They show little sequence homology with LTRs of infectious mammalian type C viruses. The primer tRNA binding site in these proviral genomes is related to the rat glutamic acid tRNA. The second structure contained sequences related to the LTR-like element of the full-length proviral genome but was not associated with retroviral *gag*-, *pol*-, or *env*-related sequences. This type of genetic element, which is flanked by direct repeat sequences, has been called a "solitary LTR" (Steele et al. 1984). It probably reflects the excision of viral-related sequences by homologous recombination between the 5' and 3' proviral LTRs. The third structure appears to be a truncated 6-kb proviral genome bounded by highly repetitive cellular DNA sequences. Nucleotide sequence analysis demonstrates that (1) the truncated proviral genome could be aligned with the region of MLV spanning the p12 *gag* gene through the *pol* gene, indicating that the 5' end of the *gag* gene and the entire *env* gene have been deleted; and (2) the repeat sequences bounding the truncated viral genome are related to the KPN-1 family of highly repetitive cellular DNA sequences.

It has been estimated (Steele et al. 1984) that as many as 35–50 copies of each family of class I retroviral-related sequences may exist. However, many of the full-length and truncate proviral genomes may have resulted from the amplification of a limited number of primary germ line integration events early in primate evolution (Steele et al. 1986). Although the mechanism by which the viral genomes were amplified is unclear, it appears to have included flanking cellular sequences. The distribution of these proviral genomes in the human genome appears to include many, if not all, of the chromosomes (Steele et al. 1986).

In separate experiments, Bonner et al. (1982) and O'Connell et al. (1984) also identified recombinant clones of human cellular DNA containing mammalian type C retroviral-related sequences. They used chimpanzee recombinant cellular DNA containing sequences related to the *pol* gene of the endogenous baboon virus as a probe (Bonner et al. 1982). Two proviral genomes designated Erv-1 and Erv-3 have been extensively characterized. The Erv-1 endogenous provirus lacks only the 5' LTR and is located on chromosome 18 (O'Brien et al. 1983). Erv-3 represents a full-length provirus and is located on chromosome 7 (O'Connell et al. 1984). Nucleotide sequence analysis has demonstrated that the *gag-pol* region Erv-1 and Erv-3 shares significant homology with the human endogenous genomes isolated by Martin et al. (1981b). Although Erv-1 and Erv-3 are members of this same class of human endogenous proviruses, they possess three distinguishing characteristics (O'Connell and Cohen 1984): (1) their LTRs share little sequence homology with each other or with those described by Steele et al. (1984); (2) the architecture of the LTRs differs from that of other retroviruses with respect to the size of the U-3, R, and U-5 regions; and (3) the primer tRNA binding site corresponds to the mouse arginine tRNA.

Class II Herv Genomes

Recombinant clones of human cellular DNA containing MMTV-related sequences have been isolated and characterized (Callahan et al. 1982, 1985; May et al. 1983; Deen and Sweet 1986). These sequences are organized within 9-kb full-length endogenous proviral genomes. More recently, a Syrian hamster IAP genome was used as a probe to isolate related recombinant clones of human cellular DNA (Ono 1986). Analysis of these endogenous viral genomes has shown that they are very similar to those identified with the MMTV genome. This result, however, is not surprising, since the MMTV and IAP *pol* genes share significant nucleotide sequence homology (Chiu et al. 1984; Ono et al. 1985, 1986; L. Johnson et al., in prep.). There are approximately 50 copies of this family of endogenous proviral genomes in human cellular DNA (Deen and Sweet 1986; Ono et al. 1986; T.M. Horn et al., in prep.). Restriction enzyme mapping and heteroduplex analysis have shown significant levels of sequence divergence among members of the family with the frequent occurrence of deletions or insertions of sequences (Westley and

May 1984; Ono et al. 1986; T.M. Horn et al., in prep.). These endogenous proviral genomes are located on many, if not all, of the human chromosomes. Two of them, HLM-2 and HLM-25, have been shown to be located on chromosomes 1 and 5, respectively (Horn et al. 1986).

Blot hybridization and nucleotide sequence analysis of two representative proviral genomes, designated Herv K (Ono et al. 1986) and HLM-2 (L. Johnson et al., in prep.), have revealed several characteristic features of the class II proviral genomes. The 1-kb LTR shares little sequence homology with other retroviral LTRs and, unlike the MMTV LTR, contains no significant open reading frame that could encode a protein. The primer tRNA binding site, like that of type D retroviruses, is complimentary to the 3' 18 nucleotides of the lysine tRNA, having a CUU anticodon. The polypurine tract adjacent to the 3' LTR is most similar (16 of 18 base pairs) to that of MMTV. The major region of nucleotide and peptide sequence homology (60% and 40–60%, respectively) spans the protease and *pol* genes of type A (IAP), type B (MMTV), and type D (Mason pfizer monkey virus, MPMV) retroviruses. Although there is no nucleotide sequence homology with the *env* genes of MMTV or MPMV, there is amino acid sequence homology with MMTV gp52 (18%) and gp36 (27%) envelope proteins. When the putative human peptide sequences are compared with those of gp52 and gp36 on the basis of chemically similar amino acids in the sequences, the homology increases to 42% and 50%, respectively.

Recently, May and Westley (1986) have isolated a recombinant clone (NMWV4) containing a novel member of this family of proviruses. The proviral genome contains MMTV *pol*-related sequences but is truncated (5.4 kb). It has two additional distinguishing features. The LTR is 424 bp, contains short stretches of homology with the MMTV, human T-lymphotropic virus (HTLV), and bovine leukemia virus LTRs, but no homology with the Herv K18 or HLM-2 LTRs. The primer tRNA binding site is complimentary (17 of 18 base pairs) with the 3' end of the lysine tRNA having a UUU anticodon. This is the same primer tRNA used by MMTV and human immunodeficiency virus.

In contrast to the class II viral-related structural genes, the LTR-like sequences are reiterated 1000 times in human cellular DNA (T.M. Horn et al., in prep.). In general, most of these sequences appear to be unassociated with viral structural genes. Preliminary analysis of two recombinant clones containing only HLM-2 LTR-related sequences revealed that they are adjacent to or interrupt members of the *kpn*-1 family of repetitive cellular DNA sequences (L. Johnson et al., in prep.). Nucleotide sequence analysis showed that, in two recombinant clones, the solitary LTRs are in pairs separated by 8 kb of unrelated cellular sequences and are in the same transcriptional orientation (L. Johnson et al., in prep.). In each case, the same tRNA primer binding site, as found in full-length proviral genomes, was located 3' to the 5' LTR. Examination of Southern blots of restricted cellular DNA suggests that many of the solitary LTRs have been amplified in the human genome. Like the viral structural genes, the solitary LTRs appear to be distributed throughout the cellular genome (Horn et al. 1986).

Expression of Herv Genomes

The presence of multiple copies of endogenous proviral genomes in human cellular DNA raises the possibility that at least some of these genomes are biologically active. In fact, immunological studies using heterologous antisera prepared against either the feline type C RD114 endogenous retroviral *gag* p30 protein (Suni et al. 1981) or an unadecapeptide deduced from the p30 *gag* gene of Erv-1 (Vahari et al. 1985) react with normal placenta, blighted ova, hydatidiform and destructive moles, and choriocarcinomas. No other adult or embryonic tissue reacted with these sera. Similarly, human breast tumors and the T47D breast tumor cell line have been reported to react with antisera raised against the MMTV *env* gp52 protein (Keydar et al. 1984; Segev et al. 1985).

These observations have been reinvestigated using recombinant DNA probes corresponding to various regions of both classes of human endogenous proviral genomes. Human class I retroviral RNA transcripts have been observed in placenta, spleen, colon carcinomas, and a breast carcinoma (Rabson et al. 1983; Gattoni-Celli et al. 1986). In these tissues, LTR (3.6 kb) and LTR-*env* (3.0, 1.7, and 0.6 kb) RNA species were detected. In the acute lymphocytic leukemia cell line, 8402 (not infected with HTLV), a 6.8-kb species of *gag-pol-env*-related RNA was detected. However, nucleotide sequence analysis of partial cDNA clones of the 3.0- and 1.7-kb RNA species showed that they contained in-frame termination codons (Rabson et al. 1985). Thus, neither could encode a full-length *env* protein.

Information on the expression of class II endogenous proviral genomes is only beginning to emerge. In one study of poly(A)$^+$ RNA from 20 primary breast tumors, a 6.7-kb RNA species containing *gag-pol-env*-related sequences was detected (L. Johnson et al., in prep.). This species of RNA did not react with the HLM-2 LTR. In another study, the human epithelial-2, HMT-2 (malignant melanoma), KATO III (gastric carcinoma), HeLa, and T47D (breast carcinoma) tissue culture cell lines were found to express an 8.8-kb species of RNA reacting with *gag, pol,* and *env* sequences (Ono et al. 1987). In addition, progesterone was reported to enhance the expression of the 8.8-kb RNA species in the T47D tissue culture cell line. Since T47D cells have been reported to react with antisera prepared against MMTV *env* gp52 protein (Segev et al. 1985), it is surprising that no LTR-*env* species of RNA was detected.

DISCUSSION

Two classes of Herv genomes have been distinguished on the basis of the relative homology their *pol* genes share with either infectious mammalian type C viruses or type A, B, and D retroviruses. Genetic interactions between the progenitors of the different retroviral genera (types A, B, C, and D) have contributed to the evolution of infectious retroviruses. The association of two different LTRs and tRNA PBSs with common viral structural genes in each of the two classes of human endogenous proviral genomes suggests that their respective progenitors were the result of a similar process.

Several lines of evidence suggest that many of the endogenous proviral genomes are the result of retroviral infection early in human evolution. Comparison of Old World primates and human DNAs reveals common patterns of restriction fragments containing viral-related sequences. In addition, the class I Erv-1 proviral genome located on human chromosome 18 is also present in chimpanzee cellular DNA (O'Brien et al. 1983). Similarly, the class II HLM-2 proviral genome of chromosome 1 is present in chimpanzee, gorilla, and orangutan DNAs (T.M. Horn and R. Mariani-Costantini, in prep.). These observations suggest that many of the class I and II proviral genomes have been genetically transmitted for at least 17 million years.

During primate evolution, endogenous proviral genomes, like other cellular genes, have been subjected to genetic alterations such as point mutations, small deletions, and insertions. These changes probably account, in part, for the extensive nucleotide sequence divergence observed within each class of proviral genomes. In addition, both classes of proviral genomes have undergone a process of partial excision, which in some cases was followed by amplification of the remaining viral sequences (class I truncated genomes, class II solitary LTRs). Similarly, certain class I full-length proviral genomes and flanking cellular sequences have been amplified. The process of proviral amplification in the germ line has evidently resulted in their dispersion throughout the human genome. In considering the mechanism by which the class I truncated genomes and class II solitary LTRs arose and were amplified, it is pertinent that each are within or are associated with the *kpn*-1 family of highly repetitive cellular DNA sequences. These genetic events probably occurred early in human evolution, since the pattern of viral-related restriction fragments and their level of amplification does not vary significantly in DNA from unrelated individuals.

Expression of class I and class II proviral genomes has been detected in normal tissues, several types of tumors, and tumor-derived tissue culture cell lines. Although the inability to isolate corresponding infectious retroviruses may reflect the absence of sensitive tissue culture cell lines, other observations suggest that many of the proviral genomes are either defective or unavailable for expression. Nucleotide sequence analysis of recombinant cDNA revealed several mutations that would preclude the complete expression of viral proteins. In addition, evidence is lacking for the amplification or acquisition of new proviral sequences in DNA from tissues or tissue culture cell lines expressing retroviral RNA or proteins.

These considerations, in addition to the absence of any compelling epidemiological evidence, suggest that infectious retroviruses corresponding to the endogenous proviral genomes are probably not widespread in the human population. Nevertheless, several interesting questions regarding class I and II proviral genomes remain unanswered: (1) the molecular basis for the regulation of their expression, (2) their potential involvement in mutations associated with inherited diseases, including cancer, and (3) the possibility that they can act as insertion mutagens in a manner similar to that observed for IAP genomes in mice (Kuff, this volume), or *kpn*-1 repetitive elements (Singer et al., this volume).

REFERENCES

Barbacid, M., L.K. Long, and S.A. Aronson. 1980. Major structural proteins of type B, type C, and type D oncoviruses share interspecies antigenic determinants. *Proc. Natl. Acad. Sci.* **77:** 72.

Bonner, I.I., C. O'Connell, and M. Cohen. 1982. Cloned endogenous retroviral sequences from human DNA. *Proc. Natl. Acad. Sci.* **79:** 4709.

Callahan, R., W. Drohan, S. Tronick, and J. Schlom. 1982. Detection and cloning of human DNA sequences related to the mouse mammary tumor virus genome. *Proc. Natl. Acad. Sci.* **79:** 5503.

Callahan, R., E.L. Kuff, K.K. Lueders, and E. Birkenmeier. 1981. Genetic relationship between the *Mus* cervicolor M432 retrovirus and *Mus musculus* intracisternal type A particle. *J. Virol.* **40:** 901.

Callahan, R., I.M. Chiu, J.F.H. Wong, S.R. Tronick, B. Roe, S.A. Aaronson, and J. Schlom. 1985. A new class of endogenous human retroviral genomes. *Science* **228:** 1208.

Chiu, I.M., R. Callahan, S.R. Tronick, J. Schlom, and S.A. Aaronson. 1984. Major *pol* gene progenitors in the evolution of oncoviruses. *Science* **223:** 364.

Coffin, J. 1982. Endogenous viruses. In *Molecular biology of tumor viruses,* 2nd edition: *RNA tumor viruses* (ed. R. Weiss et al.), p. 1109. Cold Spring Harbor Laboratory, Cold Spring Harbor, New York.

Cohen, M., N. Rice, R. Stephens, and C. O'Connell. 1982. DNA sequence relationship of baboon endogenous virus genome to the genomes of other type C and type D retroviruses. *J. Virol.* **41:** 801.

Copeland, N.G., K.W. Hutchison, and N.A. Jenkins. 1983. Excision of the DBA ectotropic provirus in dilute coat-color revertants of mice occurs by homologous recombination involving LTRs. *Cell* **33:** 379.

Daniel, M.D., N.W. King, N.L. Letvin, R.D. Hunt, P.K. Sehgal, and R.C. Desrosiers. 1984. A new type D retrovirus isolated from macaques with an immunodeficiency syndrome. *Science* **223:** 602.

Deen, K.C. and R.W. Sweet. 1986. Murine mammary tumor virus *pol*-related sequences in human DNA: Characterization and sequence comparison with the complete murine mammary tumor virus *pol* gene. *J. Virol.* **57:** 422.

Devare, S.G., R.E. Hanson, J.R. Stephenson. 1978. Primate retroviruses: Envelope glycoproteins of endogenous type C and type D viruses possess common interspecies antigenic determinants. *J. Virol.* **26:** 316.

Gattoni-Celli, S., K. Kirsch, S. Kalled, and K.J. Isselbacher. 1986. Expression of type C related endogenous retroviral sequences in human colon tumors and colon cancer cell lines. *Proc. Natl. Acad. Sci.* **83:** 6127.

Horn, T.M., K. Huebner, C. Croce, and R. Callahan. 1986. Chromosomal locations of members of a family of novel endogenous human retroviral genomes. *J. Virol.* **58:** 955.

Jenkins, N.A., N.G. Copeland, B.A. Taylor, and B.K. Lee. 1981. Dilute (d) coat color mutation of DBA/25 mice is associated with the site of integration of an ecotropic MuLV genome. *Nature* **293:** 370.

Keydar, I., T. Ohno, R. Nayak, R. Sweet, F. Simoni, F. Weiss, S. Karby, K. Mesa-Tejada, and S. Spiegelman. 1984. Properties of retrovirus-like particles produced by a human breast carcinoma line: Immunological relationship with mouse mammary tumor virus proteins. *Proc. Natl. Acad. Sci.* **81:** 4188.

Martin, M.A., T. Bryan, T.F. McCutchan, and H.W. Chan. 1981a. Detection and cloning of murine leukemia virus related sequences from African green monkey liver DNA. *J. Virol.* **39:** 835.

Martin, M.A., T. Bryan, S. Rasheed, and A.S. Khan. 1981b. Identification and cloning of endogenous retroviral sequences present in human DNA. *Proc. Natl. Acad. Sci.* **78:** 4892.

Marx, P.A., D.H. Maul, K.G. Osborn, N.W. Lerche, P. Moody, L.J. Lowenstine, R.V. Henrickson, L.O. Arthur, R.V. Gilden, M. Gravell, W.T. London, J.L. Sever, J.A. Levy, R.J. Munn, and M.B. Gardner. 1984. Simian AIDS: Isolation of a type D retrovirus and transmission of the disease. *Science* **223:** 1083.

May, F.E.B. and B.R. Westley. 1986. Structure of a human retroviral sequence related to mouse mammary tumor virus. *J. Virol.* **60:** 743.

May, F.E.B., B.R. Westley, H. Rochefort, E. Buetti, and H. Diggelmann. 1983. Mouse mammary tumor virus related sequences are present in human DNA. *Nucleic Acids Res.* **11:** 4127.

O'Brien, S.J., T.I. Bonner, M. Cohen, C. O'Connell, and W.G. Nash. 1983. Mapping of an endogenous retroviral sequence to human chromosome 18. *Nature* **303:** 74.

O'Connell, C. and M. Cohen. 1984. The long terminal repeat sequences of a novel human endogenous retrovirus. *Science* **226:** 1204.

O'Connell, C., S.J. O'Brien, W.G. Nash, and M. Cohen. 1984. ERV3, a full length human endogenous provirus: Chromosomal localization and evolutionary relationships. *Virology* **138:** 225.

Ono, M. 1986. Molecular cloning and long terminal repeat sequences of human endogenous retroviral genes related to type A and B retroviral genes. *J. Virol.* **58:** 937.

Ono, M., M. Kawakami, and H. Ushikubo. 1987. Stimulation of expression of human endogenous retrovirus genome by female steroid hormones in human breast cancer cell line T47D. *J. Virol.* **61:** 2059.

Ono, M., T. TeYasunaga, T. Miyata, and H. Ushikubo. 1986. Nucleotide sequence analysis of human endogenous retrovirus genome related to the mouse mammary tumor virus genome. *J. Virol.* **60:** 589.
Ono, M., H. Toh, T. Miyata, and T. Awaya. 1985. Nucleotide sequence of the Syrian hamster intracisternal A-particle gene: Close evolutionary relationship of type A particle gene to type B and D oncovirus genes. *J. Virol.* **54:** 764.
Rabson, A.B., P.E. Steele, C.F. Garon, and M.A. Martin. 1983. mRNA transcripts related to full-length endogenous retroviral DNA in human cells. *Nature* **306:** 604.
Rabson, A.B., Y. Hamagishi, P.E. Steele, M. Tykocinski, and M.A. Martin. 1985. Characterization of human endogenous retroviral envelope RNA transcripts. *J. Virol.* **56:** 176.
Repaske, R., R.R. O'Neill, P.E. Steele, and M.A. Martin. 1983. Characterization and partial nucleotide sequence of endogenous type C retrovirus segments in human chromosomal DNA. *Proc. Natl. Acad. Sci.* **80:** 678.
Sarkar, N. 1980. Type B virus and human breast cancer. In *The role of viruses in human cancer* (ed. G. Giraldo and E. Beth), vol. 1, p. 207. Elsevier, New York.
Segev, N., A. Hizi, F. Kirenberg, and I. Keydar. 1985. Characterization of a protein released by the T47D cell line immunologically related to the major envelope protein of mouse mammary tumor virus. *Proc. Natl. Acad. Sci.* **82:** 1531.
Steele, E., A.B. Rabson, T. Bryan, and M.A. Martin. 1984. Distinctive termini characterize two families of human endogenous retroviral sequences. *Science* **225:** 943.
Steele, P.E., M.A. Martin, A.B. Rabson, T. Bryan, and S.J. O'Brien. 1986. Amplification and chromosomal dispersion of human endogenous retroviral sequences. *J. Virol.* **59:** 545.
Stephenson, J.R., S. Hino, E.W. Garrett, and S.A. Aaronson. 1976. Immunological cross reactivity of Mason-Pfizer monkey virus with type C RNA viruses endogenous to primates. *Nature* **261:** 609.
Stromberg, R., R.E. Benveniste, L.O. Arthur, H. Rabin, W.E. Giddens, H.E. Ochs, W.R. Morton, and C.C. Tsai. 1984. Characterization of exogenous type D retrovirus from a fibroma of a macaque with simian AIDs and fibromatosis. *Science* **227:** 289.
Suni, J., T. Wahlstrom, and A. Vaheri. 1981. Retrovirus p30 related antigen in human syncytiotrophoblasts and IgG antibodies in cord blood sera. *Int. J. Cancer* **28:** 559.
Teich, N. 1982. Taxonomy of retroviruses. In *Molecular biology of tumor viruses,* 2nd edition: *RNA tumor viruses* (ed. R. Weiss et al.), p. 25. Cold Spring Harbor Laboratory, Cold Spring Harbor, New York.
Vaheri, A., J. Sumi, A. Narvanen, R. Pakkanen, and T. Wahlstrom. 1985. Activation of retroviral genes in human reproductive tissues and tumors. In *Retroviruses and human pathology* (ed. R.C. Gallo et al.), p. 275. Humana Press, Clifton, New Jersey.
Weiss, R. 1982. The search for human RNA tumor viruses. In *Molecular biology of tumor viruses,* 2nd edition: *RNA tumor viruses* (ed. R. Weiss et al.), p. 1205. Cold Spring Harbor Laboratory, Cold Spring Harbor, New York.
Westley, B. and F.E.B. May. 1984. The human genome contains multiple sequences of varying homology to mouse mammary tumor virus DNA. *Gene* **28:** 221.

Inducers/Regulators of Transposable Element Expression and Transposition: Host Effects

Concerted Transposition in *Drosophila melanogaster*

NICKOLAI A. TCHURIKOV,* TATIANA I. GERASIMOVA,[‡]
SOFIA G. GEORGIEVA,[‡§] LEV J. MIZROKHI,[†] PAVEL G. GEORGIEV,[†‡]
AND YURII V. ILYIN[†]
*Department of Nucleic Acids Biosynthesis
[†]Department of Genome Mobility
Institute of Molecular Biology
[‡]Mobile Elements Group
N.I. Vavilov Institute of General Genetics
[§]Department of Molecular Genetics
N.K. Koltsov Institute of Developmental Biology
U.S.S.R. Academy of Sciences
Moscow B-334, Union of Soviet Socialist Republics

OVERVIEW

We studied the molecular nature of mutations occurring during transposition explosions in *Drosophila melanogaster*. Besides mobile element excisions and long terminal repeat (LTR)-targeted reinsertions, deletion mutations of single-copy gene sequences and their reversions to the wild type were observed. The characteristic sets of mutation changes were shown to appear independently in certain strains. Such simultaneous multiple transposition events were demonstrated in the cut (*ct*) and white (*w*) loci. The possible mechanisms of the described events are discussed.

INTRODUCTION

The mobility of dispersed repetitive sequences in the eukaryotic genome was first suggested by demonstrating their variable localization in *D. melanogaster* chromosomes (Ilyin et al. 1978). However, the investigation of the transposition process and of its control was progressing rather slowly because experimental systems, in which such transpositions occurred permanently at a relatively high rate, were not available. During the last few years, we were studying transpositions in *D. melanogaster* strains with prolonged genomic instability. The original strain, ct^{MR2}, carried a mutation in the *ct* locus induced by an insertion of mobile element mdg4 (gypsy). A number of derivative strains were obtained from the ct^{MR2}. They contained novel mutations in the *ct* locus and/or mutations in several other loci (Gerasimova 1981; Gerasimova et al. 1984b).

Transposition explosions have been discovered in these strains. They occur with a rate of about 10^{-3} and represent multiple changes in the location of mobile elements belonging to different classes (mdg, or copia-like elements, P element, FB elements, and retrotransposons), which take place in one and the same germ cell. Transposition explosions frequently lead to multiple mutagenesis. In the great majority of germ cells, however, even single transposition events cannot be detected (Gerasimova et al. 1984b, 1985).

In the course of successive transposition explosions, mobile elements rather often move back to the same places of the genome where they have previously been located (Gerasimova et al. 1985). This phenomenon designated as transposition memory at least in several cases can be attributed to the fact that one LTR of an mdg element is being retained upon mdg excision. Later on, such a solo LTR serves as a target site for mdg reinsertion (Mizrokhi et al. 1985).

This paper presents the recent results of molecular analysis of the events occurring in the course of transposition explosions. Besides mdg excision and LTR-targeted reinsertions, several other events were observed. These include specific deletions of single-copy gene sequences that may revert to the wild type. We also present examples of concerted changes in the distribution of mobile elements characteristic of certain strains. Possible mechanisms of these events are discussed, such as the involvement of cellular systems of homologous recombination and gene conversion.

RESULTS

All Events Taking Place in the Course of Transposition Explosions Are Simultaneous

Certain data support the idea that all transposition events detected by in situ hybridization upon a transposition explosion occur simultaneously in a particular germ cell during a single cell cycle (Gerasimova et al. 1984a, 1985). The most clear-cut experimental results were obtained by analyzing the offspring of mutated and nonmutated brothers from individual crosses. An example of such an experiment is given in Table 1.

Flies for individual crosses were taken from the unstable strain, cm^+ct^+ (homozygous for the X chromosome, cm-carmine), which was generated from the $cm^{MR19}ct^{MRpN19}$ strain as a result of double reversion. The offspring of one individual cross contained five males with the parent phenotype, six males with a mutation in the ct locus (ct^{MR19}), and two males with mutations in the white (w) locus (w^{MR19}). The clusters of mutants suggest that the mutations occurred at the premeiotic stage.

Each of the thirteen males was crossed individually with females carrying attached X chromosomes to yield thirteen substrains, and the X chromosomes of males were analyzed by in situ hybridization. One can see that the

Table 1
Distribution of MDG Elements in the X Chromosomes of the Parental Strain cm^+ct^+ and Progeny Strains Obtained from Nonmutated and Mutated Brothers in the Individual Cross

Strain	MDG1						MDG2											MDG3		MDG4	copia		No. of transposition events
	3 C	4 D	4 F	9 A	19 E	20	1 B	3 C	3 D	4 A	4 D	5 B	6 C	11 D	18 E	19 E	20	13 A	20	7 B	1 C	5 A	
Parent																							
cm^+ct^+	+		+			+		+		+		+			+	+	+	+			+		—
Progeny																							
N1	+		+			+		+	+	+		+			+	+	+	+			+		0
cm^+ct^+ N2	+		+			+		+	+	+		+			+	+	+	+			+		0
N3	+		+			+		+	+	+		+			+	+	+	+			+		0
N4	+		+			+		+	+	+		+			+	+	+	+			+		0
N5	+		+			+		+	+	+		+			+	+	+	+			+		0
ct^{MR19} N1	+			+	+	+	+	+			+			+			+	+	+	+	+		13
N2		+		+	+	+	+						+					+	+		+	+	13
N3	+			+	+	+	+	+		+	+	+	+	+			+	+	+	+	+		13
N4	+			+	+	+	+	+		+	+	+	+	+			+	+	+	+	+		13
N5	+			+	+	+	+	+		+	+	+	+	+			+	+	+	+	+		13
N6	+			+	+	+	+	+		+	+	+	+	+			+	+	+	+	+		13
w^{MR19} N1	+		+			+	+	+	+	+		+		+			+	+			+	+	10
N2	+		+			+	+	+	+	+		+		+			+	+			+	+	10

five tested mobile elements in the X chromosome of the offspring from five nonmutated brothers have exactly the same distribution as in the parents. However, the localization of mdg elements in the X chromosome is dramatically changed in the offspring of ct^{MR19} and w^{MR19} males. On the other hand, no diversity can be detected while comparing the progeny of six different ct^{MR19} males or of two different w^{MR19} males. This and similar experiments demonstrate that all changes in the localization of mobile elements occur simultaneously in one germ cell between two cell divisions.

Transposition Explosions and Multiple Changes within the *ct* Locus

Originally, transposition explosions were demonstrated either genetically or by in situ hybridization. We will show here how multiple changes in the location of mobile elements can be detected even within the individual genetic *ct* locus using blot analysis and molecular cloning.

The *ct* locus of *D. melanogaster* was cloned by Tchurikov et al. (1986) and somewhat earlier by Jack (1985) by means of chromosome walking (Bender et al. 1983). Tchurikov et al. (1986) started to walk from the gene *H55* (Tchurikov et al. 1982) located in the 7B subdivision of the X chromosome (Fig. 1). The borders of the *ct* locus are located at about −150 kb and +20 kb of the walk (*0—H55* gene).

The majority of *ct* mutations are housed in a relatively short segment of the locus, from −125 kb to −115 kb (Fig. 1 and 2). All the mututations with a phenotype similar to ct^{MR2} depend on the mdg4 insertion at −117.5 kb. The mutations of the ct^{MRpN} phenotype (wings with small excisions), which independently appeared many times in the ct^{MR2} strain and its derivatives, contain (in addition to mdg4 at −117.5 kb) another mobile element, jockey, inserted into mdg4. Jockey transposes to the same region of mdg4, but it can be located in either of the two possible orientations (Fig. 2).

More than one transposition event can be observed frequently within either the *ct* locus or even its small part. The first example is a mutation change, $ct^{MRpN19} \rightarrow ct^{MR-P}$, that was phenotypically realized by a change in the wing shape. This was the result of an excision of mdg4 with jockey from the position −118 kb and an insertion of the mdg element roo (B-104, mdg5) into the position −123 kb (Fig. 2).

Interestingly, the ct^{MR-P} mutation was identical with the collection ct^n mutation both in phenotypical and molecular terms (a roo insertion). At least in our case, the specificity of a roo insertion depended on the presence of roo LTR at −123 kb in ct^{MRpN19} as well as in the progenitor strain ct^{MR2}. Thus, the mechanism of transposition memory realized here was the same as that described previously for mdg4 reinsertion (Mizrokhi et al. 1985). It seems to

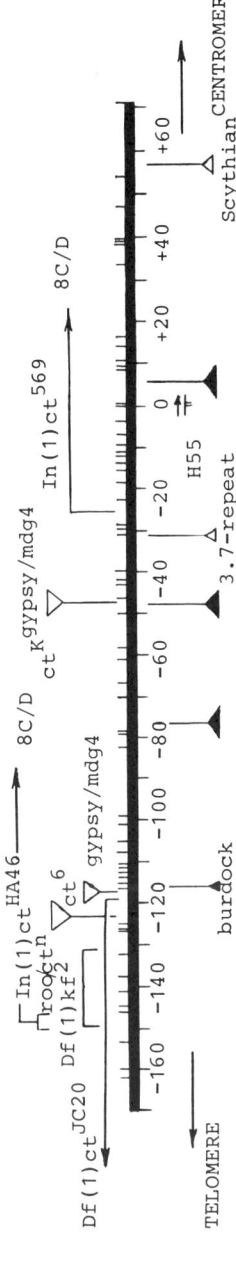

Figure 1

Cloning of the *ct* locus of *D. melanogaster*. The *Eco*RI restriction map of the area including the *ct* locus is presented. The positions of collection mutations induced by insertions of mobile elements, deletions, and inversions are shown above. The mobile repetitive elements present in the Oregon RC strain, but not in other strains, are indicated below; (▲) insertions that appear in all *ct* lethals.

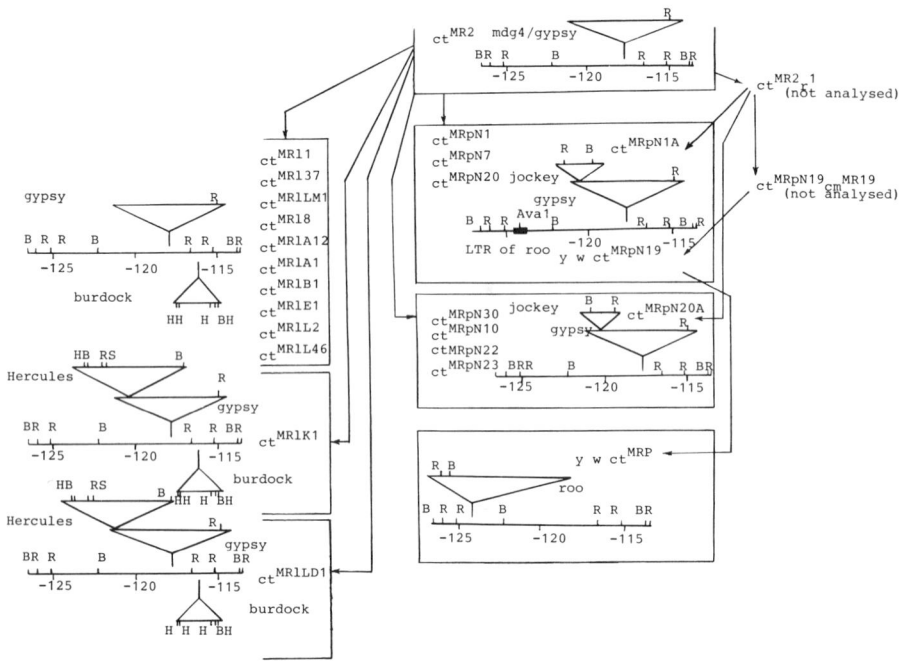

Figure 2

Transposition events that occur in the area −125 to −115 kb of the *ct* locus in the course of transposition explosions. Cloning, restriction mapping, and blot hybridization were performed with the mutants designated in the blocks.

depend on the homologous recombination of a particular mdg element with its solo LTR retained in the genome.

Thereafter, we analyzed a series of lethal mutations in the *ct* locus (ct^l) appearing independently in the ct^{MR2} strain. Unexpectedly, similar insertions were detected in all the twelve cases analyzed (Fig. 2). mdg4 was retained in all the cases, but twice a novel mobile element 7 kb long, designated as Hercules, was inserted into mdg4 at the same position but in two different sites.

In addition, all twelve lethals contained four other insertions at −116 (mobile element burdock), −80, −48, and +5 kb (Fig. 1 and 2). It remains unclear which of these insertions is responsible for the lethal effect. It should be pointed out that the area of active transcription is located at least at the +5-kb region.

Most interesting is that there may be up to five simultaneous insertions in a rather short genetic area (∼240 kb) and from two to three simultaneous transposition events even in a shorter ∼10-kb segment. Thus, it is possible to detect transposition explosions directly at the molecular level. Possibly, the *ct* locus in the ct^{MR2} family is a hot spot for transposition events.

Deletion Mutations and Their Reversions

Not all events occurring during transposition explosions can be attributed to the transposition of mobile elements. Unexpectedly, a number of mutations in the *w* locus were found to depend on a deletion of the genetic material. The pedigree of several *w* mutations appearing in the ct^{MR2} family is shown in Figure 3a. One can see that many of these *w* mutations are capable of reversing to the wild type.

We have cloned and analyzed the structure of genomic segments containing the *w* locus from the strains Oregon RC (wild type), y^{MR19}(yellow) $w^{MR19}ct^{MRpN19}$, and $y^+w^+ct^+sn^{L2}$ (*sn*, singed) (Fig. 3a) using the probe of the *w* locus kindly provided by Dr. G. Rubin.

The restriction map of the *w* locus in the Oregon RC strain is identical with that previously described (O'Hare et al. 1984) (Fig. 3b). In contrast to that, the $y^{MR19}w^{MR19}ct^{MRpN19}$ strain contains a 462-bp deletion in the 700-bp *Pvu*II-*Sac*I fragment. The borders of the deletion were determined by sequencing as shown in Figure 3c. The deletion includes coding sequences at the end of the second exon and at the beginning of the third exon and also induces a frameshift. Therefore, it should completely inactivate the gene. On the other hand, one cannot find any prominent homologies at the ends of the deletion that would account for its appearance.

The restriction map and hybridization properties of the *w* locus obtained from the w^+ revertant are indistinguishable from those of the wild type (Fig. 3b). Thus, the deletion is completely repaired by an unknown mechanism (see Discussion section).

Among the twelve independent *w* mutations analyzed, six were represented by the same ~460-bp deletion, as followed from the results of blot analysis (Fig. 3d). All of them were capable of reverting to the wild type. The Southern blots of genomic DNA from the revertants were the same as those from the wild type. Larger deletions occupying the entire *w* locus were found in three *w* mutations, and unidentified DNA, 0.9 kb long, was inserted into the 1.5-kb *Sal*I fragment in three cases. The latter two groups of *w* mutations did not reverse to the wild type.

The repetitive appearance of similar deletions and their reversions cannot be interpreted in terms of the contamination or internal heterogeneity of X chromosomes within the strains because there are different recessive marker mutations. Therefore, one has to suggest that the material of the *w* locus was not lost in the deletion mutants but was transferred to some other parts of the genome, i.e., to either autosomes or extrachromosomal DNA. The material

A

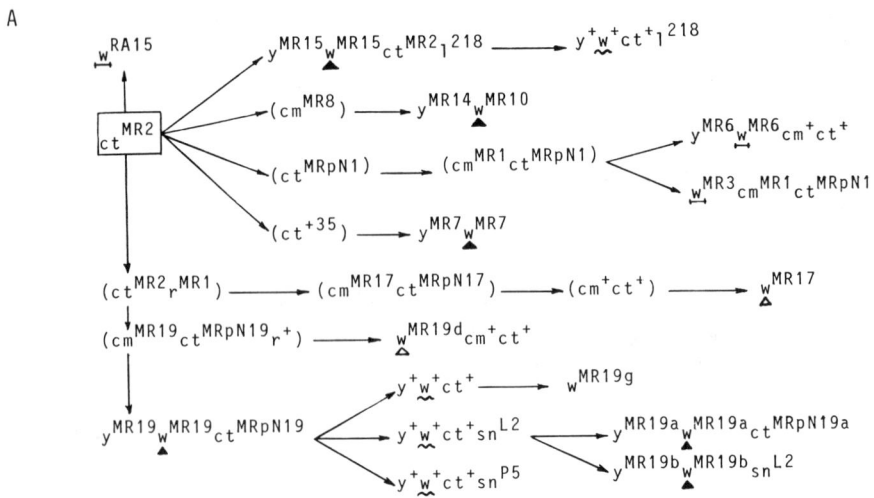

B

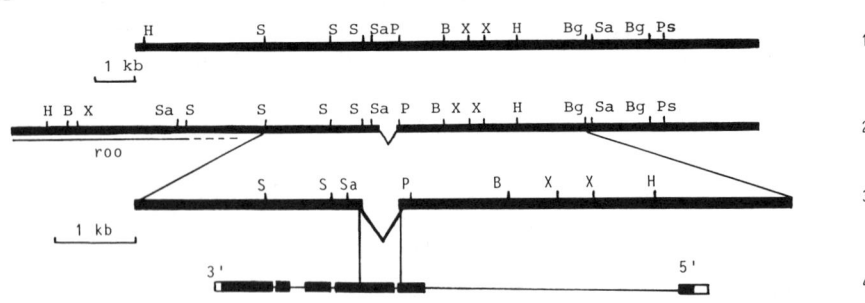

C

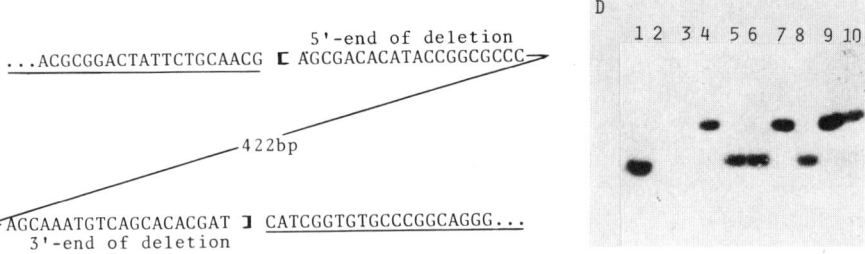

Figure 3 (See facing page for legend.)

was diluted when a homozygous strain was being obtained (and therefore was not detectable on a Southern blot) but still was present in the population giving rise to w^+ reversions.

Thus, besides simple transpositions, several more complex events may occur in the course of transposition explosions in unstable strains of *D. melanogaster*.

Concerted Transpositions in the ct^{MR2} Family of Strains

The intriguing though yet not understood feature of transposition explosions is the frequent appearance of a specific set of genomic changes, which is more or less characteristic of either a particular strain or a certain number of strains.

One such observation has already been mentioned above. The appearance of ct^1 mutations in the ct^{MR2} strain is always associated with four similar insertions in different parts of the *ct* locus (Fig. 1).

A striking example of specific concerted transpositions was obtained in the case of *w* mutations. All the six independent ~460-bp deletions in the *w* locus were accompanied by some other changes in the X chromosome. Firstly, a mobile element roo appeared from 1.0 to 2.5 kb downstream from the *w* locus, as followed from the analysis of the corresponding clone (Fig. 3b) and from the blot analysis of genomic DNA from different *w* mutants (not shown). Secondly, the *y* locus always got a mutation (Fig. 3a), which probably resulted from an mdg2 insertion (as suggested by in situ hybridization experiments with *y* mutants and y^+ revertants). Thus, four different changes occurred together six times. Still, the X chromosomes with *w* deletions were not

Figure 3

Deletion mutants and their repair in the *w* locus. (*A*) Pedigree of the *w* mutations appearing in the ct^{MR2} family of the strains: (▲) ~460-bp *w* deletion; (△) insertion into the *w* locus; (—) deletion of the entire *w* locus. (*B*) Restriction maps of the cloned area of the *w* locus: (H) *Hin*dIII; (S) *Sal*GI; (Sa) *Sac*I; (P) *Pvu*II; (B) *Bam*HI; (X) *Xho*I; (Bg) *Bgl*II; (Ps) *Pst*I. (1) Oregon RC and $y^+w^+ct^+sn^{L2}$ strains; (2) $y^{MR19}w^{MR19}ct^{MRpN19}$ (the deletion and roo insertion are indicated); (3) the region of the *w* locus with a deletion (2) at a higher scale; (4) Map of the *w* transcript (data from O'Hare et al. 1984): (■) coding sequences; (□) nontranslatable ones; (—) introns. (*C*) Nucleotide sequence at the borderlines of the 462-bp deletion in the *w* locus. (*D*) Genomic blots of several strains containing *w* mutations and w^+ reversions. Genomic DNA was isolated from the following strains: (1) $y^{MR7}w^{MR7}$; (4) ct^{+35}; (5) $y^{MR19a}w^{MR19a}ct^{MRpN19a}$; (6) $y^{MR14}w^{MR10}$; (8) $y^{MR19b}w^{MR19b}sn^{L2}$ (deletion in the 1.9-kb fragment); (7) $cm^{MR19}ct^{MRpN19}r^+$; (9) $y^+w^+ct^+sn^{L2}$; (10) $w^{MR19d}cm^+ct^+$ (normal 1.9-kb fragment); (2) $w^{MR3}cm^{MR1}ct^{MRpN1}$; (3) $y^{MR6}w^{MR6}cm^+ct^+$ (entire 1.9-kb fragment is deleted as well as other parts of the *w* locus). The DNA was restricted with *Bam*HI and *Sal*GI restriction endonucleases and probed with 1.9-kb *Bam*HI-*Sal*GI wild-type restriction fragment (see *B*).

identical. In two cases, X chromosomes did not possess any other visible mutations. In two cases, they carried a ct^{MRpN}-mutation; in one *sn* and in one $ct^{MR2}l$ (Fig. 3a). The same conclusion can be drawn from in situ hybridization experiments (not shown).

It is noteworthy that the w^+ reversions in these cases again were accompanied by several other events, i.e., a loss of the mdg element roo, a y^+ reversion, and an mdg2 excision from the 1AB region.

In situ hybridization experiments provide several other examples (see, for example, Gerasimova et al. 1985).

DISCUSSION

Transposition explosions were considered earlier to be a result of transposition activation by an unknown inducer with the P element as a stimulating component. However, the evidence accumulated within the last few years—in particular, the molecular analysis of events taking place in transposition explosions—indicates that their mechanism is more complex.

First of all, in contrast to hybrid dysgenesis, the process involves all the known types of mobile elements such as mdg, P and FB elements, and retroposons. Moreover, as shown above, other events, including the formation of deletions and their repair, may take place. All these events occur simultaneously at the premeiotic stage of germ cell development. Probably, some cellular factors, rather than the specialized transposases of mobile elements, are involved.

It should be pointed out that the number of genomic changes is very high (at least several dozens), and in spite of that, the integrity of chromosomes is not damaged. One the other hand, transpositions of the P element in the course of hybrid dysgenesis frequently lead to chromosomal rearrangements and even to complete "pulverization" of chromosomes.

The appearance of specific sets of mutations and characteristic insertions is another important feature of transposition explosions. Some examples were presented above. In a particular strain, such specific sets of events were repeated in several transposition explosions.

In other words, these characteristic sets of events are in some way encoded. Sometimes such an encoding may be explained by the conservation of LTR used for mdg targeting, as was shown for $ct^+ \rightarrow ct$ (Mizrokhi et al. 1985) and $ct^{MRpN} \rightarrow ct^{MRP}$ (this paper) mutations. This mechanism, however, does not explain many other cases of transposition described above and, in particular, specific deletions and their reversions. To explain these phenomena, a hypothesis was proposed by P. Georgiev (unpubl.) according to which nonhomologous chromosomes interact in transposition explosions to induce the transpositions of mobile elements and sometimes of single-copy sequences to nonhomologous chromsomes or extrachromosomal DNA with the aid of mechanisms similar to gene conversion. Thus, the specificity of transposition at the level of a whole chromosome may be encoded. Interestingly, the properties of autosomes significantly influence the specificity of changes in the localization of mobile elements in the X chromosome. Of course, this hypothesis is not yet well defined and remains quite speculative. Its testing is under way now.

REFERENCES

Bender, W., M. Akam, F. Karch, P. Beachy, M. Peifer, P. Spierer, E. Lewis, and D. Hogness. 1983. Molecular genetics of the bithorax complex in *Drosophila melanogaster*. *Science* **221**: 23.

Gerasimova, T.I. 1981. Genetic instability of the *cut* locus of *Drosophila melanogaster* induced by the *MRh12* chromosome. *Mol. Gen. Genet.* **184**: 544.

Gerasimova, T.I., L.J. Mizrokhi, and G.P. Georgiev. 1984a. Transposition bursts in genetically unstable *Drosophila melanogaster*. *Nature* **309**: 714.

Gerasimova, T.I., L.V. Matjunina, Y.V. Ilyin, and G.P. Georgiev. 1984b. Simultaneous transposition of different mobile elements: Relation to multiple mutagenesis in *Drosophila melanogaster*. *Mol. Gen. Genet.* **194**: 517.

Gerasimova, T.I., L.S. Matjunina, L.J. Mizrokhi, and G.P. Georgiev. 1985. Successive transposition explosions in *Drosophila melanogaster* and reverse transpositions of mobile dispersed genetic elements. *EMBO J.* **4**: 3773.

Ilyin, Y.V., N.A. Tchurikov, E.V. Ananiev, A.P. Ryskov, G.N. Yenikolopov, S.A. Limborska, N.E. Maleeva, V.A. Gvozdev, and G.P. Georgiev. 1978. Studies on the DNA fragments of mammals and *Drosophila* containing structural genes and adjacent sequences. *Cold Spring Harbor Symp. Quant. Biol.* **42**: 959.

Jack, J.W. 1985. Molecular organization of the *cut* locus of *Drosophila melanogaster*. *Cell* **42**: 869.

Mizrokhi, L.J., L.A. Obolenkova, A.F. Priimägi, Y.V. Ilyin, T.I. Gerasimova, and G.P. Georgiev. 1985. The nature of unstable insertion mutations and reversions in the *cut* locus of *Drosophila melanogaster*: Molecular mechanism of transposition memory. *EMBO J.* **4**: 3781.

O'Hare, K., C. Murphy, R. Levis, and G.M. Rubin. 1984. DNA sequence of the *white* locus of *Drosophila melanogaster*. *J. Mol. Biol.* **180**: 437.

Tchurikov, N.A., A.K. Naumova, E.S. Zelentsova, and G.P. Georgiev. 1982. A cloned unique gene of *Drosophila melanogaster* contains a repetitive 3'-exon whose sequence is present at the 3'-ends of many different mRNAs. *Cell* **28**: 365.

Tchurikov, N.A., A.L. Kenzior, N.V. Trubetskaya, N.I. Barbakar, T.K. Johnson, and G.P. Georgiev. 1986. Molecular cloning of the locus *cut* of *Drosophila melanogaster*. *Proc. Acad. Sci. USSR* **286**: 1257.

Transposons as Probes of Regulatory Machinery

PAUL M. BINGHAM, TZE-BIN CHOU, AND ZUZANA ZACHAR
Department of Biochemistry
State University of New York
Stony Brook, N.Y. 11794

OVERVIEW

$su(w^a)$, suppressor-of-white-apricot, is a retrotransposon insertion allele-specific suppressor locus in *Drosophila*. As such, $su(w^a)$ is a presumptive regulatory gene. We review here our studies indicating that $su(w^a)$ expression is autoregulated at the level of splicing. Specifically, the $su(w^a)$ protein product apparently feeds back to block splicing of the first two of the seven introns whose removal is required to produce the message coding for this protein. Implications of these and related observations for mechanisms of regulation of metazoan gene expression are discussed. Furthermore, we briefly review prospects for progress resulting from using transposon parasites to probe the regulatory machinery of their cellular hosts.

INTRODUCTION

Retrotransposons are developmentally programmed, somatically transcribed polymerase II transcription units. This developmental programming apparently involves parasitism of host regulatory information (for recent discussions, see Varmus 1983; Winston et al. 1984; Zachar et al. 1985; Parkhurst and Corces 1986). Retrotransposon parasites are, thus, attractive cases for analysis of the regulatory machinery of their cellular hosts.

In practice, retrotransposon transcription units have an advantage over host genes for such studies. In *Drosophila*, retrotransposon insertion causes a large fraction of spontaneous mutations in host genes (Zachar and Bingham 1982). A subset of these insertion alleles couple expression of the host gene and the inserted retrotransposon so as to generate convenient "reporters," allowing genetic screens for host genes regulating retrotransposon expression (for example, see Zachar et al. 1985; Parkhurst and Corces 1986). Among host genes so identified are allele-specific suppressors and enhancers that interact quite specifically with individual retrotransposon families.

The strict allele-specificity of these suppressors and enhancers is informative. It argues that they do not influence reporter phenotypes for generalized or trivial reasons. Rather, this specificity indicates that suppressors and enhancers are likely to be involved in processes idiosyncratic to individual retrotransposons. These, presumably, include developmental programming of expression.

Motivated by these considerations, we have studied one of these loci in *Drosophila*, $su(w^a)$. The hypomorphic mutant phenotype of the white-apricot (w^a) allele is specifically suppressed by $su(w^a)$, resulting from insertion of the copia retrotransposon into the second intron of the white (w) locus. This copia insertion contains a transcript terminus formation site at which $\sim 95\%$ of w^a transcripts are terminated; some or all of the remaining $\sim 5\%$ of w^a transcripts are processed at the conventional w splice sequences to produce a wild-type mature w message accounting for the low level of w^+ function of the allele (see Zachar et al. 1985 and references therein). Thus, w^a is a reporter for at least two RNA processing events: splicing of the second w^a intron and polyadenylated terminus formation in the copia transposon. Mutational inactivation of $su(w^a)$ increases the levels of wild-type transcripts from w^a, presumably by influencing one or both of these RNA processing events (Zachar et al. 1985.)

Results of our analysis indicate that $su(w^a)$, which apparently regulates a w^a RNA processing event in *trans*, apparently regulates its own expression at the level of RNA processing (Chou et al. 1987; Zachar et al. 1987). We review these studies here and discuss some implications of the study of allele-specific suppressors for analysis of metazoan gene expression.

RESULTS

Transcription of $su(w^a)$

$su(w^a)$ transcripts show a complex developmental pattern (Chou et al. 1987; Zachar et al. 1987). During the first several hours of development, a 3.5-kb polyadenylated RNA is the primary product. Beginning at around cellular blastoderm (~ 3 hours postoviposition at 25°C), the 3.5-kb RNA is largely replaced by two larger RNAs (4.4 kb and 5.2 kb in size). Trace amounts of the 3.5-kb RNA continue to be made throughout postcellular development.

We have characterized the structures of these three transcript classes (Chou et al. 1987), and these structures are diagrammed in Figure 1. The three $su(w^a)$ transcript classes begin and end at the same site and differ only in pattern of splicing. Seven introns are removed to produce the 3.5-kb RNA. The 4.4-kb RNA is produced when the first of these seven is not removed and from the 5.2-kb RNA when the first and second are not removed. As a result of their origins in the failure to remove introns, we refer to the 4.4-kb and 5.2-kb RNAs as "blocked forms."

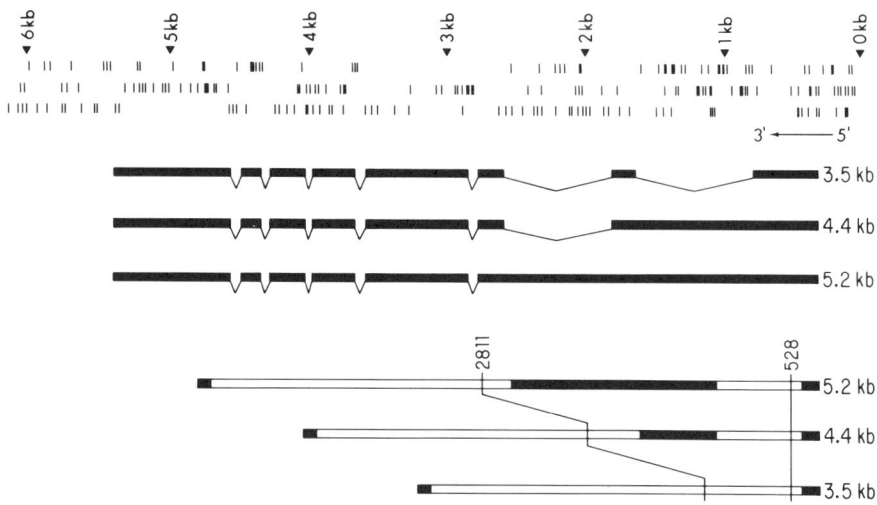

Figure 1
Structures of the three major $su(w^a)$ transcripts based on S_1 protection and cDNA sequence results (Chou et al. 1987). Shown at top is a conventional ORF diagram of the su(w^a) sequence and the standard coordinate grid used to describe the sequence. The three reading frames (5' to 3', right to left; defined beginning at base 1 of the sequenced interval) are indicated as three lines of symbols (top line is frame 1, middle is frame 2 and bottom is frame 3). A vertical line is placed at each translation termination codon in the reading frame. Thus, extended segments of uninterrupted potential peptide coding sequences are indicated by intervals free of vertical lines. Immediately below the ORF diagram are the structures of the three mature $su(w^a)$ transcripts (■)exons; (—) introns; transcripts oriented 5' to 3', right to left). At bottom are the structures of the three mature $su(w^a)$ transcripts diagrammed with introns removed. These are drawn to the same scale as the upper diagrams and are aligned at their 5' ends. The open portions of the bars indicate ORFs longer than 450 bases. The positions of the first in-frame AUG in each ORF (AUGs at coordinates 528 and 2811, respectively) are indicated. The 3.5-kb RNA contains a single ORF extending from an AUG at coordinate 528 through a UGA at coordinate 5364. This extended ORF remains separated in two or three noncontinuous components when one or both of the first two splices fails to occur, resulting in the 4.4-kb and 5.2-kb mature RNAs.

The Decision to Express $su(w^a)$ Is Made at the Posttranscriptional Level

The structure of the 3.5-kb RNA is that of a relatively conventional messenger RNA (Chou et al. 1987; Fig. 1). The bulk of its length consists of a long uninterrupted segment of peptide coding sequence (open reading frame, ORF). This long ORF is flanked by relatively short 5' and 3' presumptive untranslated segments. In contrast, the structures of the 4.4-kb and 5.2-kb RNAs are not those of conventional messenger RNAs. If these two larger RNAs function as messenger RNAs, each must either contain a very short translated ORF near its 5' end with a very long 3' untranslated region or a long 3′ translated ORF with a 5′ untranslated segment well in excess of 1.5-kb (Chou et al. 1987; Fig. 1).

These observations suggested to us that the 3.5-kb RNA might be the $su(w^a)$ message and that the 4.4-kb and 5.2-kb RNAs might merely be by-products of posttranscriptional repression of production of the 3.5-kb RNA, this in spite of the status of the 4.4-kb and 5.2-kb RNA as the preponderant species throughout most of development.

We tested this hypothesis by introducing frameshift mutations into peptide coding sequences of the first, second, fourth, and eighth exons of the 3.5-kb RNA (Zachar et al. 1987; see Fig. 1 for reference). These frameshift mutations all inactivate $su(w^a)^+$ genetic function, and those tested (first, second, and fourth exons) are allelic to one another. Thus, all of these frameshifts apparently affect the same extended block of peptide coding sequence. They do so only in the 3.5-kb RNA; they affect separate blocks of coding sequence in the 4.4-kb and 5.2-kb RNAs.

We conclude on this basis that the 3.5-kb RNA is a message coding for a protein essential for $su(w^a)^+$ function. These experiments failed to detect any messenger RNA function for the 4.4-kb and 5.2-kb RNAs, thus supporting the proposal that these larger RNAs are by-products of posttranscriptional repression of production of the 3.5-kb RNA (Zachar et al. 1987).

Evidence That Posttranscriptional Regulation of $su(w^a)$ Is Autogenous

The results so far reviewed indicate that the 3.5-kb RNA codes for the $su(w^a)$ protein and production of this protein is repressed (partially or entirely) throughout postcellular blastoderm development by preventing occurrence of one or both of the first two splices necessary to produce the 3.5-kb RNA. We have investigated the effects of mutational inactivation of $su(w^a)$ protein coding sequences on the pattern of $su(w^a)$ transcripts (Zachar et al. 1987). These studies indicate that the posttranscriptional repression of production of the 3.5-kb RNA is autogenous. Specifically, we find that inactivation of peptide coding sequences of the 3.5-kb RNA causes persistent expression of the 3.5-kb RNA and failure to undergo the cellular blastoderm transition to production of the 4.4-kb and 5.2-kb RNAs (blocked forms). Moreover, we find that a wild-type $su(w^a)$ allele can act in *trans* to cause efficient production of the blocked transcript forms from a $su(w^a)$ allele whose peptide coding sequences have been mutationally inactivated and which would thus not normally produce the blocked forms (Zachar et al. 1987).

These various results lead us to propose the following model for $su(w^a)$ regulation (Chou et al. 1987; Zachar et al. 1987; Fig. 2). The $su(w^a)$ promoter fires either continuously or sporadically independently of the $su(w^a)$ protein.

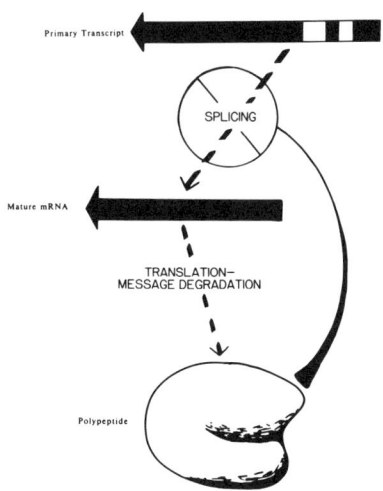

Figure 2
Model for autogenous regulation of $su(w^a)$. The $su(w^a)$ primary transcript is processed to produce a mature, functional messenger RNA (the 3.5-kb RNA [*thick center arrow*]) in the absence of effective levels of the $su(w^a)$ protein. This functional message is translated to produce the $su(w^a)$ protein and is also rapidly degraded. The $su(w^a)$ protein represses further production of the $su(w^a)$ messenger RNA by blocking the removal of one or both of the first two introns from the continuously produced $su(w^a)$ primary transcript (regulated introns [□]) leading to the accumulation of the incompletely processed 4.4-kb and 5.2-kb RNAs.

In the absence of the $su(w^a)$ protein, the primary transcript is spliced seven times to produce the 3.5-kb functional $su(w^a)$ messenger RNA. As the $su(w^a)$ protein (translation product of the 3.5-kb RNA) accumulates, this protein represses its own further synthesis by preventing further production of the 3.5-kb messenger RNA by blocking removal of one or both of the first two of the seven $su(w^a)$ introns (Fig. 2). As the 3.5-kb RNA apparently has relatively short half-life (< one-half hour), this autogenous feedback circuit is capable of precise, temporally responsive control of the levels of $su(w^a)$ protein (Zachar et al. 1987; Fig. 2).

DISCUSSION

Selected General Implications of $su(w^a)$ Autoregulation

The promise of *Drosophila* as a powerful molecular genetic experimental system has begun to come to fruition during the last several years. One unexpected insight gained from this new progress is that genes may be quite commonly turned on and off at the level of splicing. (This is in addition to the extensively documented production of related families of proteins by alternative splicing [for review, see Breitbart et al. 1987].) A few tens of cases of regulated decisions to express genes in *Drosophila* are currently understood in sufficient detail that decisions to express made at the level of splicing might have been detected. Of these at least three clearly defined cases of such regulation have been recognized. In addition to our analysis of $su(w^a)$ reviewed above, these include the transformer (*tra*) gene of the sex determination cascade (Boggs et al. 1987) and the P element (Laski et al. 1986). In light of this frequency of known cases, one would expect many others to be discovered in the next several years both in arthropods and, presumably, elsewhere.

A related point concerns the definition of messenger RNAs. Traditionally, the preponderant, stable, discreet, polyadenylated RNA product of a gene was presumed to be its messenger RNA. However, we note that the 4.4-kb and 5.2-kb $su(w^a)$ RNAs are, by far, the preponderant $su(w^a)$ transcripts throughout postcellular blastoderm development, and yet these RNAs are probably merely by-products of posttranscriptional regulation. Likewise, the nonfunctional form of the *tra* locus transcript (the male form) and the presumably nonfunctional form of the P element transcript are the preponderant transcripts under the appropriate conditions (Laski et al. 1986; Boggs et al. 1987). In general, caution in classifying a polyadenylated RNA as a messenger RNA is indicated. This is especially the case when that RNA in question has an unusual structure (for example, an unusually long presumptive 5' untranslated region).

It is also important to note that these three cases of regulation at the level of splicing were recognized only because the RNA by-products of the regulation were relatively stable. When the by-products of such regulation are unstable, the process will generally escape detection because of the prevalence of approaching study of regulation through analysis of steady-state RNA populations. The profound prevalence of introns in metazoan genes is quite tantalizing in this context. In view of these considerations, we speculate that expression of many (and possibly most) intron-containing genes may be regulated in part by controlling occurrence of one or more splicing events.

Prospects for Progress in $su(w^a)$ Analysis

Numerous cases of production of families of related protein products by alternative splicing and several cases of outright control of expression at the level of splicing are known (see above). However, in none of these cases has the detailed molecular mechanism of regulation been established. As a consequence of the autogenous nature of its splicing regulation, $su(w^a)$ represents a uniquely accessible case for detailed analysis of a *trans*-acting splicing regulator, the $su(w^a)$ protein, and its regulated RNA, the $su(w^a)$ primary transcript.

Implications of Allele-specific Suppressors as Experimental Systems

The analysis of transposon transcription and of allele-specific suppressor loci clearly indicates that most *Drosophila* transposons are transcribed in somatic tissues in a developmentally programmed fashion in response to host regulatory cues (for brief review, see Parkhurst and Corces 1986; Chou et al. 1987). As most or all *Drosophila* transposons are strict germline parasites, this presents a paradox. Although this apparent paradox remains to be unambiguously resolved, it is reasonable to suppose that it results from fundamentally combinatorial control of transposon expression. Under these circumstances, activation of a transposon promoter is only one of several conditions that must be fulfilled for effective expression of transposon genes, and this activation occurs in a large subset of tissues including the germline and various somatic tissues. Full expression might require other factors such as a germline-specific splicing pattern as is observed in the case of P (see above). On this hypothesis, allele-specific suppressor loci represent genes coding for a subset of the circuitry-regulating expression of these germline parasites.

Although the analysis of allele-specific suppressor loci is still in its infancy, the results of the studies completed to date suggest that it will be unusually informative (for brief review, see Parkhurst and Corces 1986). Among other things, these early analyses suggest that suppressor loci are essential genes and that the various suppressor loci currently known are part of one or a small number of interacting regulatory circuits. As a result of the properties of transposon-insertion alleles as reporters, these loci represent one of a very small number of accessible opportunities to apply to metazoans the genetically driven approaches to analysis of regulatory circuits that have been so successful in various microbial systems.

ACKNOWLEDGMENT

This work was supported by National Institutes of Health grant GM-320003 to P.M.B.

REFERENCES

Boggs, R.T., P. Gregor, S. Idriss, J.M. Belote, and M. McKeown. 1987. Regulation of sexual differentiation in *D. melanogaster* via alternative splicing of RNA from the transformer gene. *Cell* **50:** 739.

Breitbart, R.E., A. Andreadis, and B. Nadal-Ginard. 1987. Alternative splicing: A ubiquitous mechanism for the generation of multiple protein isoforms from single genes. *Annu. Rev. Biochem.* **56:** 467.

Chou, T.-B., Z. Zachar, and P.M. Bingham. 1987. Developmental expression of a regulatory gene is programmed at the level of splicing. *EMBO J.* **6:** 4095.

Laski, F.A., D.C. Rio, and G.M. Rubin. 1986. Tissue-specificity of *Drosophila* P element transposition is regulated at the level of mRNA splicing. *Cell* **44:** 7.

Parkhurst, S.M. and V.G. Corces. 1986. Retroviral elements and suppressor genes in *Drosophila*. *Bioessays* **5:** 52.

Varmus, H.E. 1983. Retroviruses. In *Mobile genetic elements* (ed. J. Shapiro), p. 411. Academic Press, New York.

Winston, F., D.T. Chaleff, B. Valent, and G.R. Fink. 1984. Mutations affecting Ty-mediated expression of the HIS4 gene of *Saccharomyces cerevisiae*. *Genetics* **107:** 179.

Zachar, Z. and P.M. Bingham. 1982. Regulation of white locus expression: The structure of mutant alleles at the white locus of *D. melanogaster*. *Cell* **30:** 529.

Zachar, Z., T.-B. Chou, and P.M. Bingham. 1987. Evidence that a regulatory gene autoregulates splicing of its transcript. *EMBO J.* **6:** 4105.

Zachar, Z., D. Davison, D. Garza, and P.M. Bingham. 1985. A detailed developmental and structural study of the transcriptional effects of the insertion of the copia transposon into the white locus of *Drosophila melanogaster*. *Genetics* **111:** 495.

Molecular Basis of Transposable Element-induced Mutations in *Drosophila melanogaster*

PAMELA K. GEYER,* MELVIN M. GREEN,[†] AND VICTOR G. CORCES*
*Department of Biology
The Johns Hopkins University
Baltimore, Maryland 21218
[†]Department of Genetics
University of California at Davis
Davis, California 95616

OVERVIEW

In order to understand the mechanism underlying the mutagenic effect of the gypsy transposable element, we carried out a detailed structural and functional analysis of the yellow (*y*) locus in wild-type flies, in the gypsy-induced allele y^2, and several phenotypic revertants. The complex pattern of temporal and spatial expression of the yellow gene is controlled by four distinct and independent regulatory elements. Sequence analysis of y^2 revertants suggests that the phenotype of the gypsy-induced allele is not due to the integration of the transposon into regulatory sequences, or distancing of these sequences from the TATA box, but rather to the interaction between transcriptional regulatory elements located in gypsy and the yellow control regions responsible for the proper expression of this gene in the affected tissues.

INTRODUCTION

Insertion of transposable elements is a major cause of spontaneous mutations in *Drosophila melanogaster*. The mechanisms by which an element alters gene expression will depend upon several factors such as the location of the insertion within the gene, the manner in which the target gene is regulated, and the structural and transcriptional properties of the mobile element.

Several classes of transposable elements have been identified in *Drosophila* (for reviews, see Rubin 1983; Finnegan and Fawcett 1986). The members of one of these classes show structural characteristics similar to those of vertebrate retroviral proviruses, suggesting that these elements are members of the eukaryotic family of retrotransposons. The gypsy transposable element is a member of this class. This transposon is approximately 7.5 kb in length and is flanked by two 0.5-kb long terminal repeats (LTRs) that contain signals for transcription initiation and termination. It is transcribed into a 6.5-kb polyadenylated RNA which encodes proteins with similarities to the retroviral protease, endonuclease, and reverse transcriptase (Marlor et al. 1986). Accumulation of the gypsy transcript is temporally regulated, being maximally expressed during the pupal stages of development (Parkhurst and Corces 1985). The insertion of this transposon has been found to be associated with a number of spontaneous mutations whose phenotypes can be reversed by mutations at unlinked suppressor loci (Modolell et al. 1983; Parkhurst and Corces 1986a).

The mechanisms of gypsy mutagenesis and phenotypic suppression are not clearly understood. To examine these processes, we have chosen to study the y locus (1–0.0) and the spontaneous mutant allele y^2 that results from the insertion of the gypsy element near the gene. The yellow locus is necessary for proper pigmentation of adult and larval cuticular structures. This gene has been cloned and a 1.9-kb RNA identified as the yellow transcript (Biessman 1985; Campuzano et al. 1985; Chia et al. 1986; Parkhurst and Corces 1986b). This RNA accumulates during late embryogenesis and in early larval development, declines during late larval stages and puparium formation, and increases again approximately 50 hours after pupation. The phenotype of flies carrying the y^2 mutant allele is such that coloration in larval structures and adult bristles are unaffected by gypsy integration, whereas pigmentation in all other adult structures is mutant. Analysis of RNA accumulation in y^2 mutant flies shows that the level of the 1.9-kb transcript is unchanged in the larval stage of development but is substantially reduced during pupation (Parkhurst and Corces 1986b). Several possible explanations can be proposed as to how the gypsy element exerts its mutagenic effect on the yellow gene. For example, the observed phenotype could result from integration of the transposon into an intron present only in the pupal transcript; alternatively, the transposon could be integrated into regulatory sequences necessary for pupal but not larval expression, or transcription of the gypsy element could interfere with the ability of the yellow gene to be transcribed during the pupal stages of development. To approach these issues, we have carried out a structural and functional analysis of the yellow locus in wild-type flies, in the gypsy-induced y^2 allele, and in some of its phenotypic revertants.

RESULTS

The first step in our analysis was to determine the limits of the yellow transcription unit by comparing DNA sequence data obtained from genomic and larval and pupal cDNA clones. The results of this investigation are shown in Figure 1. The yellow transcription unit is composed of two exons separated by a large 2720-bp intron (Geyer et al. 1986). Both the transcription start site and the splice pattern are identical for the larval and pupal RNAs, indicating that the yellow locus encodes a single transcript that is differentially expressed

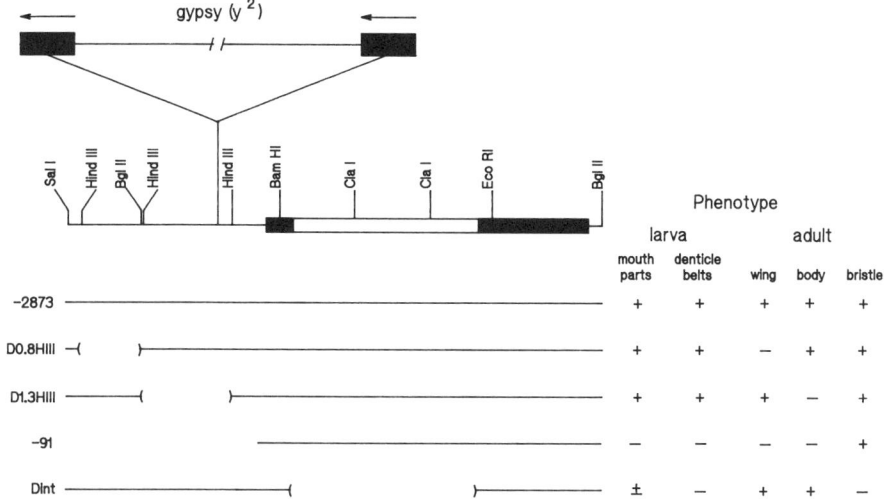

Figure 1
The structure of the yellow gene. The upper part of the figure shows a restriction map of the yellow locus as defined by the sequences that are necessary and sufficient to obtain phenotypic rescue of null y^- flies after P element-mediated transformation. The insertion of the gypsy element relative to the yellow gene is also diagrammed. The lower half of the figure indicates the structure and phenotypic effect of the different plasmids used in the transformation experiments described in the text. The blank spaces between brackets signify deleted sequences.

during the development of the fly. Sequence analysis of genomic clones from the y^2 allele shows that the gypsy insertion point is located 700 bp upstream of the transcription initiation site, such that the transcription of the element is opposite to that of the yellow gene (Fig. 1). Since the gypsy is inserted outside of the yellow coding region, its mutagenic effect cannot be explained as a consequence of pupal-specific disruption of RNA processing or transcription termination.

To understand the molecular basis of the tissue-specific gypsy-induced adult phenotype, we carried out a detailed functional analysis of the yellow promoter. DNA sequences controlling various patterns of the temporal and spatial transcription of the yellow gene were identified by studying the phenotypic effect of in vitro deleted plasmids after P element-mediated transformation into *Drosophila* preblastoderm embryos. A 7.7-kb *Sal*I-*Bgl*II fragment (see Fig. 1) containing the yellow coding region and 2.8 kb of 5' and 0.1 kb of 3' sequences was inserted into the transformation vector, Carnegie 20, which contains the rosy$^+$ (ry^+) (xanthine dehydrogenase) gene as a selectable marker to identify transformed flies (Rubin and Spradling 1982). This plasmid rescues the mutant phenotype of larval structures (mouth hooks and denticle belts) and adult tissues (wings, body, and bristles) when used to transform embryos carrying a deletion of the yellow and achaete loci (y^-, ac^-) or ry^-. Deletions were then made in different regions of the 7.7-kb SalI-BglII fragment; the altered plasmids were injected into y^-, ac^-; ry^- embryos and the resulting phenotype of the transformed adults was analyzed. Flies carrying the plasmid deleted between −2690 and −1853 bp from the start of transcription (pD0.8HIII in Fig. 1) have wild-type pigmentation in larval and adult structures with the exception of the wings. The orientation of these sequences can be inverted with respect to the yellow coding region without affecting the expression in the wings (Geyer and Corces 1987). Deletion of DNA sequences located between −1853 and −495 bp from the transcription initiation site (pD1.3HIII in Fig. 1) results in wild-type coloration of larval structures as well as the wings and bristles of the adult flies, but the body of the adult animals shows mutant pigmentation. Again, the orientation of these sequences can be inverted with no effect on the transcription of the yellow gene. The expression of this locus, during the early stages of development, is only affected when sequences very close to the yellow promoter are deleted (see construct −91 in Fig. 1). Finally, sequences responsible for the proper coloration of the adult bristles could be located mainly in the intron region (pDint in Fig. 1). Deletion of these sequences also resulted in a larval phenotype with mutant denticle belts and mouth parts that showed a coloration intermediate between that of yellow and wild-type flies (Geyer and Corces 1987).

These results suggest that the temporal and spatial expression of the yellow locus is regulated by at least four different and separate elements that control the transcription of this gene during development in a tissue-specific fashion. It is interesting to note that the sequences responsible for yellow expression in the wings and body of the adult flies are located upstream of the insertion site of the gypsy element in the y^2 allele. It is these structures that show abnormal pigmentation in y^2 adult flies, whereas the expression of the yellow locus in structures that show a wild-type phenotype is controlled by sequences located downstream from the gypsy insertion site in y^2. These results suggest that the gypsy element might be inserted into sequences important for the proper expression of the yellow locus in the affected tissues, or that, as a consequence of the insertion, sequences responsible for this expression lie 7.5 kb further from the promoter in the mutant than in the wild-type locus and are, therefore, unable to function properly at this distance.

To test these various possibilities, we carried out a molecular analysis of ten complete revertants of y^2. These revertants arose spontaneously and were found to fall into two different classes. Class A revertants appeared as single events occurring in the progeny from either a y^2 male or female, whereas class B revertants always arose as a cluster (15–30 individuals) of phenotypically wild-type flies from a mutant mother. DNA from the yellow locus of both classes of revertants was cloned and sequenced in order to study the molecular basis of the reverted phenotype. Class A mutations contained a normal yellow locus with the exception that a single 0.5-kb LTR was then present at the original insertion point of the gypsy element in the y^2 mutation (Fig. 2A). These revertants probably arise by homologous recombination between the two LTRs of gypsy. The fact that they are phenotypically wild type while still containing 0.5 kb of gypsy sequences suggests that the gypsy element in the y^2 allele is not inserted into a region necessary for proper transcription of the locus (Geyer et al. 1986). Class B revertants showed a much more complicated structure. Sequence analysis showed that this group contained the 3' LTR of the gypsy element and half of the 5' LTR (nucleotides 1–163); the rest of the gypsy element had been deleted and replaced by a novel ca. 8-kb element that contains a poly(A) tail at its 3' end (Fig. 2B). The fact that these revertants are phenotypically wild type, but still contain at least 8 kb of DNA sequences inserted in the same place where the gypsy element is located in y^2, suggests that gypsy does not cause the mutant phenotype of y^2 by merely increasing the distance between the yellow promoter and sequences responsible for its proper expression in the tissues of the wing and adult body (P. Geyer et al., in prep.).

DISCUSSION

Transposable elements can cause mutations by a variety of mechanisms. They can produce a mutant phenotype by inserting into the coding region of a gene and causing the synthesis of an altered protein or by inserting and disrupting

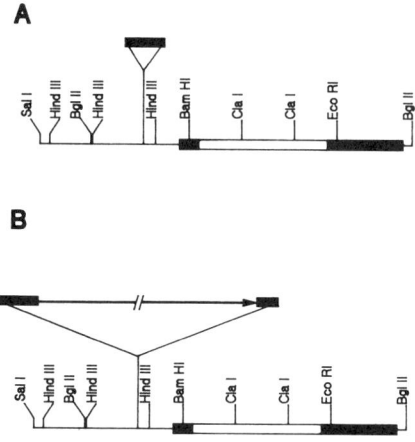

Figure 2
The structure of y^2 revertants. This figure shows the structure of the yellow locus and inserted sequences in various types of revertants of gypsy-induced mutations. Figure 2A shows revertants containing a single gypsy LTR at the insertion site (class A revertants), whereas Figure 2B indicates the organization of class B revertants in which deletions of gypsy sequences and insertion of new transposons have taken place (see text for further reference).

sequences important for proper transcription initiation, RNA processing, or transcription termination. In addition, retrotransposons carry with them their own transcription signals that can interact with regulatory sequences of adjacent genes affecting their expression and causing a mutant phenotype (Rubin 1983; Finnegan and Fawcett 1986). The results presented here suggest that the latter is the mechanism by which the gypsy element causes mutations in the yellow gene.

The yellow locus shows a complex pattern of developmental and tissue-specific expression. Sequences located close to the TATA box and in the large intron are responsible for controlling yellow locus transcription in the mouth parts and denticle belts during the larval stages of development and in the bristles in the adult animals. On the other hand, the proper coloration of the wings and body of adult flies is regulated by enhancer-like elements that reside upstream of the gene, between -0.4 and -2.8 kb (Geyer and Corces 1987). Since the insertion of the gypsy element in the y^2 allele only results in a mutant phenotype in the wings and body color of the adults, it follows that the mutagenic effect of gypsy is the consequence of its interaction with these two enhancer elements.

The analysis of phenotypic revertants of y^2 sheds some light on the molecular basis of this interaction. All revertants analyzed still contain gypsy sequences at the same insertion point as y^2, indicating, as a first possibility, that the mutant phenotype is not a consequence of insertional disruption of regulatory sequences. The second possibility, that gypsy insertion results in an increased distance between the TATA box and the control elements responsible for yellow gene transcription in the wings and body of the adult, is not supported by the structure of class B revertants. These complete revertants contain new transposable elements in addition to gypsy LTRs, bringing the control elements even further away from the yellow promoter.

In view of these findings, we propose that gypsy exerts an active mutagenic effect on genes it disrupts by the interaction of gypsy regulatory sequences with the transcriptional controlling elements of the affected gene. This interaction between regulatory sequences is most likely transcription-contingent, and, since the gypsy element is preferentially transcribed during the pupal stages of development (Parkhurst and Corces 1985), this could explain the apparent specificity in the pupal-dependent phenotypes of the mutations induced by gypsy. This type of mutagenic effect is probably not restricted to gypsy, but it might also be employed by other transposable elements. The fact that *Drosophila* retrotransposons show different patterns of developmental expression (Parkhurst and Corces 1987) adds a new dimension to the specificity of their mutagenic capabilities. Furthermore, this hypothesis offers a logical framework to explain the effect of second-site suppressor mutations on gypsy-induced alleles, since these suppressor loci encode products that control the transcription of the gypsy element (Parkhurst and Corces 1986b,c). The lack of the suppressor gene-encoded proteins will have an effect on the expression of gypsy and, therefore, on the transcription-dependent interaction between gypsy regulatory regions and promoter sequences of the mutated gene. The disruption of this interaction ultimately results in the normal expression of the gene and a wild-type phenotype.

ACKNOWLEDGMENTS

This work was supported by Public Health Service grant GM-32036 from the National Institutes of Health. P.K.G. was supported by National Research Service Award GM-11156 from the National Institutes of Health.

REFERENCES

Biessman, H. 1985. Molecular analysis of the yellow gene (*y*) region of *Drosophila melanogaster. Proc. Natl. Acad. Sci.* **82:** 7369.

Campuzano, S., L. Carramolino, C.V. Cabrera, M. Ruiz-Gomez, R. Villares, A. Boronat, and J. Modolell. 1985. Molecular genetics of the *achaete-scute* gene complex of *D. melanogaster. Cell* **40:** 327.

Chia, W., G. Howes, M. Martin, Y.B. Meng, K. Moses, and S. Tsubota. 1986. Molecular analysis of the *yellow* locus of *Drosophila. EMBO J.* **5:** 3597.

Finnegan, D.J. and D.H. Fawcett. 1986. Transposable elements in *Drosophila melanogaster. Oxf. Surv. Euk. Genes* **3:** 1.

Geyer, P.K. and J.G. Corces. 1987. Separate regulatory elements are responsible for the complex pattern of tissue specific developmental transcription of the *yellow* locus in *Drosophila melanogaster. Genes Dev.* **1:** 996.

Geyer, P.K., C. Spana, and V.G. Corces. 1986. On the molecular mechanism of gypsy-induced mutations at the *yellow* locus of *Drosophila melanogaster. EMBO J.* **5:** 2657.

Marlor, R.L., S.M. Parkhurst, and V.G. Corces. 1986. The *Drosophila melanogaster* gypsy transposable element encodes putative gene products homologous to retroviral proteins. *Mol. Cell. Biol.* **6:** 1129.

Modolell, J., W. Bender, and M. Meselson. 1983. *Drosophila melanogaster* mutations suppressible by the *suppressor of hairy-wing* are insertions of a 7.3 kilobase mobile element. *Proc. Natl. Acad. Sci.* **80:** 1678.

Parkhurst, S.M. and V.G. Corces. 1985. *Forked*, gypsys and suppressors in *Drosophila. Cell* **41:** 429.

———. 1986a. Retroviral elements and suppressor genes in *Drosophila. Bioessays* **5:** 52.

———. 1986b. Interactions among the gypsy transposable element and the yellow and the suppressor of hairy-wing loci in *Drosophila melanogaster. Mol. Cell. Biol.* **6:** 47.

———. 1986c. Mutations at the suppressor of forked locus increase the accumulation of gypsy-encoded transcripts in *Drosophila melanogaster. Mol. Cell. Biol.* **6:** 2271.

———. 1987. Developmental expression of *Drosophila melanogaster* retrovirus-like transposable elements. *EMBO J.* **6:** 419.

Rubin, G.M. 1983. Dispersed repetitive DNAs in *Drosophila*. In *Mobile genetic elements* (ed. J.A. Shapiro), p. 329. Academic Press, New York.

Rubin, G.M. and A.C. Spradling. 1982. Genetic transformation of *Drosophila* with transposable element vectors. *Science* **218:** 348.

Regulation of Expression and Transposition of Murine Endogenous Retroviral Elements

MICHAEL C. WILSON, PAUL F. POLICASTRO, AND MERETE FREDHOLM
Research Institute of Scripps Clinic
Department of Molecular Biology
La Jolla, California 92037

OVERVIEW

The majority of murine leukemia virus (MLV)-related sequences in the murine genome are defective for production of infectious virus but are transcribed in a variety of tissues. Their transcription is governed by at least two *trans*-acting genes, *Gv-1* and *Gv-2*, as well as tissue specific factors. Some of these endogenous retroviral transcripts contain deletions of *gag-pol* and *env* coding sequence. Oligonucleotide probes spanning these deletions have been used to both isolate *Gv-1* responsive genes from murine strain 129 and demonstrate the mobility of these elements among different strains of inbred mice. A target sequence for *trans* regulation within the long terminal repeats (LTR) of these endogenous sequences, but not in infectious virus, is postulated. We propose that the movement of these sequences, mediated by transcription and retrotransposition, may represent mutagenic events within the murine genome.

INTRODUCTION

Retroviral related sequences are endogenous to the genome of mice and to most mammalian genomes including that of human. These elements may constitute as much as 1% of genomic DNA, but for the most part do not give rise to infectious viral particles. In the mouse, one class of retroviral sequence is highly homologous to, but can be distinguished from, infectious type C MLV. These endogenous elements are present as 12–30 copies per haploid genome (O'Neill et al. 1986; for review, see Stoye and Coffin 1985) in all inbred and wild subspecies of *Mus musculus* but are absent, in distant murine species such as *Mus cooki*, *Mus caroli* and *Mus pahari* (Schmidt et al. 1984; Kozak and O'Neill 1986). Thus, as with the retroviral-like transposable elements of yeast, maize, and *Drosophila*, their multiplicity and potential for movement within the genome suggests that these murine endogenous retroviral elements can act as insertional mutagens, altering the expression of cellular genes. Important precedents for this mutagenic role are the activation of certain cellular proto-oncogenes after adjacent integration of recombinant mink cells focus-forming (MCF) virus during retroviral mediated leukemogenesis in mice (Corcoran et al. 1984; Cuypers et al. 1984; Selton et al. 1984) and of murine mammary tumor virus (MMTV), a type B retrovirus during mammary tumorigenesis (Peters et al. 1983; Nusse et al. 1984). Similarly, the insertion of a complete ecotropic type C virus in DBA/2 mice at the dilute locus alters gene(s) mediating coat color (Jenkins et al. 1981; Strobel et al., this volume). Reversion of this mutation can result from excision of retroviral sequence by homologous recombination through the LTR (Copeland et al. 1983).

In certain strains of inbred mice at least a subset of endogenous, noninfectious retroviral sequences is actively transcribed resulting in a tissue-specific accumulation of polyadenylated transcripts (Levy et al. 1982, 1985a). This provides the potential for reverse transcription of these transcripts into circular, double stranded DNA forms, as generated in a typical retroviral life cycle, and integration at new sites possibly within or adjacent to genes in the genome. The regulation of endogenous retroviral gene expression was initially described by Boyse and colleagues as the G_{IX} system, in which the appearance of a MLV-related thymocyte antigen was mapped to the IX linkage group of the murine genome (Stockert et al. 1971). Subsequent studies defined G_{IX} regulation as involving two independent loci *Gv-1* and *Gv-2*. The *Gv-1*^{a} positive allele is codominant in penetrance, whereas the *Gv-2*^{a} positive allele has a dominant effect on G_{IX} expression. *Gv-2* has been mapped to mouse chromosome 7, but *Gv-1* has eluded classical genetic mapping techniques, showing equal linkage at a distance of 35–50 cM from *Fv-1* on CH4 and *H-2* on CH17; linkage to adjacent CH4 marker *Gpd-1* or the *H-2* associated *Tla* locus, however, is not observed (for review, see Boyse 1977). To define the regulation conferred by *Gv-1*, Boyse and colleagues developed a congenic partner to the highly expressive G_{IX^+} 129 strain by introducing the G_{IX} negative *Gv-1*^{b} allele in a cross with C57BL/6 (Stockert et al. 1975). Here we describe the isolation of a subset of the endogenous retroviral elements present in the 129 congenic strains and our findings to date on their mobility and expression, both of which are proposed to be under control of the *Gv-1* and *Gv-2* loci.

RESULTS

Gv-1 is proposed to act on endogenous retroviral gene expression in *trans*. This model was based on the observed coordinate transcriptional regulation of a number of polyadenylated RNA transcripts in congenic 129 strains bearing alternative alleles at *Gv-1* (Levy et al. 1982). As illustrated in Figure 1, RNA from various tissues hybridized with an endogenous retroviral probe demonstrates that multiple RNA transcripts are present at high relative

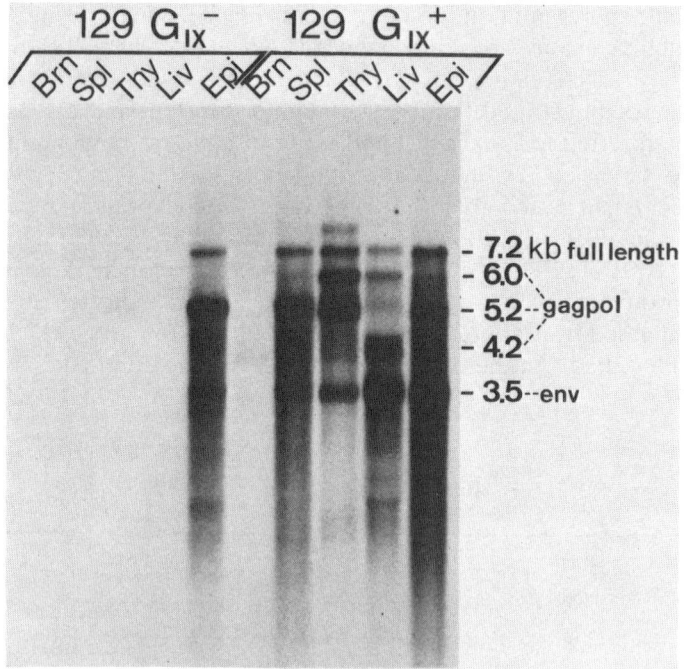

Figure 1
Regulation of endogenous retroviral transcripts in 129 G_{IX^+} and G_{IX^-} strains. Polyadenylated RNA of different tissues (2.5 µg, except for epididymus where 1 µg was applied) was fractionated on a 0.9% formaldehyde-agarose gel, transferred to nitrocellulose, and probed with a ^{32}P-labeled cDNA containing endogenous *pol* and *env* sequence. (Brn) brain; (Spl) spleen; (Thy) thymus; (Liv) liver; (Epi) epididymus.

abundance in G_{IX^+} animals compared to the 129 G_{IX^-} strain. Over exposure of the hybridization shows that transcripts are present in 129 G_{IX^-} tissues at a 10–20-fold lower abundance (not shown). The major exception to the apparent coordinate regulation is the abundant level of a 5.2-kb transcript and to a lesser extent a 3.0-kb transcript 129 G_{IX^-} epididymis. S_1 nuclease protection assays suggest, in fact, that this 5.2-kb transcript is not expressed in 129 G_{IX^+} epididymal tissue (Levy et al. 1985a) and may be a constitutively transcribed product of a retroviral element contributed by the C57BL/6 genome used to provide the Gv-1^b allele in the 129 G_{IX^-} strain.

The retroviral transcripts differentially expressed in the two partner 129 strains range in size from 7.2 kb to 3.0 kb in length. Both the 7.2- and 3.0-kb transcripts contain envelope (*env*) coding sequence, consistent with a structure of full length and *env*-coding spliced mRNAs typical of infectious retrovirus. In contrast, the intermediate 4.0–6.0-kb transcripts, which show a tissue specific representation, do not contain *env* probe hybridizing sequence, but do contain *gag* and U3 LTR sequence (Levy et al. 1985a). The nucleotide sequence of cDNA clones to two of these transcripts, from spleen and epididymus, demonstrated that they contain a different and extensive deletion of *env*, referred to as *del env*-1 and *env*-2 and that the breakpoints of both mapped to seven bp direct repeats found in nondeleted cDNA sequences (Fig. 2). This suggested that the transcripts were expressed by retroviral elements deleted through homologous recombination, rather than resulting from an atypical splicing event of full-length transcripts. Analysis of S_1 nuclease protection experiments (Levy et al. 1985a) indicates at least two more *env* deleted transcripts, diagrammed in Fig. 2, for which cDNA clones were not identified. As these *gag-pol* transcripts, deleted of *env*, are coordi-

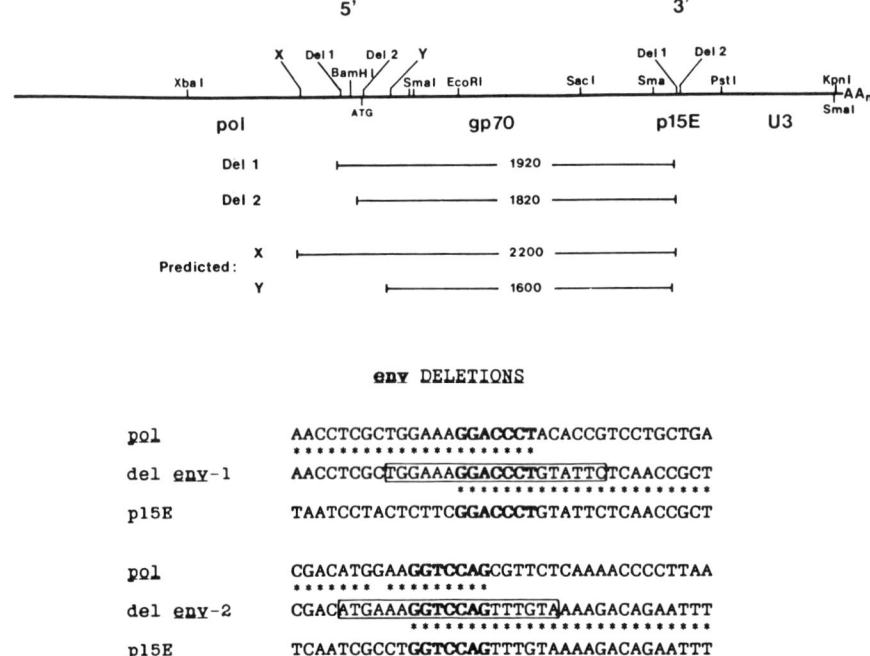

Figure 2

Deletions of *env* sequence are mediated by 7-bp direct repeats. The top schematic represents the position of the endpoints of two cloned *env*-deleted transcripts, *del env*-1 and -2, as well as additional *env* deletions, x and y, determined by S_1 analysis (Levy et al. 1985a) for which no cloned cDNAs are available. The lower portion compares the sequence of the deleted transcripts with that determined from an intact sequence (Levy et al. 1985b) to illustrate the 7-bp repeat (in bold type face) at the breakpoints of *del env*-1 and -2 loci. Sequence homology is indicated by asterisks and the boxed regions indicate the sequence used to generate *env*-1 and *env*-2 specific oligonucleotides used in Fig. 3.

nately regulated by *Gv-1*, their respective genomic sequences could be considered functional targets of *Gv-1* and, thus, be distinguished from the multitude of potentially quiescent, *Gv-1* nonresponsive retroviral elements.

To identify such *Gv-1* responsive genomic sequences in various strains of mice, oligonucleotide probes spanning these deletion breakpoints were hybridized to Southern blots of *Eco*RI digested DNA under conditions that distinguish between full-length and partial homology. As summarized in Figure 3 the 129 G_{IX^+} and G_{IX^-} congenic partner strains contain an identical complement of four *del env*-1 loci and a single *env*-2 locus, based on the size of fragments generated by *Eco*RI cleavage in flanking nonretroviral DNA. Interestingly, the strain 129/SvJ contains an additional *del env*-1 hybridizing sequence, whereas the independent lines 129/J and 129/Rr appear identical to the 129 congenic lines raised by Boyse. Inbred strains 101 and LP/J, closely related to 129, contain at least one hybridizing fragment comigrating with

		env Del-1 loci					env Del-2 loci			additional gag-pol Del loci	
		3.1	3.2	11.1			15.3				
129+		*13.5*		*10*	9	8.8	8.9			7.4	6.6
129-		*13.5*		*10*	9	8.8	8.9			7.4	6.6
129/J		*13.5*		*10*	9	8.8	8.9			7.4	6.6
129/SvJ		*13.5*	12	*10*	9	8.8	8.9			7.4	6.6
129/Rr		*13.5*		*10*	9	8.8	8.9			7.4	6.6
101	18	-		10	9.4	-		9.6	8.9	7.4	*13.5*
LP/J	18	*13.5*		-		9		9.6	8.9		
A/J	23	-		-		9	20	10.5	8.9		
BALB	23	-	12	-		9	20	10.5	-		8.1
C3H		-		-		9	20		-		8.1
C57		-	12	-		-	22		-		8
DBA		-		-		-			-	25	8.1
AKR		-		-		-	20	10.5			*13.5*

Figure 3

*Eco*RI fragments in kb recognized by hybridization with *del env*-1 and -2 and *del gag-pol* oligonucleotide probes. Strain specific distribution of *env* and *gag-pol* deleted endogenous retroviral sequences demonstrates their mobility. Oligonucleotide probes depicted in Fig. 2 that detect *env*-1 and *env*-2 deletions and another that was made to the *gag-pol* deletion found in the *env*-1, 3.1 isolate, were used as probes to blots of *Eco*RI digested genomic DNAs of various inbred strains. Independent loci were scored by the length of the hybridizing *Eco*RI fragment as indicated in kilobases. A dash indicates the lack of those elements (A, 3.1; B, 3.2; C, 11.1; and D, 15.3) which are found in the 129 genome and whose cloned structure is shown in Fig. 4. Additional loci for which clones are not available lie outside the vertical dotted lines, and include additional *gag-pol* deleted structures not bearing *env*-1 and *env*-2 hybridizable sequence (left columns) and *env* deleted sequences not recognized by the *gag-pol* deletion probe, i.e., those not depicted in italics.

those found in the 129 genome. An additional *env*-1 locus is present in both strains, within an 18-kb *Eco*RI fragment, and strain 101 contains a locus in a 9.4-kb fragment. The single *env*-2 locus of 129, present on a 8.9-kb *Eco*RI fragment, is also found in related 101 and LP/J strains as well as in the more distant A/J strain. The 101 and LP/J strains also carry a 9.6-kb *env*-2 fragment. Strain 101 carries an additional 7.4-kb *env*-2 not found in other strains. More distantly related inbred strains contain other *env*-1 loci with *Eco*RI fragments of 23 kb and 12.5 kb and *env*-2 fragments of 22, 20, and 10.5 kb. Thus, between strains these specific *env*-deleted retroviral elements are not fixed but appear to be mobile elements within the murine genome.

To isolate these potential targets for *Gv-1* regulation, genomic DNA *Eco*RI fragments were cloned from 129 G_{IX^+} DNA using the respective oligonucleotide probes. Since the deletion of *env* sequence results in the loss of an internal *Eco*RI site common to full length endogenous retroviral sequence (Hoggan et al. 1982), these genomic clones bear both 5' and 3' LTR sequences and flanking DNA as expected (see Fig. 4). Restriction enzyme map and nucleotide sequence analyses of the LTRs indicated that the *env*-1

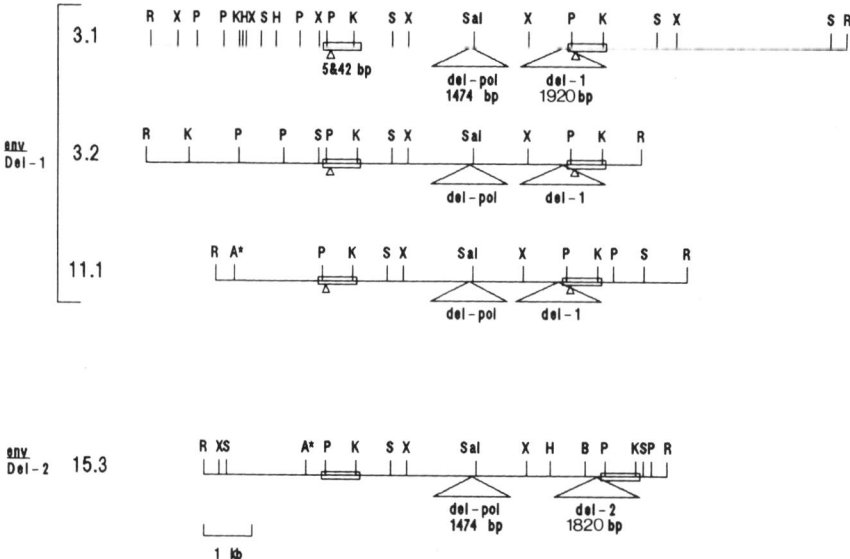

Figure 4

Restriction maps of *del env*-1 and *env*-2 proviral loci isolated from the 129 genome. The loci were isolated from an EMBL-4 library of complete *Eco*RI digested 129 G_{IX^+} genomic DNA by screening with oligonucleotide probes, indicated in Fig. 2 (in prep.). The approximate position and extent of *env* and *gag-pol* (*del-pol*) and U3 deletions are indicated. The 5' and 3' LTRs are represented as boxes. The position of restriction sites are indicated for enzymes *Eco*RI (R), *Xba*I (X), *Pst*I (P), *Kpn*I (K), *Hind*III (H), *Sac*I (S), *Ava*II (A), *Bam*HI (B) and *Sal*I (Sal).

and *env*-2 loci each represent one each of the two divergent polytropic and modified polytropic classes of endogenous retroviral elements. This classification is based on distinctive *env* coding regions, which includes a 9 amino acid deletion in gp70, and a cosegregating *Hin*dIII site at the 3' end of *pol* in intact modified polytropic as compared to polytropic elements (Stoye and Coffin 1987). The two classes are, however, otherwise closely related, suggesting that they have diverged recently after differentiating from more distantly related ecotropic and xenotropic infectious proviral structures.

Unexpectedly, sequence analysis indicates that both *env*-1 and *env*-2 retroviral elements contain yet another major deletion of *pol* sequence. Comparison of the endogenous sequence just upstream of a conserved *Sal*I site demonstrates that precisely the same 1474-bp deletion, spanning from $p10^{gag}$ to the middle of the reverse transcriptase coding region, was present in *env*-1 and *env*-2 loci (P.F. Policastro et al., in prep.). As found for the *env* deletions, a 7-bp sequence at *gag-pol* deletion breakpoints is present as a pair of direct repeats in the AKV genome, indicating the deletion occurred through homologous recombination. Since *env*-1 and *env*-2 are likely to be derived from full-length proviruses, each from one of the classes of endogenous elements, a simple model of divergence, first resulting in the *gag-pol* deletion followed by independent *env* deletions, cannot be proposed. Rather, it is more likely that these *env*-deleted sequences have undergone recombination possibly as DNA forms during retrotransposition.

Hybridization of an oligonucleotide probe of the sequence spanning the *gag-pol* deletion endpoints demonstrates that in 129 strains there are additional 7.4 and 6.6 kb *Eco*RI fragments bearing this deletion that do not contain the *env*-1 or *env*-2 deletion (see Fig. 3). The genome of other strains contain yet additional *gag-pol* deleted loci not bearing *env* deletions detected by oligonucleotide probes to del *env*-1 or *env*-2. At this time it cannot be distinguished whether this is due to nucleotide sequence divergence at deletion breakpoints, resulting in insufficient homology for hybridization, or conservation of *env* sequence at these specific sites.

As shown in Figure 3 endogenous retroviral elements present in 129 strains are often not present in other inbred mice. If this was due to the loss of the individual element by homologous recombination mediated through the LTR, then a residual or solitary copy of the LTR ought to remain in the genome of these strains. Copies of such "solo LTRs" are found in the *M. musculus* genome indicating that this is a common mechanism in generating the variation of endogenous retroviral loci between strains (Wirth et al. 1983). To investigate the polymorphism of *env*-1 and *env*-2 loci, regions of unique flanking sequence were used as probes in genomic DNA blots to test for the potential presence of a solo LTR in strains not bearing the endogenous *env*-deleted retroviral sequence. As illustrated in Figure 5, the size of the

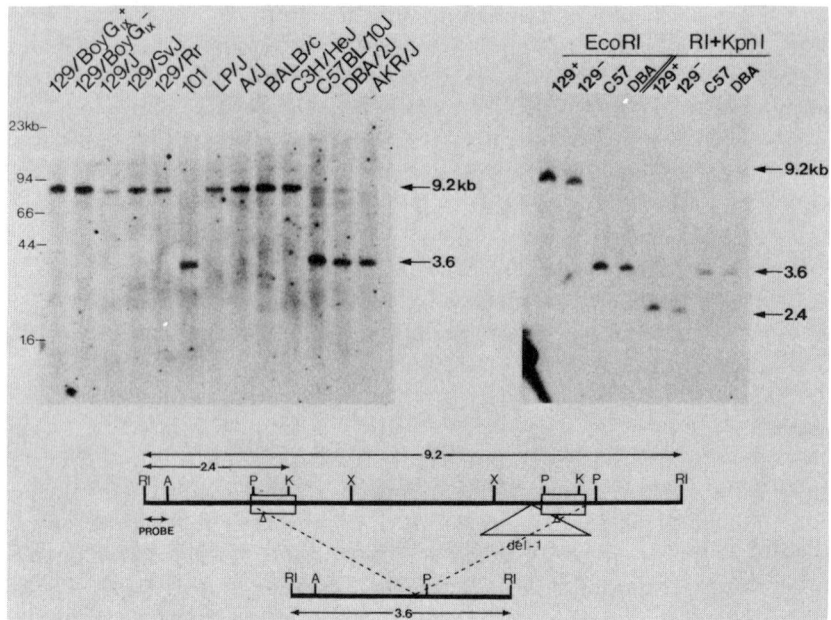

Figure 5
Lack of retroviral locus 11.1 in strain 101, C57BL/6 DBA/2, and AKR is not due to excision of the proviral sequence by homologous recombination. Genomic DNAs were digested with either EcoRI or EcoRI and KpnI as indicated and probed with the 5' flanking fragment defined by EcoRI and AvaII sites. Presence of the retroviral insertion produces a EcoRI fragment of 9.2 kb and a EcoRI/KpnI fragment of 2.4 kb (where KpnI cleaves within the LTR sequence). Absence of the 11.1 locus or of a residual LTR, potentially left by homologous recombination, results in an EcoRI fragment of 3.6 kb that is not cut by KpnI (left panel strains, C57 and DBA).

EcoRI fragment with del env-1 11.1 flanking sequence in strains 101, C3H, DBA/2, and AKR, and the lack of a characteristic KpnI site retained in such solo LTRs, is inconsistent with the presence of a solitary LTR sequence. Similar results have been obtained with probes to flanking regions of the other 129 del env-1 and the env-2 loci (M. Fredholm et al., in prep.). This evidence strongly supports the relatively recent insertion of these deleted endogenous retroviral elements during the derivation of inbred laboratory mice. If a retrotransposon mediated event is indeed responsible for this apparent movement, the establishment of these sequences in the germline must occur either through an abortive infection, perhaps facilitated by infectious virus, or by intracellular retrotransposition during gametogenesis.

The Potential Site of G_{IX} Regulation

The LTR of infectious retrovirus has been demonstrated to confer both host- and tissue-specific regulation, presumably at the level of transcription (see

Varmus 1982). These regulatory sequences are considered to reside in the U3 region where recent studies on infectious MLV have demonstrated the binding of both constitutive and tissue specific factors (Speck and Baltimore 1987). As described previously the endogenous retroviral LTRs contain U3 sequence homologous to the enhancer region of ecotropic or xenotropic virus, but are distinguished by the presence of a 192-bp insertion (Khan and Martin 1983; Ou et al. 1983). As diagramed in Figure 6 this insert is flanked by 6-bp direct repeats in the LTR of transcribed endogenous sequences. This 6-bp sequence is present as imperfect tandem repeats in the LTR of MCF 247, a recombinant leukemogenic virus and NZB xenotropic virus, but as a single copy in ecotropic AKV MLV. The 192-bp insert is also present in approximately 500 copies in the *M. musculus* genome, presumably as a result of numerous excision events of retroviral elements by homologous recombination (Wirth et al. 1983). It is intriguing, however, that within such residual solo LTRs the upstream enhancer sequence is often highly diverged to the extent of nonhomology, whereas the 192-bp sequence shares greater than 95% identity with that sequence in the LTR of intact or *env*-deleted endogenous retroviral elements (Schmidt et al. 1984; Levy et al. 1987). It is also

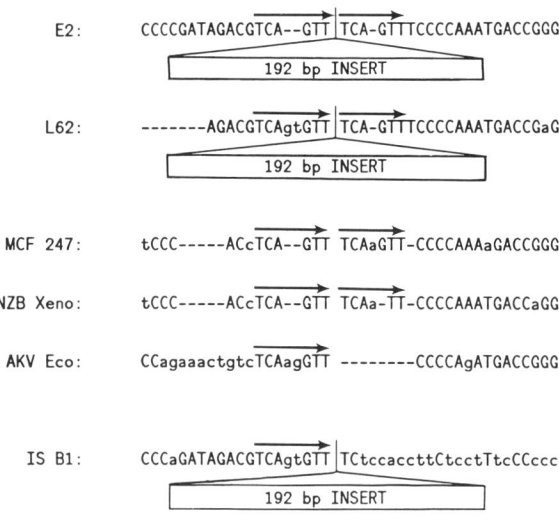

Figure 6

Six bp repeats flank a middle repetitive DNA element in endogenous but noninfectious viral LTRs. A schematic representation of the insertion of a 192-bp sequence in the LTR of endogenous elements is compared with infectious retroviral LTRs: the insertion element (□) and the flanking repeat TCAGTT (→). Identical sequence between the endogenous E2 (*env*-1) and L62 (*del env*-2) cDNA transcripts (Levy et al. 1987) and that of recombinant MCF247 (Kelley et al. 1983) and NZB xenotropic (O'Neill et al. 1985) and AKV ecotropic (Herr 1984) virus and a solo LTR, IS B1 (Wirth et al. 1983) is indicated by capital letters, while small letters represent nonidentical sequence. The dashes represent spaces introduced to maximize sequence alignment.

interesting that the 192-bp sequence lacks or is deficient in CpG residues, sites of potential regulatory methylation, while not particularly of low G + C content, as indicated by the frequency of GpC residues (data not shown). In contrast, upstream enhancer homologous and the downstream promoter regions (containing the obligatory CAAT and TATA box signal sequences) of both endogenous and infectious viral LTRs contain a significant number of potential CpG methylation sites. In preliminary studies, we have observed that either the enhancer homolog or the 192-bp insert sequence when inserted downstream from the enhancer sequence of the Moloney MLV (Mo-MLV) LTR inhibits the transcriptional activity of that promoter in murine cells not expressing their own complement of endogenous retroviral transcription units (M. Lucas and M. Wilson, unpubl.). This provides support for the idea that the 192-bp insert as well as the upstream enhancer homolog do provide some function, possibly as negative regulatory elements, in governing transcription from endogenous LTRs.

DISCUSSION

The structure of MLV related endogenous retroviral elements is compared to that of a typical infectious MLV genome in Figure 7. Although several distinctive features serve to distinguish these endogenous elements, these differences do not preclude their transcription or their potential for genetic movement through retrotransposition. This includes an *env* region in nondeleted transcripts that differs significantly from either ecotropic or xenotropic sequence but still potentially encodes *env* gene products gp70 and p15E (Khan 1984; Levy et al. 1985b). Evidence suggests, in fact, that the 5' region of endogenous *env* is incorporated into an ecotropic parent virus during the de novo generation of MCF recombinant viruses and contributes to their leukemogenic potential by altering the host range and cellular recognition of these viruses (for review, see Stoye and Coffin 1985). A region of nonhomology to infectious virus is also common to endogenous elements at the 5' region of *gag* sequence. Although this precludes the expression of a minor *gag* product gPr80^{gag}, it does allow for the potential downstream initiation of the major *gag* and *gag-pol* precursor polypeptides (Colicelli and Goff 1986). Most important, however, to the proposed retrotransposition of these elements is that these endogenous retroviral elements bear a sequence complementary to glutamine tRNA, rather then proline tRNA (Nikbakht et al. 1985), the typical primer utilized by infectious type C-MLV for the initiation of second strand synthesis during a retroviral replication cycle. Incorporation of either the nonhomologous *gag* region and glutamine tRNA primer binding site (PBS) into the genome of ecotropic Mo-MLV, however, has been shown not to significantly affect the replication of such a recombinant virus in cultured

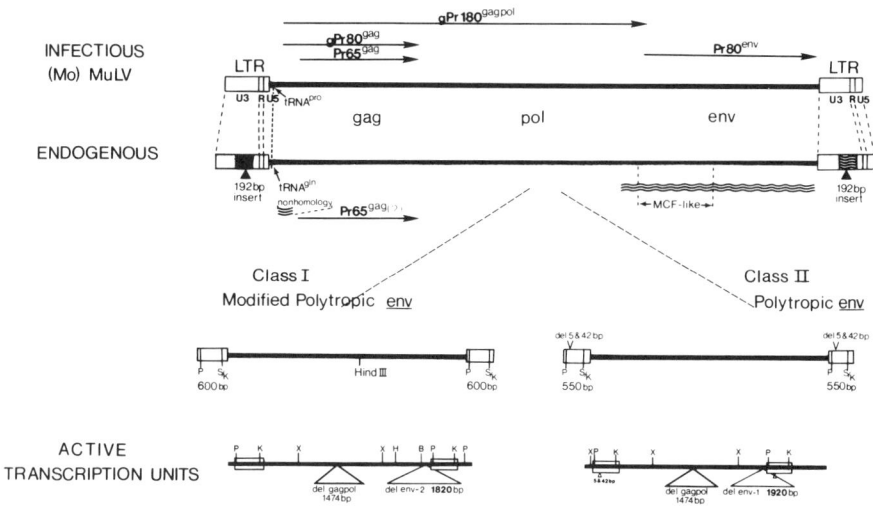

Figure 7
Relationship of *Gv-1* responsive retroviral transcriptional units to infectious MLV. Comparison of endogenous retroviral elements with the genome of infectious MLV. The regions of nonhomology between the sequences are indicated as wavy lined areas and are described in the text. The two classes of endogenous elements, modified polytropic (containing an internal *Hin*dIII site) and polytropic structures, are depicted as described by Stoye and Coffin 1987. The *gag-pol-* and *env*-deleted proviral structures of *del env*-1 and *env*-2 are represented below as *Gv-1* responsive, transcribed genes.

cells (Colicelli and Goff 1986) and, therefore, is compatible with mechanisms mediating retrotransposition.

The potential for transcriptional regulation of retroviral-like elements has several consequences for their action as mutagenic agents. First, since transcription is essential to retroviral mediated transposition, the mechanism that regulates transcriptional activity of these elements would in turn govern the frequency of genetic movement. Second, active transcription may be required, once integration has occurred, for mutagenesis to be activated. For example, analogous to *Gv-1* and *Gv-2* regulation of murine endogenous retroviral elements, the suppressor of hairy wing mutation is an independent locus that exerts its effect by down regulation of a gypsy retroviral-like element at the affected locus (see Corces, this volume). Similarly, the glucocorticoid responsiveness of the MMTV LTR (Ringold et al. 1975) when integrated at the *int-1* (Nusse et al. 1984) or the *int-2* locus (Peters et al. 1983) may help determine the tumorigenic potential of that mutagenic event. The intriguing possibility that regulation of endogenous retroviral genes by *Gv-1* and *Gv-2* may play a role in such mechanisms of mutagenesis, however, has not been directly addressed. This is in part because of the difficulty of distinguishing *Gv-1* responsive from inactive endogenous retroviral loci. With the identification of at least two, and perhaps four, *Gv-1* responsive genes and flanking cellular DNAs, we are now in the process of defining the sequences conferring *trans* activation. The identification of the *Gv-1* responsive element of the endogenous LTR will allow the utilization of selectable gene markers linked to the regulatory sequence for the isolation and characterization of the *trans*-acting *Gv-1* locus.

ACKNOWLEDGMENTS

We wish to thank Dr. John Elder for discussions and Drs. Jonathan Stoye and John Coffin for relating results prior to publication. We acknowledge the help of Michelle Dietrich in preparing the manuscript. This paper is publication number 4915MB of the Research Institute of Scripps Clinic. The work is supported in part by National Institutes of Health grant CA-33730 (M.C.W.), postdoctoral fellowship CA-07788 (P.F.P.), and a Danish National Research Council (SJVF) grant (M.F.)

REFERENCES

Boyse, E.A. 1977. The G_{IX} system in relation to C-type viruses and heredity. *Immunol. Rev.* **33**: 125.

Colicelli, J. and S.P. Goff. 1986. Isolation of a recombinant murine leukemia virus utilization of a new primer tRNA. *J. Virol.* **57**: 37.

Copeland, N.G., K.W. Hutchinson, and N.A. Jenkins. 1983. Excision of the DBA ecotropic provirus in dilute coat-color revertants of mice occurs by homologous recombination involving viral LTRs. *Cell* **33**: 379.

Corcoran, L.M., J.M. Adams, A.R. Dun, and S. Cory. 1984. Murine T-lymphomas in which the cellular *myc* oncogene has been activated by retroviral insertion. *Cell* **37**: 113.

Cuypers, H.T., G. Selton, W. Quint, M. Zijlstra, E.R. Maandag, W. Boelens, P. van Wezenbeck, C. Melief, and A. Berns. 1984. Murine leukemia virus-induced T-cell lymphomagenesis integration of proviruses in a distinct chromosomal region. *Cell* **37**: 141.

Herr, W. 1984. Nucleotide sequence of AKV murine leukemia virus. *J. Virol.* **43**: 413.

Hoggan, M.D., C.E. Buckler, J.F. Sears, H.W. Chan, W.P. Rowe, and M.A. Martin. 1982. Internal organization of endogenous proviral DNAs of xenotropic murine leukemia viruses. *J. Virol.* **43**: 8.

Kelley, M., C.A. Holland, M.L. Lung, S.K. Chattopadhyay, D.R. Lowy, and N.H. Hopkins. 1983. Nucleotide sequence of the 3' end of MCF 247 murine leukemia virus. *J. Virol.* **45**: 291.

Jenkins, N.A., N.G. Copeland, B.A. Taylor, and B.K. Lee. 1981. Dilute(d) coat colour mutation of DBA/2J mice is associated with the site of integration of an ecotropic MuLV genome. *Nature* **293**: 370.

Khan, A.S. 1984. Nucleotide sequence analysis establishes the role of endogenous murine leukemia virus DNA segments in formation of recombinant mink cell focus-forming murine leukemia viruses. *J. Virol.* **50:** 864.

Khan, A.S. and M.A. Martin. 1983. Endogenous murine leukemia proviral long terminal repeats contain a unique 190-base-pair insert. *Proc. Natl. Acad. Sci.* **80:** 2699.

Kozak, C.A. and R.R. O'Neill. 1986. Xenotropic and MCF related retroviral genes in wild mice. *Curr. Top. Microbiol. Immunol.* **127:** 349.

Levy, D.E., R.A. Lerner, and M.C. Wilson. 1982. A genetic locus regulates the expression of tissue specific mRNAs from multiple transcription units. *Proc. Natl. Acad. Sci.* **79:** 5823.

―――――. 1985a. The Gv-1 locus coordinately regulates the expression of multiple endogenous murine retroviruses. *Cell* **41:** 289.

―――――. 1985b. Normal expression of polymorphic endogenous retroviral RNA containing segments identical to mink cell focus-forming virus. *J. Virol.* **56:** 691.

Levy, D.E., R.H. McKinnon, M.N. Brolaski, J.W. Gautsch, and M.C. Wilson. 1987. The 3' long terminal repeat of a transcribed yet defective endogenous retroviral sequence is a competent promoter of transcription. *J. Virol.* **61:** 1261.

Nikbakht, K.N., C.-Y. Ou, L.R. Boone, P.L. Glover, and W.K. Yang. 1985. Nucleotide sequence analysis of endogenous murine leukemia virus-related proviral clones reveals primer-binding sites for glutamine tRNA. *J. Virol.* **54:** 889.

Nusse, R., A. van Ooyen, D. Cox, T. Kai, Y. Fung, and H. Varmus. 1984. Model of proviral activation of a putative mammary oncogene (int-1) on mouse chromosome 15. *Nature* **307:** 131.

O'Neill, R.R., C.E. Buckler, T.S. Theodore, M.A. Martin, and R. Repaske. 1985. Envelope and long terminal repeat sequences of a cloned infectious NZB xenotropic murine leukemia virus. *J. Virol.* **53:** 100.

O'Neill, R.R., A.S. Khan, M.D. Hoggan, J.W. Hartley, M.A. Martin, and R. Repaske. 1986. Specific hybridization probes demonstrate fewer xenotropic than mink cell focus-forming murine leukemia virus *env*-related sequences in DNAs from inbred laboratory mice. *J. Virol.* **58:** 359.

Ou, C.-Y., L.R. Boone, and W.E. Yang. 1983. A novel sequence segment and other nucleotide structural features in the long terminal repeat of a BALB/c mouse genomic leukemia virus-related DNA clone. *Nucleic Acids Res.* **11:** 5603.

Peters, G., S. Brookes, R. Smith, and C. Dickson. 1983. Tumorigenesis by mouse mammary tumor virus: Evidence for a common region for provirus integration in mammary tumors. *Cell* **33:** 369.

Ringold, G.M., K.R. Yamamoto, G.M. Tonkins, J.M. Bishop, and H.E. Varmus. 1975. Dexamethasone-mediated induction of mouse mammary tumor cirus RNA: A system for studying glucocorticoid action. *Cell* **6:** 299.

Schmidt, M., K. Gloggler, T. Wirth, and I. Horak. 1984. Evidence that a major class of mouse endogenous long terminal repeats (LTRs) resulted from recombination between exogenous retroviral LTRs and similar LTR-like elements (LTR-IS). *Proc. Natl. Acad. Sci.* **81:** 6696.

Selton, G., H.T. Cuypers, M. Zylstra, C. Melief, and A. Berns. 1984. Involvement of c-myc in MuLV-induced T-cell lymphomas in mice: Frequency and mechanisms of activation. *EMBO J.* **3:** 3215.

Speck, N.A. and D. Baltimore. 1987. Six distinct nuclear factors interact with the 75 base pair repeat of the moloney murine leukemia virus enhancer. *Mol. Cell. Biol.* **7:** 1101.

Stockert, E., L.J. Old, and E.A. Boyse. 1971. The G_{IX} system. A cell surface allo-antigen associated with murine leukemia virus; implications regarding chromosomal integration of the viral genome. *J. Exp. Med.* **149:** 200.

Stockert, E., E.A. Boyse, Y. Obata, H. Ikeda, N.H. Sarker, and H.A. Hoffman. 1975. New mutant and congenic mouse stocks expressing the murine leukemia virus-associated thymocyte surface antigen G_{IX}. *J. Exp. Med.* **142:** 512.

Stoye, J.P. and J.M. Coffin. 1985. Endogenous viruses. In *Molecular biology of tumor viruses*, 2nd edition: *RNA tumor viruses* (ed. R. Weiss et al.), vol. 2, p. 1109. Cold Spring Harbor Laboratory, Cold Spring Harbor, New York.

―――. 1987. The four classes of endogenous murine leukemia virus: Structural relationships and potential for recombination. *J. Virol.* **61:** 2659.

Varmus, H.E. 1982. Form and function of retroviral proviruses. *Science* **216:** 812.

Wirth, T., K. Gloggler, T. Baumruker, M. Schmidt, and I. Horak. 1983. Family of middle repetitive DNA sequences in the mouse genome with structural features of solitary retroviral long terminal repeats. *Proc. Natl. Acad. Sci.* **80:** 3327.

Genes That Affect Ty-mediated Gene Expression in Yeast

FRED WINSTON
Department of Genetics
Harvard Medical School
Boston, Massachusetts 02115

OVERVIEW

Ty and solo δ insertion mutations in the 5' noncoding regions of genes in *Saccharomyces cerevisiae* can inhibit adjacent gene expression. Frequently, this occurs because transcription signals on the Ty or δ interfere with those of the adjacent gene. *Trans*-acting suppressors of Ty and δ insertion mutations are candidates for mutations in genes that encode transcription factors that are required for the function of Ty transcription signals. We have isolated and analyzed a large number of suppressors of Ty and δ insertion mutations and have identified eight new genes. Mutations in these genes alter transcription of genes adjacent to Ty and δ insertion mutations. In addition, they frequently cause other mutant phenotypes, suggesting that they may also be important for more general cellular functions.

INTRODUCTION

Ty elements of *S. cerevisiae* are a set of retrotransposons similar to those found in many types of eukaryotic cells. Ty elements are approximately 6000 base pairs long and are flanked by long terminal repeats, δ sequences, of approximately 330 bp. These elements share several important features with other retrotransposons, including mode of transposition, genetic organization, and the ability to affect gene expression when inserted in or near a gene.

Ty insertion mutations in 5' noncoding regions make adjacent genes mutant (for review, see Roeder and Fink 1983). In some cases, the insertion mutations cause constitutive expression of the adjacent gene. In other cases, the insertion mutations inhibit expression of the adjacent gene. In all of the cases that have been examined, these effects occur at the transcriptional level. Ty insertion mutations can give rise to solo δ derivatives, presumably via δ–δ recombination. In most cases, the solo δ derivatives also cause mutant phenotypes, sometimes distinct from those of the complete Ty insertion mutations.

The mutant phenotypes caused by Ty and δ insertion mutations in 5' noncoding regions can often be attributed to interference with a gene's normal transcriptional signals by those on the Ty or δ sequence. Ty elements are transcribed from δ to δ sequence, with transcription initiating in the 5' δ sequence and terminating in the 3' δ (Elder et al. 1983). Therefore, δ sequences themselves most likely contain signals for transcription initiation and termination. Furthermore, recent work has demonstrated that the Ty internal ε region contains sequences near the 5' δ sequence that are also important for adjacent gene expression (Errede et al. 1985; Roeder et al. 1985).

Assuming that particular gene products act in *trans* at Ty transcription signals, selection for mutations that alter Ty- or δ-mediated gene expression is a selection for mutations in the genes that encode these *trans*-acting factors. By a large scale selection for suppressor mutations of insertion mutations, we have identified at least eight *trans*-acting genes (*SPT* genes, *s*uppressor of *T*y). Genetic and molecular analysis of *spt* mutants demonstrates that they have altered transcription of genes adjacent to Ty and solo δ insertion mutations. Furthermore, distinct suppression patterns and additional mutant phenotypes caused by *spt* mutations show that these genes fall into different classes and suggest that they are important for other functions in cellular growth.

RESULTS

Isolation and Analysis of *spt* mutations

To select for suppressors of insertion mutations, we initially selected His$^+$ revertants of *his4-912δ*, a solo δ insertion mutation inserted at a position -98 from the start of *HIS4* transcription and in the same transcriptional orientation at *HIS4* (Fig. 1a). This insertion mutation confers a cold-sensitive His$^+$ phenotype: His$^-$ at 23°C, weak His$^+$ at 30°, and His$^+$ at 37°. We selected for revertants of a solo δ insertion mutation because Chaleff and Fink (1980) demonstrated that revertants of the complete Ty insertion mutation *his4-912* are frequently chromosomal rearrangements, apparently by recombination between Ty elements. We reasoned that rearrangements would be less frequent as revertants of a solo δ insertion mutation.

Spontaneous His$^+$ revertants of *his4-912δ* were selected, purified, and analyzed with respect to dominance and complementation. By these tests, we identified seven complementation groups, designated *SPT1–SPT7* (Winston et al. 1984b). To examine if each complementation group represents distinct genes, representative mutants from each complementation group were crossed by each other to examine linkage relationships. By this analysis, the seven *SPT* complementation groups represent seven unlinked genes. Furthermore, for the alleles tested, *spt4 spt5*, *spt4 spt6*, and *spt5 spt6* double mutants are inviable, suggesting a possible functional interaction between these genes.

a) his4-912δ

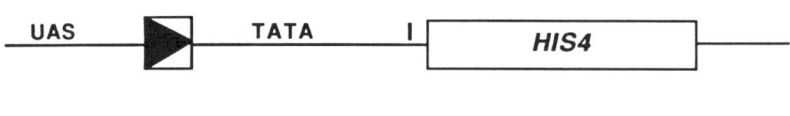

b) his4-917δ

Figure 1
(*a*) The *his4-912δ* mutation is a solo δ insertion mutation located 98 bases from the start point of *HIS4* transcription in the same transcriptional orientation as *HIS4*. The relative positions of other *HIS4* transcription signals are shown: the upstream activation sequence (UAS), the TATA region, and the start point of transcription (I). (*b*) The *his4-917δ* mutation is a solo δ insertion mutation located 7 bases from the start point of *HIS4* transcription in the opposite transcriptional orientation as *HIS4*. *his4-917δ* is located between the *HIS4* TATA region and the start point of transcription.

Among the dominant mutations isolated, we demonstrated by linkage analysis that at least some of them are *spt2* mutations. Therefore, mutation of *SPT2* gives rise to both dominant and recessive alleles.

The *spt* mutants were further analyzed for suppression of other solo δ insertion mutations and suppression of complete Ty insertion mutations at *HIS4* and *LYS2*. The results (Table 1) demonstrate that mutations in different *SPT* genes confer different suppression patterns with respect to this set of insertion mutations, suggesting that mutations in different *SPT* genes suppress by different mechanisms.

In addition to suppression of insertion mutations, *spt* mutants were analyzed for additional mutant phenotypes. From this analysis, we found that some alleles of *SPT6* are temperature sensitive for growth, mutations in *SPT3* confer sporulation and mating defects, and at least one allele of *SPT4* causes hypersensitivity to methyl methanesulfonate.

Table 1
spt Suppression Patterns

Mutant gene	his4-912δ	his4-917δ	his4-917	lys2-173R2
wild type	cs	−	−	+
spt1	+	−	+/−	+
spt2	+	−	−/+	+
spt3	+	+	+	−
spt4	+	−/+	−	+
spt5	+	−	−	+
spt6	+	−/+	−/+	+
spt7	+	+	+	−/+
spt8	+	+	+	+/−

Symbols indicate growth on media lacking histidine (or lysine for *lys2-173R2*) as follows: cs = cold sensitive; + = strong growth; +/− = intermediate growth; −/+ = weak growth; − = no growth.

Three Genes Are Required for Transcription Initiation from δ Sequences

Examination of Ty transcription in representative *spt* mutants demonstrated that only *spt3* mutations cause a defect in Ty transcription. In an *spt3* null mutant, full-length δ-δ transcription is virtually abolished. Instead, a shorter Ty transcript, whose 5' end is located approximately 800 bases 3' of the normal Ty transcription start site, is made at a low level (Winston et al. 1984a).

In addition to alteration of Ty transcription, analysis of transcription of both *his4-912δ* and *his4-917δ* (Fig. 1) in wild-type and in *spt3* mutant backgrounds reveals a similar transcriptional alteration for both insertion mutations. In an SPT^+ host, *HIS4* transcription initiates in the δ sequence in both cases. Such transcripts contain translation initiation and termination codons upstream of the normal *HIS4* translation initiation codon and are assumed to be nonfunctional, explaining the His⁻ phenotype in an $SPT3^+$ genetic background. In strains that contain *spt3* null mutations and either of these insertion mutations, the phenotype is His⁺. Northern hybridization analysis reveals that in each case the transcript from the solo δ sequence is gone or greatly reduced and, instead, a wild-type length *HIS4* transcript is made. Therefore, suppression of these two solo δ insertion mutations is through alteration of the transcription start site from within the δ sequence to a position at or near the wild-type *HIS4* initiation site.

Examination of suppression patterns reveals that mutations in *SPT3* cause a unique suppression pattern: strong suppression of *his4-917δ* and an opposite effect on *lys2-173R2*, altering the phenotype from Lys⁺ to Lys⁻ (Table 1). Taking advantage of this unique *spt3* suppression pattern, we were able to select for new *spt* mutations also involved in regulation of Ty transcription.

Beginning with a strain that contains *his4-917δ* and *lys2-173R2*, we selected for His⁺ revertants and screened among them for those with a Lys⁻ phenotype. From this search, we isolated mutations in three genes: *SPT3*, *SPT7*, and a new gene, *SPT8* (Winston et al. 1987). By Northern hybridization analysis and S1 protection experiments, we demonstrated that mutant alleles of *SPT7* and of *SPT8* cause a similar effect on Ty transcription as do those in *SPT3*. Presumably, the original allele of *SPT7* that we examined, *spt7-159* obtained as a *his4-912δ* revertant, is leaky and, therefore, does not reveal a defect in Ty transcription. In addition to defects in Ty transcription, mutations in *SPT7* and *SPT8*, like mutations in *SPT3*, cause a sporulation defect.

These results led to the model that three genes, *SPT3*, *SPT7*, and *SPT8*, are all required for transcription initiation from δ sequences (Fig. 2). In the absence of any of these three gene products, transcription fails to initiate from the δ sequence and, instead, initiates from a site farther downstream.

Analysis of *SPT6*

Analysis of the *SPT6* gene has shown that it is distinct from *SPT3*, *SPT7*, or *SPT8*. Furthermore, it is essential for growth, and it is probably involved in some fundamental aspect of gene expression. Additional evidence strongly suggests that *SPT6* interacts in a complex to carry out its function.

SPT6 is distinct from *SPT3*, *SPT7*, and *SPT8* based on suppression patterns, effects on Ty transcription, and additional mutant phenotypes. First, *spt3*, *spt7*, and *spt8* mutations suppress complete Ty insertion mutations, as

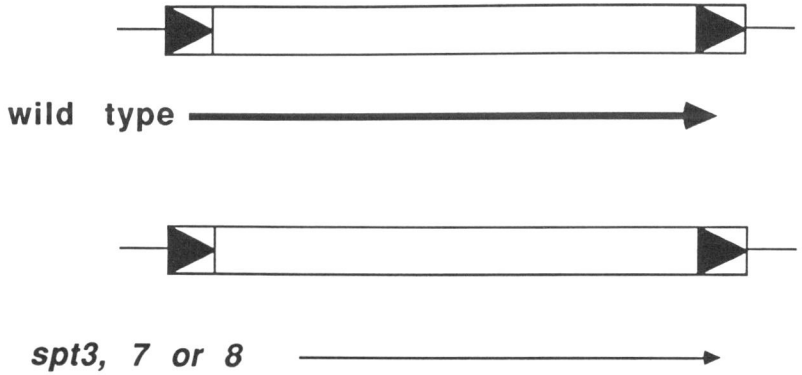

Figure 2
Ty transcription requires *SPT3*, *SPT7*, and *SPT8*. In wild-type strains, a δ-δ Ty transcript is made. In *spt3*, *spt7*, or *spt8* mutants, virtually no full-length Ty transcription is made; instead, a transcript approximately 800 bases shorter at the 5' end is made at low level.

well as solo δ insertion mutations; *spt6* mutations suppress Ty insertion mutations very weakly. Second, *spt6* mutations have no detectable effect on the level or size of Ty transcripts (Clark-Adams and Winston 1987). Third, *spt6* mutations do not have the *spt3*, *spt7*, and *spt8* mutant phenotype of a defect in sporulation. Finally, as discussed below, *SPT6* function is sensitive to gene dosage in a way not seen for *SPT3*, *SPT7*, or *SPT8*.

Suppression of solo δ insertion mutations by *spt6* mutations occurs at the transcriptional level. Examination of two insertion mutations, *his4-912δ* and *lys2-128δ* demonstrates transcriptional alterations in the presence of an *spt6* mutation, in both cases, resulting in a functional *HIS4* or *LYS2* transcript, respectively (Clark-Adams and Winston 1987). In these cases, it is not yet proved whether the transcripts made in the *spt6* mutants are the same as those made in *spt3*.

Alteration of *SPT6* gene copy number has demonstrated that *SPT6* is essential for growth and that wild-type *SPT6* function is extremely dosage-sensitive. First, haploid strains that contain an *spt6* null mutation are inviable. Second, overproduction of wild-type *SPT6* in a wild-type host results in suppression of insertion mutations. Finally, a diploid strain that contains an *spt6* null mutation on one homolog and wild-type *SPT6* on the other homolog shows suppression of insertion mutations (Table 2). Therefore, we conclude that either overproduction or underproduction of wild-type *SPT6* causes a mutant phenotype.

Genetic and physical evidence has demonstrated that *SPT6* and *SSN20* are the same gene (Clark-Adams and Winston 1987; Neigeborn et al. 1987). Of mutations that suppress the *s*ucrose-*n*onfermenting mutations, *snf2* and *snf5*, *SSN20* (*s*uppressor of *snf*) was identified (Neigeborn et al. 1986). SNF2 and SNF5 are required for transcription of *SUC2*, the structural gene for invertase (Abrams et al. 1986). The *ssn20* mutations also suppress deletions of the

Table 2
SPT6 Gene Dosage Effects

SPT6 genotype	Suppression of *lys2-128δ*[a]
SPT6+	−
spt6-140[b]	+
SPT6+/YEp24-*SPT6*+[c]	+
SPT6+/YCp50-*SPT6*+[d]	−
SPT6+/*spt6* null[e]	+

[a]Suppression of *lys2-128δ* (Clark-Adams and Winston 1987) is indicated as ability to grow on media lacking lysine; + = Lys+, − = Lys−.
[b]*spt6-140* is a temperature sensitive allele that suppresses *lys2-128δ* at 30°.
[c]This strain is a haploid with a high copy number plasmid with *SPT6*+.
[d]This strain is a haploid with a low copy number plasmid with *SPT6*+.
[e]This strain is a diploid with *SPT6*+ on one homolog and *spt6* null on the other.

SUC2 upstream activation sequence (Neigeborn et al. 1987). SPT6 and SSN20 have the same genetic map position, the cloned genes have the same restriction map, spt6 and ssn20 mutations fail to complement, and spt6 and ssn20 mutations segregate as alleles in a genetic cross. Furthermore, ssn20 mutations suppress solo δ insertion mutations, and spt6 mutations suppress snf2 (Clark-Adams and Winston 1987; Neigeborn et al. 1987). Therefore, mutations in SPT6 can suppress both cis- and trans-acting transcriptional defects in 5' noncoding regions of genes.

DISCUSSION

Selection for suppressors of insertion mutations in S. cerevisiae has identified eight unlinked genes, SPT1–SPT8. By both genetic and molecular criteria, mutations in different SPT genes cause different defects. This suggests that the selection has identified distinct functions involved in transcriptional regulation both of Ty- and δ-mediated gene expression and perhaps in other aspects of cellular transcription as well.

The SPT3, SPT7, and SPT8 genes are required for transcription initiation from δ sequences. Only mutations in these three SPT genes cause a defect in Ty transcription. Furthermore, mutations in all three confer the same suppression pattern, and all three share the same type of sporulation defect. It is tempting to speculate that these genes interact in a regulatory pathway or directly in a complex. However, it is possible that they act independently as positive activators of transcription initiation from δ sequences.

The function(s) of these genes may be similar to that of the *Drosophila* gene suppressor-of-Hairy wing, *su(Hw)*. Similar to spt3, spt7, and spt8 mutations, *su(Hw)* mutations suppress insertion mutations caused by the *Drosophila* retrotransposon gypsy at the transcriptional level (Modolell et al. 1983; Parkhurst and Corces 1986a). Also similar to spt3, spt7, and spt8 mutations, transcription of gypsy is reduced in *su(Hw)* homozygous mutant flies (Parkhurst and Corces 1986a). More detailed molecular analysis of the functions of these gene products in both organisms will be essential for understanding their degree of relatedness.

There must also be other mechanisms for suppression of retrotransposon insertion mutations, as Parkhurst and Corces (1986b) have shown that the gene suppressor-of-forked, *su(f)*, which suppresses a subset of the gypsy mutations suppressed by *su(Hw)*, actually increases transcription of gypsy elements approximately tenfold. Furthermore, other spt mutations have been identified in S. cerevisiae that suppress complete Ty insertion mutations, yet do not have any significant effect on Ty transcription (Fassler and Winston 1988).

Mutations in *SPT6* also suppress solo δ insertion mutations at the transcriptional level. *SPT6*, however, is distinct from *SPT3*, *SPT7*, and *SPT8* in many mutant phenotypes. *SPT6*, therefore, is most likely functionally unrelated to *SPT3*, *SPT7*, and *SPT8*. The fact that *spt6* mutations were also isolated as suppressors of the sucrose-nonfermenting mutations *snf2* and *snf5* strongly suggests that it may play a very general role in cellular transcription. The sensitivity of *SPT6* function to altered gene dosage indicates that it acts in a complex. Further molecular analysis of *SPT6* and its action will hopefully elucidate its role in yeast growth.

ACKNOWLEDGMENTS

This work was supported by the National Institutes of Health grant GM-32967, National Science Foundation grant DCB-84-51649, and grants from Monsanto Company, The Stroh Brewery, and Molecular Therapeutics, Inc.

REFERENCES

Abrams, E., L. Neigeborn, and M. Carlson. 1986. Molecular analysis of *SNF2* and *SNF5*, genes required for expression of glucose-repressible genes in *Saccharomyces cerevisiae. Mol. Cell. Biol.* **6:** 3643.

Chaleff, D.T. and G.R. Fink. 1980. Genetic events associated with an insertion mutation in yeast. *Cell* **21:** 227.

Clark-Adams, C.D. and F. Winston. 1987. The *SPT6* gene is essential for growth and is required for δ-mediated transcription in *Saccharomyces cerevisiae. Mol. Cell. Biol.* **7:** 679.

Elder, R.T., E.Y. Loh, and R.W. Davis. 1983. RNA from the yeast transposable element Ty1 has both ends in the direct repeats, a structure similar to retrovirus RNA. *Proc. Natl. Acad. Sci.* **80:** 2432.

Errede, B., M. Company, J.D. Ferchak, C.A. Hutchison III, and W.S. Yarnell. 1985. Activation regions in a yeast transposon have homology to mating type control sequences and to mammalian enhancers. *Proc. Natl. Acad. Sci.* **82:** 5423.

Fassler, J.S. and F. Winston. 1988. Isolation and analysis of a novel class of suppressor of Ty insertion mutations in *Saccharomyces cerevisiae. Genetics* (in press).

Modolell, J., W. Bender, and M. Meselson. 1983. *Drosophila melanogaster* mutations suppressible by the suppressor of hairy-wing are insertions of a 7.3-kilobase mobile element. *Proc. Natl. Acad. Sci.* **80:** 1678.

Neigeborn, L., J.L. Celenza, and M. Carlson. 1987. *SSN20* is an essential gene with mutant alleles that suppress defects in *SUC2* transcription in *Saccharomyces cerevisiae. Mol. Cell. Biol.* **7:** 679.

Neigeborn, L., K. Rubin, and M. Carlson. 1986. Suppressors of *snf2* mutations restore invertase derepression and cause temperature-sensitive lethality in yeast. *Genetics* **112:** 741.

Parkhurst, S.M. and V.G. Corces. 1986a. Interactions among the gypsy transposable element and the yellow and the suppressor of hairy-wing loci in *Drosophila melanogaster*. *Mol. Cell. Biol.* **6:** 47.

———. 1986b. Mutations at the suppressor of forked locus increase the accumulation of gypsy-encoded transcripts in *Drosophila melanogaster*. *Mol. Cell. Biol.* **6:** 2271.

Roeder, G.S. and G.R. Fink. 1983. Transposable elements in yeast. In *Mobile genetic elements* (ed. J.A. Shapiro), p. 300. Academic Press, New York.

Roeder, G.S., A.B. Rose, and R.E. Pearlman. 1985. Transposable element sequences involved in the enhancement of yeast gene expression. *Proc. Natl. Acad. Sci.* **82:** 5428.

Winston, F., K.J. Durbin, and G.R. Fink. 1984a. The *SPT3* gene is required for normal transcription of Ty elements in *S. cerevisiae*. *Cell* **39:** 675.

Winston, F., D.T. Chaleff, B. Valent, and G.R. Fink. 1984b. Mutations affecting Ty-mediated expression of the *HIS4* gene of *Saccharomyces cerevisiae*. *Genetics* **107:** 179.

Winston, F., C. Dollard, B.M. Malone, J. Clare, J.G. Kapakos, P. Farabaugh, and P.L. Minehart. 1987. Three genes are required for *trans*-activation of Ty transcription in yeast. *Genetics* **115:** 649.

The Genetic Control of Recombination between Repeated Sequences

RODNEY J. ROTHSTEIN, W. LANE ARTHUR, JOHN W. WALLIS, JR.,*
HANS RONNE,† AND BARBARA J. THOMAS
Department of Genetics and Development
Columbia University College of
Physicians and Surgeons
New York, New York 10032

OVERVIEW

In thinking about genome stability in eukaryotes, it is important to consider the question of the control of recombination between repeated sequences. The genomes of virtually all organisms are filled with repeated sequences (for review, see Jelinek and Schmid 1982). Thus, if there were uncontrolled recombination between these elements, deletions, inversions and translocations would be generated at high frequency. Since this is not observed, it implies that these recombination events may be suppressed. Recent observations in yeast emphasize the importance of this postulated control, since meiotic recombination between homologous sequences in nonhomologous positions (ectopic gene conversion) has been found to occur very near the levels observed for homologous recombination (Jinks-Robertson and Petes 1985, 1986; Lichten et al. 1987). In this report we discuss recombination among repeated sequences including δ-sequences and Ty elements. We will focus on the mechanism of rearrangements that use the host recombination apparatus.

INTRODUCTION

The report is divided into three sections. First, we describe evidence that heteroduplex DNA is an intermediate during direct repeat recombination. Next, we outline experiments designed to show how transcription affects recombination between repeated sequences. Finally, we will describe a gene that controls several types of recombination events between δ-sequences and Ty elements.

Present addresses: *Department of Biochemistry, St. Louis University School of Medicine, St. Louis, Missouri 63104; †Ludwig Institute for Cancer Research, Uppsala Branch, Biomedical Center, Box 595, S-75123 Uppsala, Sweden.

Banbury Report 30: Eukaryotic Transposable Elements as Mutagenic Agents
© Cold Spring Harbor Laboratory. 0-87969-230-8/88. $1.00 + .00

RESULTS

Sectored Colonies as Indicators of Heteroduplex DNA

A system was designed to detect heteroduplex intermediates in recombination if they occur during mitotic direct repeat recombination (Ronne and Rothstein 1988). We took advantage of the fact that the presence or absence of an ochre suppressor can be monitored by the color of the yeast colonies.

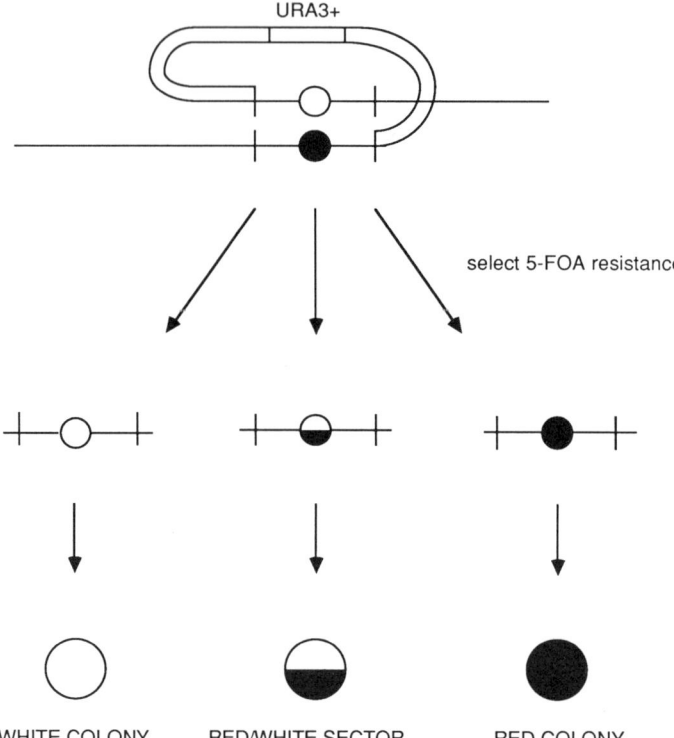

Figure 1
A direct repeat assay that detects sectored colonies. A direct repeat was created by integrating a plasmid containing the $URA3^+$ gene and a wild-type copy of the $sup3^+$ locus into the homologous chromosomal $SUP3$-o region: (●) the $sup3^+$ gene and (○) the $SUP3$-o gene. Direct repeat recombination events resulting in loss of the $URA3^+$ containing plasmid sequences are selected on 5-FOA medium (Boeke et al. 1984). Crossovers to the left of the anticodon produce red colonies, whereas crossovers to the right of the anticodon remain white. In addition to the red and white colonies, red/white sectored colonies are also associated with plasmid loss events. These are likely due to the formation of heteroduplex DNA during the excision process. Failure to repair the mismatch leads to the expression of both color types in a single colony after DNA synthesis and mitotic segregation.

Red colonies are formed in the presence of the wild-type gene, whereas white colonies form in the presence of the supressor. We placed the $URA3^+$ gene (Struhl et al. 1979; Rose et al. 1984) between a duplicated fragment containing one ochre suppressor and one wild-type tRNA gene (Fig. 1). The $URA3^+$ gene is used to select direct repeat recombinants (popouts) using the drug 5-fluoro-orotic acid (5-FOA, Boeke et al. 1984). As illustrated in Figure 1, a crossover to the left of the anticodon results in a white colony, whereas a crossover to the right of the anticodon results in a red colony. If the DNA molecules from the anticodon regions of the two tRNA genes form a heteroduplex during the excision process and the mismatch is unrepaired, then we expect that after DNA replication, the heteroduplex DNA will segregate and result in a red/white sectored colony. Using this system, we asked how often red/white sectored colonies appear after selecting for loss of $URA3^+$ (5-FOA resistance). We found that approximately 15% of the popouts give rise to sectored colonies. This result indicates that heteroduplex DNA is likely an intermediate in direct repeat recombination as implied in earlier mitotic studies of interchromosomal recombination (Wildenberg 1970; Esposito 1978; Roman and Fabre 1983) and, as has been established during meiosis, to explain postmeiotic segregation (for review, see Szostak et al. 1983).

The Association between Transcriptional Activation and Recombination

We next addressed the question of the relationship between the transcriptional state of a region and recombination as assayed by direct repeat recombination. This question is especially relevant to mobile elements, since many contain direct repeat sequences and are transcribed at a high rate (Finnegan 1985; Weiner et al. 1986). We designed a construct with direct repeats of a portion of the *GAL10* gene (St. John and Davis 1981) flanking a plasmid containing the $URA3^+$ gene. One side of the duplication contains a three-quarter length copy of *GAL10*, whereas the other has a full length copy that contains a mutation (Fig. 2). As described in the previous section, the *URA3* gene was used to select for 5-FOA resistant colonies that arise by loss of the $URA3^+$ sequences due to direct repeat recombination. The rate at which popouts occurred between repeats that are actively transcribed was compared to that for the transcriptionally inactive state. To alter the transcriptional state of the *GAL10* region, we took advantage of the fact that transcription can be controlled by two *trans*-acting mutations, *gal4* and *gal80* (Oshima 1982). The *gal4* mutation in constitutively turns off the *GAL10* gene, whereas the *gal80* mutation constitutively turns on the gene. The rate of recombination between the direct repeats, when they are transcriptionally inactive, is

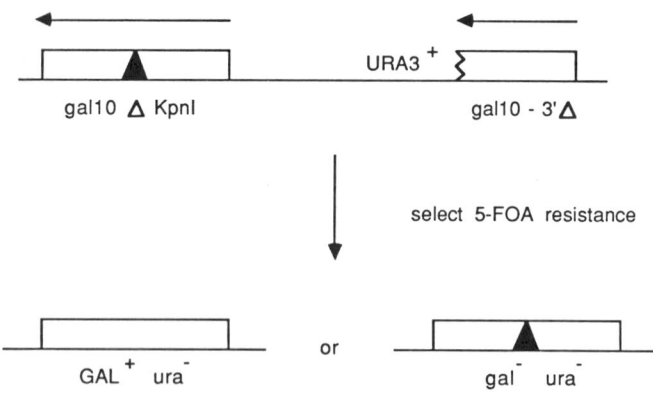

Figure 2
A direct repeat recombination assay regulated by transcriptional activation. A duplication of a portion of the *gal10* locus was created by integrating a $URA3^+$, *gal10*-containing plasmid into the chromosomal *GAL10* region. Transcription of the *gal10* repeats and the surrounding DNA can be controlled by unlinked regulatory genes. As in Figure 1, plasmid loss occurs by direct repeat recombination and can be selected on 5-FOA.

10^{-5}. When the repeats are transcribed, there is a 20- to 30-fold stimulation of the rate. To assess the effect of *cis*-acting promoter mutations on the rate of recombination, a promoter deletion mutation constructed by West et al. (1984) that abolishes transcription was substituted for the wild-type promoter. Abolishing transcription with this mutation resulted in the same level of recombination in both the *gal4* and *gal80* strains, (10^{-5}). These results suggest that transcriptional activation of the *gal10* duplication, which is mediated by the Gal4 protein, affects the rate of recombination between these direct repeats.

Genetic Control of δ and Ty recombination

The yeast retrotransposon, Ty, contains long terminal repeats called δ-sequences (for review, see Adams et al. 1987). Occasionally, the two δs flanking a Ty element recombine resulting in a solo δ in the genomic position of the original Ty. All solo δs are not identical, therefore they comprise a family of repeated sequences. We have been studying rearrangements in a region that contains 5 solo δ-sequences with no associated Ty elements (Rothstein et al. 1987). This region is shown diagrammatically in Figure 3. A genetically marked tyrosine tRNA ochre suppressor gene, *SUP4*-o, situated in the middle of these δ-sequences, deletes from the genome at high frequency (1×10^{-7} to 3×10^{-7}) (Rothstein 1979). In yeast, deletions are fairly rare, therefore we suspected recombination between these elements to be

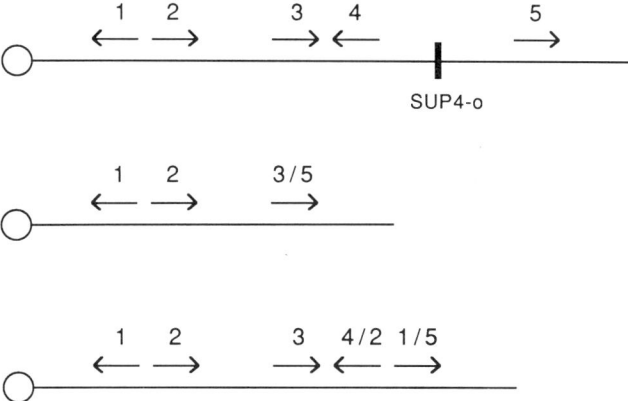

Figure 3
The *SUP4* region and two spontaneous deletion classes. The numbered arrows above the line represent the relative positions of the five δ-sequences that surround the *SUP4* gene (not drawn to scale). Two deletion classes are shown, and the position of the recombinant structure is indicated by the hybrid name given to the novel recombinant δs remaining. For example, 3/5 signifies that the δ arose via recombination of δ3 and δ5.

responsible for the high deletion frequency. Cloning and DNA sequence analysis of junction fragments from several deletions revealed that these deletions had their end points in δ-sequences. One third were due to recombination events between the two direct repeat δs (3 and 5) that flank the suppressor (Fig. 3). Interestingly, the predominant deletion class occurred by a gene conversion event in which the suppressor-containing region between δ4 and δ5 was replaced by the δ1–δ2 junction located about 3000 nucleotides downstream. These gene conversion events, defined as the nonreciprocal transfer of information from one region into another, are dependent upon general recombination genes of yeast. We discovered this by measuring the deletion frequency in a radiation-sensitive *rad52* mutant (Resnick 1969), unable to repair an HO-mediated double-strand-break (Malone and Esposito 1980) and plasmid double-strand-breaks (Orr-Weaver et al. 1981). All of the deletion classes are greatly diminished in a *rad52* mutant background.

We have recently proposed two models to account for the deletion classes that we observe at the *SUP4* locus (Rothstein et al. 1987). One model supposes that the recombination event leading to deletion occurs in G1. For the direct repeat δ-events, we imagine recombination between the directly repeated δs (3 and 5) to result in deletion of the *SUP4*-containing fragment. For the gene conversion class, we envision pairing of the two δs flanking the tRNA gene, with the inverted δs 3.0-kb upstream and the subsequent conversion of the inverted δs in place of the *SUP4* region (Fig. 4A). These post-

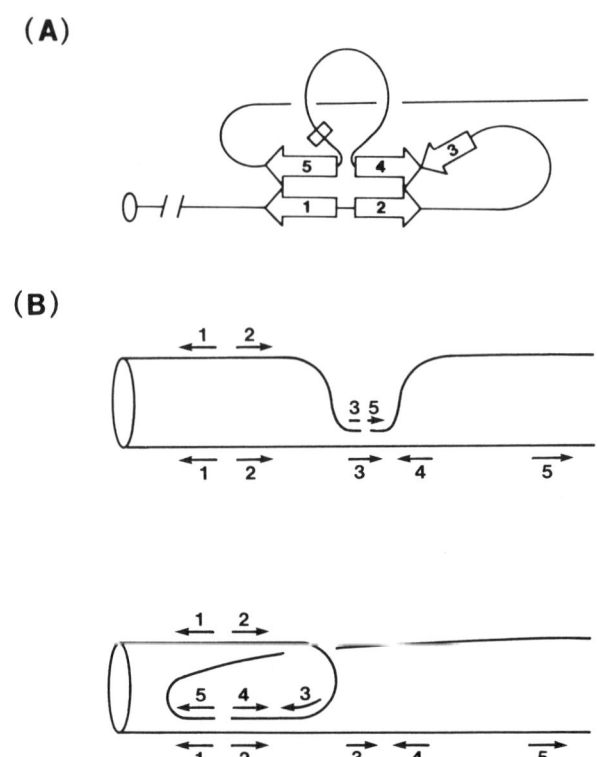

Figure 4
Pairing models for deletion formation in the *SUP4* region. (*A*) A G1 pairing model for the δ 4/2–1/5 deletion. δ4 and δ5 pair with the δ1–2 junction. Recombination resulting in the replacement of the δ1–2 junction for the *SUP4* region gives rise to the deletion. (*B*) A G2 pairing model for deletion formation. The δ3/5 deletion is generated by the pairing of δ3 and δ5 from one chromatid with either δ3 or δ5 from the sister chromatid (shown as δ3 in the figure). For simplicity, the initiating lesion is illustrated as a double-strand-break. The δ4/2–1/5 deletion results from the pairing of δ4 and δ5 with the δ1–δ2 region on the sister chromatid. Subsequent repair and resolution (Szostak et al. 1983) produces the deletion.

ulated intrastrand events account for all but one of the seven deletion classes that we observed.

In a second model, we postulate that the events leading to the deletion of the *SUP4* region occur in G2. In this model, one chromatid is preferentially used as the template for the repair of an initiating lesion(s) on the other chromatid. Depending on the pairing of the various δs, each deletion class can be generated by this model. For example, the direct repeats, δ3 and δ5, pair with either δ3 or δ5 on the sister chromatid (δ3 in Fig. 4B) and after recombination, a deletion of the *SUP4* gene results. In another example shown in Figure 4B, $\delta 4$ and $\delta 5$ pair with $\delta 2$ and $\delta 1$, respectively, on the sister chromatid and after recombination give rise to the major deletion class. Details of both the G1 and G2 models are described in Rothstein et al. (1987).

Isolation and Characterization of a Mutation That Enhances δ and Ty Recombination Events

In yeast, as in most eukaryotes, there are many repeated sequences. For example, there are at least 150 solo δs and approximately 35 Ty elements in standard laboratory strains (Cameron et al. 1979). Thus, it is remarkable that with all the potential homologous recombination possible between these repeats, yeast genomes in general remain stable. Using the power of yeast genetic methods, we searched for genes in yeast that control recombination between δ and Ty sequences. We took advantage of the suppressor system just described to look for mutations that increase the frequency of deletions at the *SUP4* locus. We isolated one such mutation after ethylmethane sulfonate mutagenesis that we named *edr1* for *e*nhanced *d*elta *r*ecombination (Rothstein 1984). This mutation causes a sixfold to tenfold increase in deletions at the *SUP4* locus. The *edr1* mutation, which behaves as a Mendelian trait, is unlinked to the *SUP4* locus. To determine the specificity of this mutation with respect to δ, Ty, and non-repeat sequence recombination, we cloned the wild copy of the gene by complementation.

We next made a series of gene disruptions of the wild-type fragment in bacteria using the Tn*3* system developed by Fred Heffron and his colleagues (Seifert et al. 1986). Fourteen independent insertions into a 3.0-kb fragment that contains *edr1*-complementing activity were shuttled back into yeast at the *EDR1* chromosomal location by homologous recombination. To determine the phenotype of these insertion mutations, each disrupting fragment was independently transformed into a wild-type diploid to create a strain with one disrupted copy and one wild-type copy of the *EDR1* region (Rothstein 1983). After sporulation, tetrads from each diploid were dissected, and in 11 cases they gave rise to two normal-sized colonies and two smaller than normal-sized yeast colonies. For the remaining three insertions, there was no change in the colony size because these do not disrupt the open reading frame of the Edr1 protein (E. Hiller et al., unpubl.). The spores giving rise to smaller colonies contained the *edr1* disruption, revealing that *EDR1* is an important gene but not an essential one. Although the two *edr1*-disrupted spore clones grow very slowly, they eventually grow into an almost normal-sized colony.

By deleting the DNA between the outermost insertions that exhibit the mutant phenotype, we created an approximately 1.5-kb deletion. This fragment was used to replace the wild-type fragment at its normal chromosomal location to create a null mutation. We found several pleiotropic phenotypes associated with the null phenotype. First, as stated earlier, *edr1*::*null*-containing cells grow about three to four times more slowly than wild-type cells. They exhibit an altered morphology that includes an abnormal budding pattern and increased cell volume. Although an *edr1*::*null* strain forms diploids at normal frequencies when mated to a wild-type strain, *edr1*::*null* homozygous diploids cannot be formed easily.

To construct homozygous diploids, we had to resort to tetraploid genetics in which two heterozygous *edr1*::*null*/ + diploids were each made into spheroplasts and fused together in the presence of polyethylene glycol (van Solingen and van der Plaat 1977). After sporulation and dissection, the resultant tetraploid generated homozygous *edr1*::*null*/*edr1*:*null* diploids in approximately 1/4 of the spore clones. Some of the homozygous diploids formed this way were heterozygous for the mating type locus, however, they failed to sporulate. Finally, diploids formed by tetraploid genetics show increased chromosomal rearrangements as measured by changes in the segregation of markers on chromosomes III and XII. In addition, changes in chromosome sizes and number are occasionally seen on chromosome separation gels (Carle and Olson 1985). An example of one such gel is shown in Figure 5.

We next inserted the *edr1*::*null* deletion into a variety of strains each containing various recombination assays to measure both δ and Ty recombination as well as non-δ-events. We assayed the frequency of deletions at *SUP4* and found a 60-fold increase compared to wild-type cells. We next assayed direct repeat δ recombination at the *his4-912* gene that was described by Fred Winston (this volume). For this assay, we used a construct of Roeder and Fink (1982) in which they cloned a $URA3^+$ gene within the Ty element of the *his4-912* Ty insertion (Fig. 6A). This construct permits the measurement of several kinds of recombination events including direct repeat δ-δ recombination, a more complicated rearrangement that simultaneously results in gene conversion of the Ty element leaving a solo δ at the *HIS4* gene and disomes of chromosome III in which one copy has lost the $URA3^+$ marked Ty (Roeder et al. 1984). We measured the effect of the *edr1*::*null* on these rearrangements and found that the δ-δ recombination event increased fivefold over wild type, whereas the other two more complicated rearrangements were both increased between 200-fold to 500-fold. Lastly, an independent Ty insertion was used to measure a Ty gene conversion event. The mutation, *his4-917*::$URA3^+$, contains a $URA3^+$ engineered into the Ty917 insert in front of *HIS4* (Roeder and Fink 1982). Unlike *his4-912*::$URA3^+$, the *his4-917*::$URA3^+$ mutation only gives rise to His$^+$ cells after the Ty is partially or completely replaced by different Ty sequences from one of the approximately

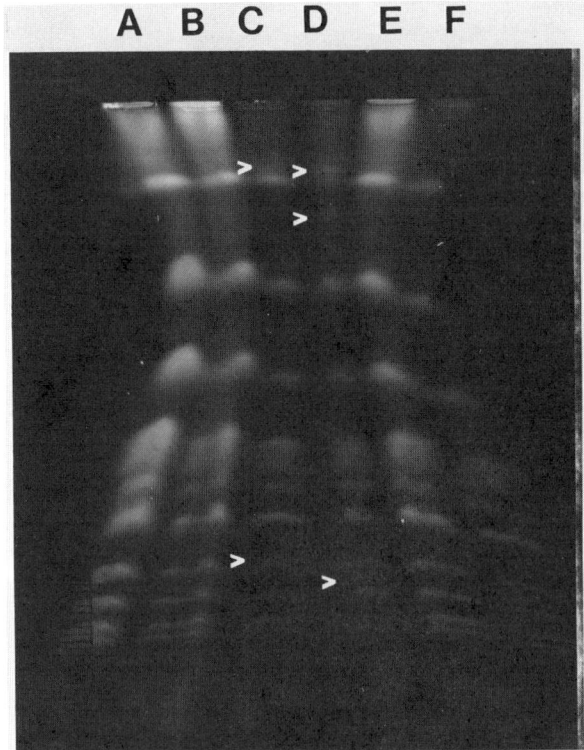

Figure 5
Chromosome separation gel of *edr1::null/edr1::null* diploids. Lanes *A*, *B*, *E*, and *F* contain isogenic parental wild-type diploids. Lanes *C* and *D* contain two subclones from the same homozygous *edr1::null* diploid. Note the appearance of new chromosome bands (→).

35 Ty elements in the genome (Roeder and Fink 1982). This assay is diagrammed in Figure 6B. We measured the frequency of this gene conversion event, and it is increased 40-fold in *edr1::null* strains.

We next examined the effect of the *edr1::null* mutation on both non-δ-direct-repeat recombination events as well as non-δ-gene-conversion events. We used an assay system similar to one we described at the beginning of this report that contained a duplication of a tRNA gene (one wild type and one suppressor) with a selectable marker in between. Loss of suppressor activity can occur in two ways: direct repeat recombination between the two tRNA segments resulting in the retention of the wild-type copy and simultaneous loss of the selectable marker (excision) or by gene conversion of the suppres-

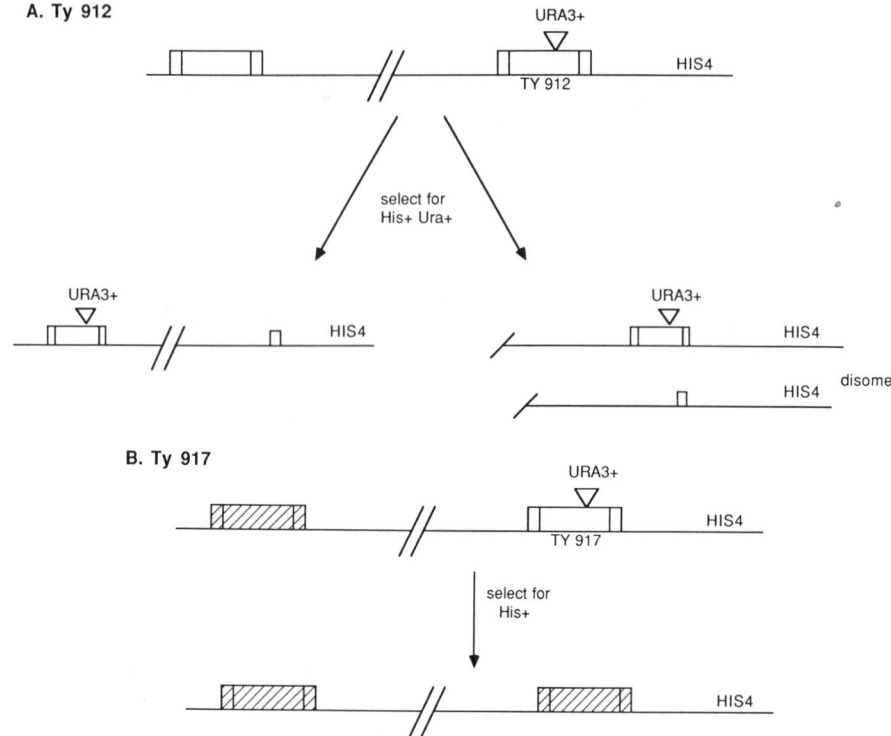

Figure 6
Ty recombination assays. (A) the $his4$-912::$URA3^+$ allele (Roeder and Fink 1982) yields His^+ Ura^+ revertants in two ways. The first is shown on the *left*. Ty912 excises from the *HIS4* region, leaving a solo δ (□), whereas the $URA3^+$ portion of Ty912 is maintained at another genomic position (usually on the same chromosome, III). The solo δ in front of the *HIS4* gene restores the His^+ phenotype. At a similar frequency, disomic strains (*right*) arise that contain one chromosome with a solo δ (thus a His^+ phenotype) and one unaltered chromosome that remains Ura^+. (*B*) The $his4$-917::$URA3^+$ allele (Roeder and Fink 1982) yields His^+ predominantly by replacement of the Ty917::$URA3^+$ sequences with Ty sequences from another position within the genome. This results in a concomitant loss of the Ura^+ information within Ty917::$URA3^+$.

sor to wild type without loss of the selectable marker. We found a twofold to fourfold increase in excision of the plasmid as well as in the gene conversion events at the duplicated *SUP3* gene in the presence of the *edr1*::*null* mutation. A similar result was found when we measured recombination between direct repeats of the *gal10* region (Fig. 2).

DISCUSSION

We are in the process of characterizing the pleiotropic phenotypes associated with the *edr1* mutation. Our first approach is to isolate suppressors of the slow growth phenotype and study their effects on the various phenotypes exhibited by *edr1* mutant strains. Preliminary results of a study in which we examined the phenotype of an *edr1* suppressor double mutant revealed that the recombination phenotype of *edr1* is not suppressed. Interestingly, the suppressor itself exhibits increased δ recombination when separated from *edr1*.

In conclusion, by focusing on δ-δ recombination and Ty-mediated gene conversion in yeast, we are discovering genes that control these events. We have isolated a gene, *edr1*, whose phenotype suggests that it is involved in the regulation of recombination between these elements. We are searching for additional genes and suppressors in order to understand the mechanism that regulates recombination between the repetitive elements scattered throughout the genome.

ACKNOWLEDGMENTS

We thank Dr. Shirleen Roeder for providing the *his4-912::URA3* and *his4-917::URA3* constructs. We also thank Gary Chrebet for excellent technical assistance, Iris Toribio for typing the manuscript and John McDonald for Figure 1. H.R. was an EMBO Long-term Fellow; J.W.W. was an American Cancer Society Fellow; and W.L.A. and B.J.T. were both supported by National Institutes of Health (NIH) grant CA-09503. R.J.R is an American Heart Association Established Investigator. This work was also supported by National Science Foundation grant DCB-8703833, NIH grants GM-34587 and CA-21111, a grant from the Irma T. Hirschl Trust, and a grant from the MacArthur Foundation.

REFERENCES

Adams, S.E., S.M. Kingsman, and A.J. Kingsman. 1987. The yeast Ty element: Recent advances in the study of a model retro-element. *Bioessays* **7**: 3.

Boeke, J.D., F. Lacroute, and G.R. Fink. 1984. A positive selection for mutants lacking orotidine-5′-phosphate decarboxylase activity in yeast: 5-fluoro-orotic acid resistance. *Mol. Gen. Genet.* **197**: 345.

Cameron, J.R., E.Y. Loh, and R.W. Davis. 1979. Evidence for transposition of dispersed repetitive DNA families in yeast. *Cell* **16**: 739.

Carle, G.F. and M.V. Olson. 1985. An electrophoretic karyotype for yeast. *Proc. Natl. Acad. Sci.* **82**: 3756.

Esposito, M.S. 1978. Evidence that spontaneous mitotic recombination occurs at the two-strand stage. *Proc. Natl. Acad. Sci.* **75**: 4436.

Finnegan, D.J. 1985. Transposable elements in eukaryotes. *Int. Rev. Cytol.* **93:** 281.
Jelinek, W.R. and C.W. Schmid. 1982. Repetitive sequences in eukaryotic DNA and their expression. *Annu. Rev. Biochem.* **51:** 813.
Jinks-Robertson, S. and T.D. Petes. 1985. High-frequency meiotic gene conversion between repeated genes on nonhomologous chromosomes in yeast. *Proc. Natl. Acad. Sci.* **82:** 3350.
———. 1986. Chromosomal translocations generated by high-frequency meiotic recombination between repeated yeast genes. *Genetics* **114:** 731.
Lichten, M., R.H. Borts, and J.E. Haber. 1987. Meiotic gene conversion and crossing over between dispersed homologous sequences occurs frequently in *Saccharomyces cerevisiae*. *Genetics* **115:** 233.
Malone, R. and R.E. Esposito. 1980. The *RAD52* gene is required for homothallic interconversion of mating types and spontaneous mitotic recombination in yeast. *Proc. Natl. Acad. Sci.* **77:** 503.
Orr-Weaver, T.L., J.W. Szostak, and R.J. Rothstein. 1981. Yeast transformation: A model system for the study of recombination. *Proc. Natl. Acad. Sci.* **78:** 6354.
Oshima, Y. 1982. Regulatory circuits for gene expression: The metabolism of galactose and phosphate. In *The molecular biology of the yeast Saccharomyces: Metabolism and gene expression* (ed. J. Strathern et al.), p. 159. Cold Spring Harbor Laboratory, Cold Spring Harbor, New York.
Resnick, M.A. 1969. Genetic control of radiation sensitivity in *Saccharomyces cerevisiae*. *Genetics* **62:** 519.
Roeder, G.S. and G.R. Fink. 1982. Movement of yeast transposable elements by gene conversion. *Proc. Natl. Acad. Sci.* **79:** 5621.
Roeder, G.S., M. Smith, and E.J. Lambie. 1984. Intrachromsomal movement of genetically marked *Saccharomyces cerevisiae* transposons by gene conversion. *Mol. Cell. Biol.* **4:** 703.
Roman, H. and F. Fabre. 1983. Gene conversion and associated reciprocal recombination are separable events in vegetative cells of *Saccharomyces cerevisiae*. *Proc. Natl. Acad. Sci.* **80:** 6912.
Ronne, H. and R. Rothstein. 1988. Mitotic sectored colonies: Evidence of heteroduplex DNA formation during direct repeat recombination. *Proc. Natl. Acad. Sci.* (in press).
Rose, M., P. Grisafi, and D. Botstein. 1984. Structure and function of the yeast *URA3* gene: Expression in *Escherichia coli*. *Gene* **29:** 113.
Rothstein, R. 1979. Deletions of a tyrosine tRNA gene in *S. cerevisiae*. *Cell* **17:** 185.
———. 1983. One-step gene disruption in yeast. *Methods Enzymol.* **101:** 202.
———. 1984. Double-strand-break repair, gene conversion and post-division segregation. *Cold Spring Harbor Symp. Quant. Biol.* **49:** 629.
Rothstein, R., C. Helms, and N. Rosenberg. 1987. Concerted deletions and inversions are caused by mitotic recombination between delta sequences in *Saccharomyces cerevisiae*. *Mol. Cell. Biol.* **7:** 1198.
St. John, T.P. and R.W. Davis. 1981. The organization and transcription of the galactose gene cluster of *Saccharomyces*. *J. Mol. Biol.* **152:** 285.
Seifert, H.S., E.Y. Chen, M. So, and F. Heffron. 1986. Shuttle mutagenesis: A method of transposon mutagenesis for *Saccharomyces cerevisiae*. *Proc. Natl. Acad. Sci.* **83:** 735.

van Solingen, P. and J.B. van der Plaat. 1977. Fusion of yeast spheroplasts. *J. Bacteriol.* **130:** 946.

Struhl, K., D.T. Stinchcomb, S. Scherer, and R.W. Davis. 1979. High frequency transformation of yeast: Autonomous replication of hybrid DNA molecules. *Proc. Natl. Acad. Sci.* **76:** 1035.

Szostak, J.W., T.L. Orr-Weaver, R.J. Rothstein, and F.W. Stahl. 1983. The double-strand-break repair model for recombination. *Cell* **33:** 25.

Weiner, A.M., P.L. Deininger, and A. Efstratiadis. 1986. Nonviral retroposons: Genes, pseudogenes, and transposable elements generated by the reverse flow of genetic information. *Annu. Rev. Biochem.* **55:** 631.

West, R.W., Jr., R.R. Yocum, and M. Ptashne. 1984. *Saccharomyces cerevisiae* GAL1-GAL10 divergent promoter region: Location and function of the upstream activating sequence UAS. *Mol. Cell. Biol.* **4:** 2467.

Wildenberg, J. 1970. The relation of mitotic recombination to DNA replication in yeast pedigrees. *Genetics* **66:** 291.

Regulation of Yeast Ty Element Transposition

JEF D. BOEKE,* DANIEL EICHINGER,* AND GERALD R. FINK[†]
*Department of Molecular Biology and Genetics
Johns Hopkins University School of Medicine
Baltimore, Maryland 21205
[†]Department of Biology
Massachusetts Institute of Technology
and Whitehead Institute for Biomedical Research
Cambridge, Massachusetts 02142

OVERVIEW

Yeast Ty elements transpose via a retrovirus-like mechanism. They are responsible for causing a variety of spontaneous mutations; systems in which the frequency of these events is greatly elevated are very useful for the study of the "retrotransposition" process. At least one gene, *SPT3*, encoded by yeast, regulates the transcription of, and hence the transposition of, Ty elements. During transposition, Ty element sequences apparently recombine with each other to generate hybrid elements. Such recombinational processes may be important in "shuffling" Ty promoters of various strengths in front of either functional or nonfunctional Ty coding sequences. We have analyzed the molecular nature of the defect in one such nonfunctional Ty element.

INTRODUCTION

In the yeast, *Saccharomyces cerevisiae*, a family of transposons called Ty elements contributes significantly to spontaneously occurring mutations. Ty insertions constitute a major fraction of mutations which result in the activation of a gene which was previously silent. This activating ability of Ty element insertions usually results from insertion of the Ty such that the direction of its transcription is opposite that of the target gene. An enhancer-like sequence within the Ty is essential for such activation (Errede et al. 1985; Roeder et al. 1985). Among spontaneously occurring "knock-out" mutations, Ty elements comprise a smaller, but still significant fraction (Eibel and Philippsen 1984; Simchen et al. 1984). These knock-out mutations consist of insertions both within structural genes and in the 5′ noncoding regions of genes.

The structures of Ty elements and their transcripts are very similar to those of retroviral proviruses (Fig. 1). Ty element transposition occurs via an RNA intermediate by a reverse transcription-like process remarkably similar to that used by vertebrate retroviruses. In yeast cells engineered to undergo very

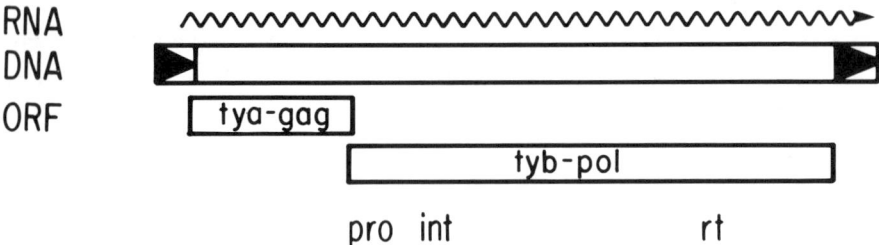

Figure 1
Anatomy of a yeast Ty element. Boxed triangle is δ or LTR sequences; central box is the coding ε region; wavy line the major transcript; lower boxes are open reading frames thought to be analogous to retroviral *gag* and *pol* genes. pro, int, and rt indicate regions of homology to retroviral protease, integrase, and reverse transcriptase functional domains.

high-frequency transposition, virus-like particles (Ty-VLPs), containing reverse transcriptase and Ty RNA, accumulate to high levels in the cytoplasm. These particles are probably intermediates in the transposition process.

The frequency of Ty element retrotransposition is quite variable and probably depends on both physicochemical as well as genetic factors (see chapters by McEntee and Bradshaw; Paquin and Moroz Williamson; Winston; all this volume). Because Ty element transposition occurs via an RNA intermediate, and because Ty-encoded gene products are required for transposition, the level of Ty transcription in the cell should be a critical determinant of the frequency of transposition. Winston and his colleagues (1984, 1987) have shown that mutations in any of several host genes can reduce or alter the transcription of Ty elements. We have studied one of these genes, *SPT3*, to determine whether it is essential for transposition. We have found that the *SPT3* gene function is indeed essential for transposition, but only at the level of Ty transcription.

By using *spt3* strains containing a *GAL1*/Ty fusion plasmid (pGTy plasmid), in which Ty transcription is directed by the *SPT3*-independent *GAL1* promoter, we have shown that Ty element transposition can be restored in the mutant strain. Moreover, using such strains we have obtained evidence that during transposition of genetically marked Ty elements, a form of homologous recombination takes place at high frequency, resulting in a considerable degree of heterogeneity in recently transposed (progeny) Ty elements.

Finally, using the pGTy plasmid system, we have shown that there are at least two classes of Ty elements in the yeast genome: functional elements, typified by Ty*H3*, and nonfunctional elements, typified by Ty*173*. The presence of multiple copies of defective Ty elements in the yeast genome suggests that Ty transposition frequency may be regulated by the amount of functional Ty mRNA produced in the cell.

RESULTS

Transposition of Yeast Ty Elements Requires a Host Function *(SPT3)*

We wanted to assay for the frequency of transposition of chromosomal (i.e., natural, nonengineered) Ty elements in the presence and absence of *SPT3* gene function. To do this, we used a plasmid bearing a promoterless *HIS3* gene from yeast as a target for transposition in a strain otherwise deleted for the *HIS3* gene; previous studies have shown that Ty insertions into this plasmid result in a His$^+$ phenotype (Scherer et al. 1982; Boeke et al. 1985, 1986). To quantitate the frequency of Ty transposition into this plasmid, we measured the frequency of His$^+$ revertants according to the (p[0]) method of Luria and Delbrück (1943), in which the frequency of His$^+$ revertants is deduced from the fraction of independently grown cultures containing no His$^+$ revertants. Because certain other events can also give rise to a His$^+$ phenotype, we then characterized the plasmids conferring the His$^+$ phenotype as to whether or not these contain a Ty insertion; the fraction of independent His$^+$ revertants caused by Ty insertion is referred to as the Ty fraction. From the product of these two numbers we obtained an estimate of transposition frequency.

The role of *SPT3* function was assessed by constructing wild-type and isogenic *spt3-101* (an in vitro-constructed frameshift allele) yeast strains by standard transplacement techniques. We then examined the transposition frequency in the isogenic *SPT3* and *spt3-101* strains bearing the *his3* promoter deletion plasmid and found that there was at least a 20-fold lower transposition frequency in the *spt3-101* strain than in the wild type (Boeke et al. 1986). This number correlates well with the relative abundance of Ty RNA in *spt3-101* and *SPT3* strains (Winston et al. 1984).

Genetic evidence suggests that *SPT3* encodes a transcriptional activator specifically required for transcription from the Ty promoter sequence. We reasoned that if the only site of action of the *SPT3* gene product were in the Ty promoter sequence, then transposition could be restored to *spt3-101* strains by introducing the pGTy*H3* plasmid, in which Ty transcription is directed by the *SPT3*-independent *GAL1* promoter. Indeed, this proved to be the case; high frequency transposition is found in *spt3* strains bearing the pGTy*H3* plasmid when grown on galactose (Boeke et al. 1986). Moreover, we could mark the Ty element in the plasmid with a synthetic *lacO* segment and determine whether transposition of chromosomal or plasmid-derived transposons were giving rise to the His$^+$ phenotype. In the *SPT3* strain, only a minority (16%) of the Tys jumping into the *his3* plasmid carried *lacO*, whereas, in the *spt3-101* strain, nearly all were derived from the pGTy*H3*-*lacO* plasmid.

Yeast Ty Elements Recombine during Retrotransposition

When transposition experiments using pGTy*H3-lacO* were carried out in *SPT3* (wild-type) strains, a curious phenomenon was noted (Boeke et al. 1985). Transposed *lacO*-marked Ty elements did not always resemble the parental Ty in the pGTy*H3-lacO* plasmid, yet they must have derived at least partially from it because they carried the *lacO* sequence. These transposed elements were found to differ at certain restriction sites such as the *Hha*I sites and the *Hin*dIII site found within the parental (Ty*H3*) sequence (Fig. 2). Importantly, these polymorphisms always reflected known polymorphisms within the population of Ty elements within yeast (i.e., these particular site differences have been noted by many workers when comparing Ty sequences). Sites that are rigorously conserved among Ty elements (such as the three *Bgl*II sites) were similarly maintained in the transposed *lacO*-marked Tys. By screening 40 *lacO*-marked Ty elements, which had transposed with three enzymes that revealed polymorphisms (*Xho*I, *Hha*I, and *Hin*dIII), we determined that at least 20% of these elements had changed in sequence during transposition. Moreover, many of these bore multiple site changes, suggesting that a recombinational mechanism generates the changes. In order to test this explicitly, we examined a similar number of *lacO*-marked elements from the *spt3-101* strain. Among these, not a single polymorphism was found (Boeke et al. 1986). This strongly suggests that transcription of chromosomal Ty elements is required in order for the polymorphisms to be generated.

A simple model that accounts for these data is presented in Figure 3. In this copy-choice model, similar to one proposed years ago to explain retroviral recombination (Coffin 1979), reverse transcriptase "jumps" between two different RNA templates (perhaps at the site of a break in the RNA), generating a reverse transcript that is a hybrid between the two parental RNA molecules. It should be pointed out that in the case of Ty, there is as yet no physical evidence that Ty RNA exists as dimers in the Ty-VLPs. Other plausible models for the generation of hybrid progeny Ty elements invoke the recombination of completed or partially completed reverse transcripts, catalyzed by the host homologous recombination system. The very high frequency of these events suggests that if this second model is correct, then Ty transposition must induce a high frequency of generalized recombination. This is not the case: mitotic recombination frequencies in heterozygous diploids undergoing high-frequency transposition is not elevated much, if at all, above the normal rate (J. Boeke, unpubl.). Furthermore, certain other events resembling homologous recombination apparently take place within the Ty-VLP, suggesting a Ty-specific mechanism (Xu and Boeke 1987). In any case, transcription of chromosomal Tys is required to generate hybrid progeny Tys.

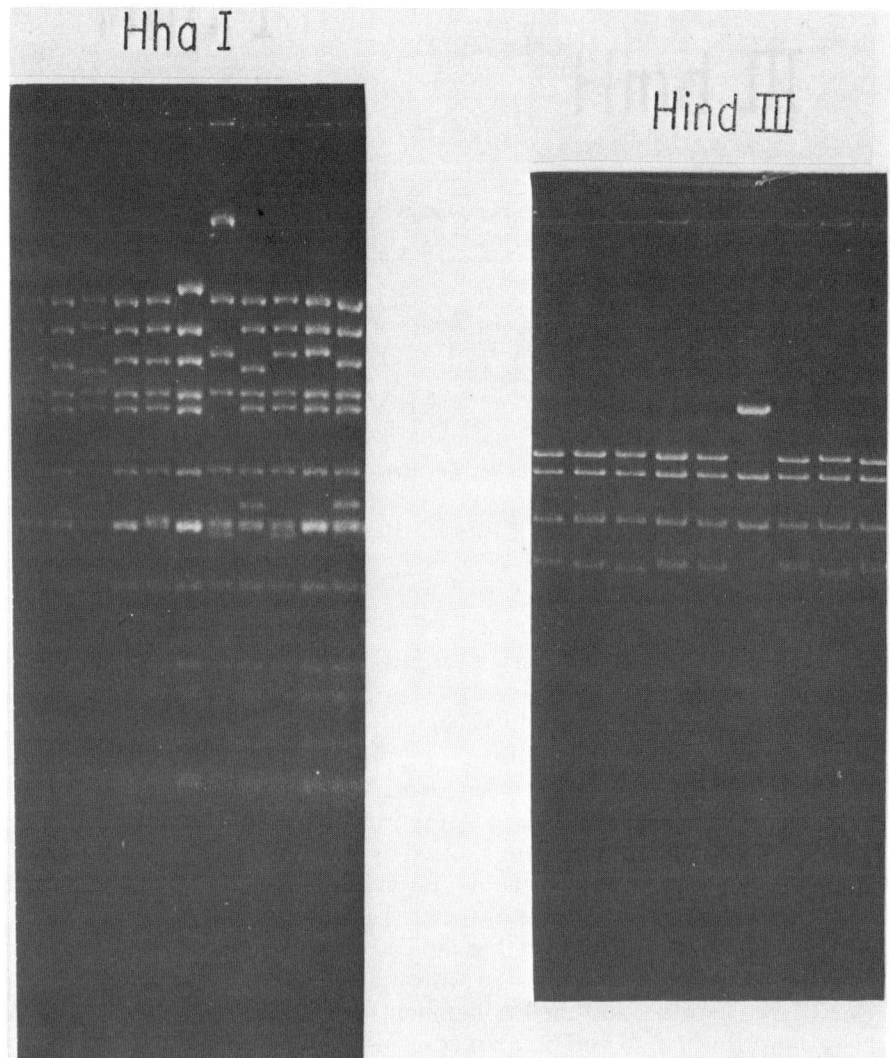

Figure 2
Restriction site changes in transposed Ty elements. Ty elements that are marked with *lacO* and had transposed into a *his3* promoter deletion plasmid were digested with the enzymes indicated. Besides the slight variation in molecular weight of plasmid junction fragments (bands *3* and *7*, *Hha*I; *1* and *4*, *Hin*dIII), two cases of loss of an internal site (lane *7*, *Hha*I; lane *6*, *Hin*dIII), and one case of a small internal insertion (lane *6*, band *1*, *Hha*I) may be seen.

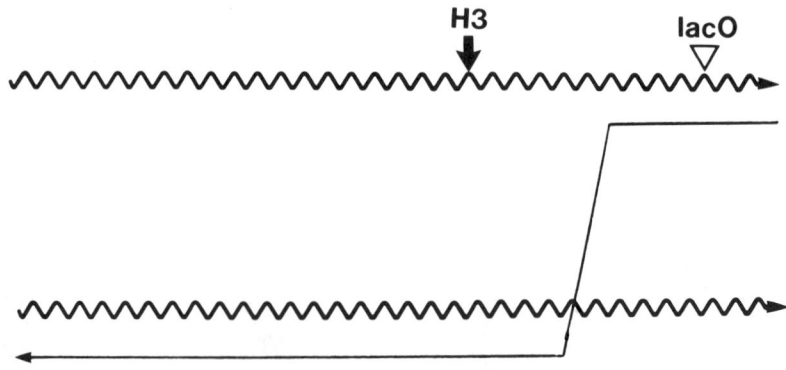

Figure 3
A model for the generation of a polymorphic progeny Ty. If more than one Ty RNA (wavy lines) can interact during reverse transcription (Ty DNA, straight line), in some cases the RNA will derive from the pGTy*H3-lacO* plasmid, in which the Ty contains a *Hin*dIII site (*H3*, ↓) and a *lacO* sequence. Many chromosomal Ty transcripts lack the *Hin*dIII site and none of them contain *lacO*. If a plasmid-derived RNA and a chromosomally derived RNA can serve as templates for the same reverse transcriptase, a template jump by the enzyme as indicated would give rise to recombinant progeny Ty elements. The terminal sequences are omitted for clarity.

Amount of RNA Overproduction from pGTy Plasmids Does Not Account for Observed Increase in Transposition Frequency

When pGTy*H3* plasmids or their derivatives are introduced into yeast cells, the transposition rate, as assayed by a variety of techniques, increases by 20–100-fold over the naturally occurring frequency (i.e., the amount of transposition seen in the absence of the plasmid or on noninducing media) (Boeke et al. 1985, 1986; Fink et al. 1986). Moreover, the amount of Ty-VLP in the cell, the amount of reverse transcriptase activity, and the amount of a *TYB*-encoded 90-kD protein are induced by similar amounts (Garfinkel et al. 1985). This is consistent with the notion that the VLP, reverse transcriptase, and 90-kD protein play a role in the transposition process.

We wished to determine whether the extent of overproduction of Ty RNA in pGTy transformants correlated with the overproduction of transposition events and Ty-VLPs and their components. Transcripts isolated from yeast cells transformed with both unmarked pGTy*H3* plasmid and pGTy*H3* carrying a 1.7-kb *neo* gene insert (enabling clear visualization of the plasmid-derived transcripts) were studied by Northern blotting. As can be seen from Figure 4, the extent of overproduction is two- to at most threefold above the amount of Ty transcript seen in uninduced cells. Thus, the amount of overtranscription is insufficient to explain the amount of transposition observed.

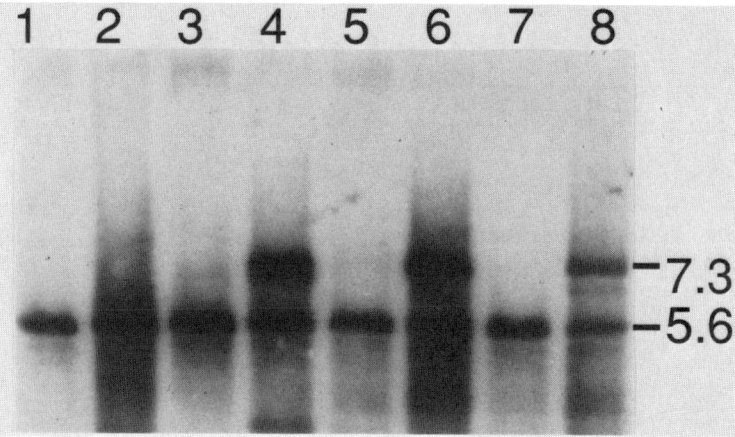

Figure 4
RNA blot analysis of yeast cells containing marked Ty plasmids. RNA was prepared from yeast transformants grown on glucose (*1,3,5,7*) or galactose (*2,4,6,8*). Transformants contain pGTy*H3* unmarked (*1* and *2*), or marked at three different positions within the Ty sequence by the insertion of a *neo* gene (*3-8*). Plasmid-derived and chromosomally derived transcripts can be distinguished by their molecular weight in *3-8*. The probe was Ty*1*.

Defective and Functional Ty Elements in the Yeast Genome

The plasmid pGTy*H3* clearly contains a fuctional Ty element; various assays have shown that this element, even when genetically marked, will transpose to new locations at high frequency. A very similar *GAL1*/Ty fusion, constructed using a different Ty element called Ty*173* (Simchen et al. 1984), behaves quite differently. Basically, this plasmid, pGTy*173*, which is virtually identical to pGTy*H3* except for the Ty sequences, does not support high levels of transposition. In order to localize the defect in this Ty element, various hybrid pGTy-*lacO* plasmids were constructed by "cutting and pasting" with restriction enzymes. These hybrid plasmids contain varying amounts of pGTy*H3* and pGTy*173*. Yeast transformants carrying these hybrid plasmids (as well as the parental plasmid) were assayed for transposition by measuring the amount of transposition of the *lacO* marker segment into genomic DNA in a period of five days at 22°C on galactose-containing (i.e., inducing) medium. The extent of such transposition is determined by segregating off the pGTy-*lacO* plasmid from a number of randomly selected colonies, preparing genomic DNA from those colonies, and Southern blotting of the genomic DNA using a *lacO* probe to determine the extent of transposition. As summarized in Figure 5, the defect in Ty*173* can be localized to

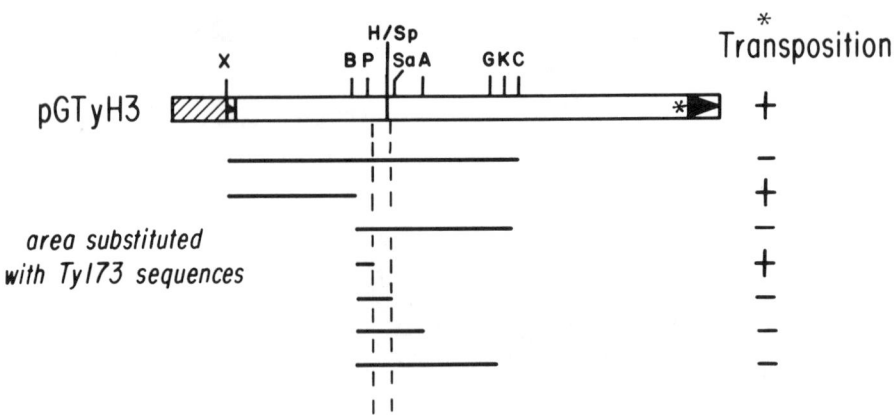

Figure 5
The *GAL1*/Ty*H3* fusion (open box) from pGTy*H3*. Asterisk indicates the *lacO* marker. Hatched box is the *GAL1* promoter. Other symbols are as in Fig. 1. Restriction sites are abbreviated as follows: (X) *Xho*I; (B) *Bst*EII; (P) *Pst*I; (H/Sp) *Hha*I/*Sph*I; (Sa) *Sal*I; (A) *Ava*II; (G) *Bgl*II; (K) *Kpn*I; (C) *Cla*I. The effect of substituting various subfragments of Ty*173* for their homologs in Ty*H3* on transcription of the *lacO* sequences is indicated.

within a 300-bp region in the *tyb* open reading frame. This region was sequenced in both Tys. Three sequence changes were seen, two of which do not result in a change in coding (both are third position base substitutions). The third results in a Leu → Ile missense mutation (Fig. 6). Using oligonucleotide-directed mutagenesis, we confirmed that this missense mutation alone, like the mutant fragments from Ty*173*, reduce transposition frequencies by 60-fold in the above assay.

It was of interest to determine how many of the 30–40 chromosomal copies of Ty elements contained the relevant Ty*173* mutation (i.e., the Leu → Ile change). Oligonucleotides complementary to the mutant and wild-type re-

```
                       2100
                        |
TyH3      CAT CGA ATG CTT GCG CAT GCC
          His Arg Met Leu Ala His Ala

Ty173     CAT CGA ATG ATT GCA CAT GCC
          His Arg Met Ile Ala His Ala
```

Figure 6
Sequence of the Ty*173* mutation. The Leu → Ile mutation at position 2101 is responsible for the inability of Ty*173* to transpose.

gions were synthesized, end-labeled with ^{32}P, and used as hybridization probes. In two yeast strains tested, five and seven copies of Ty*1* hybridized to the "mutant" probe and not to the wild-type probe. Thus, it appears that the Ty*173* mutation is widely distributed among Ty elements in the yeast genome (J.D. Boeke et al., in prep.).

DISCUSSION

It would make sense that Ty element transposition, a potentially deleterious process, might be tightly regulated by the host cell, which apparently plays a major role at least in the regulation of transcription of Ty elements. We have shown that the *SPT3* gene function, which is required for Ty element transcription, is also essential for Ty transposition. However, this requirement can be circumvented by the use of *GAL1*/Ty fusion constructs in which transcription is regulated not by *SPT3*, which apparently acts only at δ promoters, but by galactose. It has been shown that at least two other yeast genes, *SPT7* and *SPT8*, encode functions similar to *SPT3*; namely, *spt7* and *spt8* mutant strains behave like *spt3* strains with regard to Ty transcription. Presumably, Ty transposition is also eliminated in *spt7* and *spt8* mutant strains (Winston et al. 1987). However, these *SPT* gene products appear to be *positive* regulators of Ty transcription and do not appear to down-regulate Ty transcription or transposition. The composition of the mating-type (*MAT*) locus also regulates the level of Ty element transcription, which is high in mating-proficient (**a** or α) strains, and reduced three- to tenfold in **a**/α strains (Elder et al. 1983). However, the amount of Ty RNA in the cell is prodigious, estimated to be from 5% to 10% of the poly(A)$^+$ RNA. How then might Ty transposition be regulated?

Naturally, translational control may play a role in regulating Ty transposition. In fact, Ty mRNA is reported to be a poor substrate for in vitro translation systems. Similarly, the efficiency of readthrough from the *TYA* reading frame into the *TYB* reading frame determines the efficiency of *TYB* expression (Clare and Farabaugh 1985; Mellor et al. 1985) which is critical to transposition because *TYB* encodes reverse transcriptase (Garfinkel et al. 1985; Adams et al. 1987). The detailed mechanism of this frameshifting is as yet unknown, but may well be an important juncture for regulation of transposition.

One regulatory mechanism that we are investigating is the possibility of regulation by the balance between defective and functional elements among the population of elements in a genome. We have shown that the yeast genome contains many copies of nonfunctional Ty elements. Since we could assay for only one particular cause of nonfunction, namely the Ty*173* mutation, it may be that many nonfuctional copies, bearing different lesions, exist in the genome. For example, rearranged variants of Ty are often found (Kingsman et al. 1981; Errede et al. 1986). Thus, it may be that one way to regulate Ty element transposition is by adjusting the amount of functional Ty mRNA in the cell. This could be accomplished by shuffling δ promoters of varying strengths with functional or nonfunctional Ty elements. Homologous recombination events such as gene conversion, mediated by the host homologous recombination system or the more novel and potentially reverse transcriptase-mediated type of recombination alluded to above, probably occur

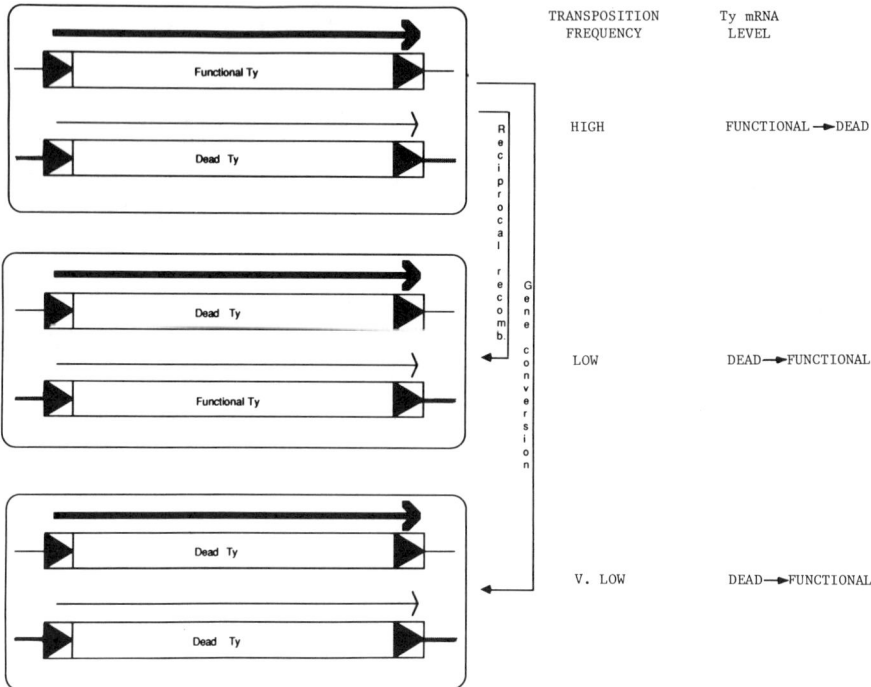

Figure 7

The promoter shuffling model for regulation of Ty transposition. An oversimplified yeast genome containing two Ty elements is indicated by the ovals. The uppermost oval represents a genome in which an actively transcribed (*upper*) Ty element contains functional sequences, whereas an inactively transcribed one (*lower*) contains nonfunctional coding sequences. Two types of homologous recombination events whose net effect would be to greatly decrease the frequency of transposition are indicated (*lower ovals*). The amplification of "dead" sequences by reverse transcriptase mediated processes such as those shown in Fig. 3 are also possible. At least three factors may influence the transcriptional competence of a given Ty promoter: the sequence of the δ itself (these are known to be very variable); the sequence in the enhancer region (Roeder et al. 1985); and the sequence context of the δ promoter (i.e., the sequences in the flanking DNA). Heavy arrows indicate active transcription (strong δ promoter or enhancer); thin arrows indicate inactive transcription (inactive δ promoter or enhancer).

frequently in individuals among populations of cells. Naturally, cells undergoing an "optimal" frequency of Ty element transposition (which is presumably rather low) will have a selective advantage and will tend to accumulate. Such a "promoter-shuffling" model of Ty element transposition (Fig. 7) could explain the state of affairs that we have observed, namely, a high level of Ty transcription in the absence of a high level of Ty element transposition.

REFERENCES

Adams, S.E., J. Mellor, K. Gull, R.B. Sim, M.F. Tuite, S.M. Kingsman, and A.J. Kingsman. 1987. The functions and relationships of Ty-VLP proteins in yeast reflect those of mammalian proteins. *Cell* **49**: 111.

Boeke, J.D., C.A. Styles, and G.R. Fink. 1986. *Saccharomyces cerevisiae* SPT3 gene is required for transposition and transpositional recombination of chromosomal Ty elements. *Mol. Cell. Biol.* **6**: 3575.

Boeke, J.D., D.J. Garfinkel, C.A. Styles, and G.R. Fink. 1985. Ty elements transpose through an RNA intermediate. *Cell* **40**: 491.

Clare, J. and P. Farabaugh. 1985. Nucleotide sequence of a yeast Ty element: Evidence for an unusual mechanism of gene expression. *Proc. Natl. Acad. Sci.* **82**: 2829.

Coffin, J.M. 1979. Structure, replication, and recombination of retroviral genomes: Some unifying hypotheses. *J. Gen. Virol.* **42**: 1.

Eibel, H. and P. Philippsen. 1984. Preferential integration of yeast transposable element Ty1 into a promoter region. *Nature* **307**: 386.

Elder, R.T., E.Y. Loh, and R.W. Davis. 1983. RNA from the yeast transposable element Ty1 has both ends in the direct repeats, a structure similar to retrovirus RNA. *Proc. Natl. Acad. Sci.* **80**: 2432.

Errede, B., M. Company, and R. Swanstrom. 1986. An anomalous Ty1 structure attributed to an error in reverse transcription. *Mol. Cell. Biol.* **6**: 1334.

Errede, B., M. Company, J.D. Ferchak, C.A. Hutchinson, and W.S. Yarnell. 1985. Activation regions in a yeast transposon have homology to mating type control sequences and to mammalian enhancers. *Proc. Natl. Acad. Sci.* **82**: 5423.

Fink, G.R., J.D. Boeke, and D.J. Garfinkel. 1986. The mechanism and consequences of retrotransposition. *Trends Genet.* **2**: 118.

Garfinkel, D.J., J.D. Boeke, and G.R. Fink. 1985. Ty element transposition: Reverse transcriptase and virus-like particles. *Cell* **42**: 507.

Kingsman, A.J., R.L. Gimlich, L. Clarke, A.C. Chinault, and J. Carbon. 1981. Sequence variation in dispersed repetitive sequences in Saccharomyces cerevisiae. *J. Mol. Biol.* **145**: 619.

Luria, S.E. and M. Delbrück. 1943. Mutations of bacteria from virus sensitivity to virus resistance. *Genetics* **28**: 491.

Mellor, J., S.M. Fulton, M.J. Dobson, W. Wilson, S.M. Kingsman, and A.J. Kingsman. 1985. A retrovirus-like strategy for the expression of a fusion protein encoded by yeast transposon Ty1. *Nature* **313**: 243.

Roeder, G.S., A.B. Rose, and R.E. Perlman. 1985. Transposable elements involved in the enhancement of yeast gene expression. *Proc. Natl. Acad. Sci.* **82:** 5428.

Scherer, S., C. Mann, and R.W. Davis. 1982. Reversion of a promoter deletion in yeast. *Nature* **298:** 815.

Simchen, G., F. Winston, C.A. Styles, and G.R. Fink. 1984. Ty-mediated gene expression of the LYS2 and HIS4 genes of *Saccharomyces cerevisiae* is controlled by the same SPT genes. *Proc. Natl. Acad. Sci.* **81:** 2431.

Winston, F., K.J. Durbin, and G.R. Fink. 1984. The SPT3 gene is required for normal transcription of Ty elements in *S. cerevisiae*. *Cell* **39:** 675.

Winston, F., C. Dollard, E.A. Malone, J. Clare, J.G. Kapakos, P. Farabaugh, and P.L. Mineheart. 1987. Three genes required for trans-activation of Ty element transcription in yeast. *Genetics* **115:** 649.

Xu, H. and J.D. Boeke. 1987. High frequency deletion between homologous sequences during retrotransposition of Ty elements in *Saccharomyces cerevisiae*. *Proc. Natl. Acad. Sci.* **84:** 8553.

Inducers/Regulators of Transposable Element Expression and Transposition: Genetic Mutator Systems

Regulatory Aspects of the Expression of P-M Hybrid Dysgenesis in *Drosophila*

MARGARET GALE KIDWELL
Department of Ecology and Evolutionary Biology
University of Arizona
Tucson, Arizona 85721

OVERVIEW

The degree of mobility of P elements in *Drosophila melanogaster* appears to be largely dependent on P-element-encoded properties. Intact, autonomous elements encode a *trans*-acting transposase and at least some classes of truncated or deleted elements apparently code for regulatory molecules. Transposition is restricted to the germline and is controlled at the level of mRNA splicing. Environmental variables, such as temperature and aging, have important effects on the level of expression of specific phenotypic traits. There is some evidence for variability in host functions affecting P element mobility, but the ease of interspecific transfer suggests that, if such functions exist, they must have been highly conserved during the evolution of the genus *Drosophila*. Site specificity for P element insertion occurs at several levels, including those of the nucleotide and the individual gene. Secondary mutations usually occur at least one to two orders of magnitude more frequently than primary insertion mutations. Studies of the evolution and population biology of P elements are providing valuable insights into the role of element-encoded regulatory functions in natural and laboratory populations.

INTRODUCTION

P elements constitute one of several structural classes of mobile elements that are found in *D. melanogaster* (Rubin 1983). Members of the P family are homologous in sequence but heterogeneous in size (O'Hare and Rubin 1983). Based on structure, P elements, can be divided into two types, complete or autonomous elements, sometimes called P factors, and defective, nonautonomous elements (Fig. 1). Complete elements are 2.9 kb in length and have 31-bp inverted terminal repeats. They have four open reading frames (ORFs), all of which are required to encode a transposase enzyme (Karess and Rubin 1984). Nonautonomous elements are smaller, vary in size, and are often derived from autonomous elements by internal deletions. The frequency of deletion formation is high under conditions of active transposition

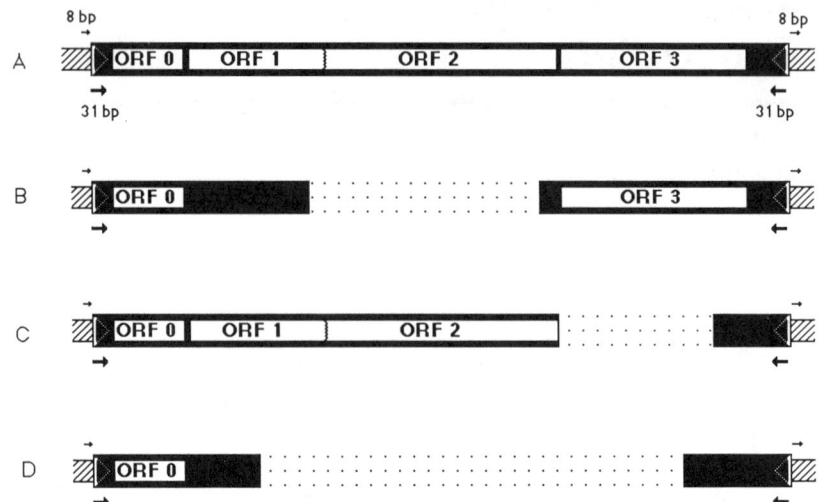

Figure 1
Structures of P elements. (*A*) A 2907-bp autonomous P element (P factor); (*B–D*) Examples of P elements that have sustained internal deletions; deleted fragments are indicated by stippling. The deleted elements shown in (*C*) and (*D*) may code for different types of regulatory molecules (see text). Both complete and deleted elements have 31-bp inverted terminal repeats (triangles not drawn to scale) and initiate a duplication of 8 bp of host DNA on insertion (indicated by small arrows). Undisrupted ORFs are shown unshaded. Only the autonomous element has the coding capacity for the transposase enzyme.

(e.g., Daniels et al. 1985b). Autonomous elements are able to catalyze their own transposition, as well as that of nonautonomous elements. An 8-bp duplication of host DNA is generated at the site of P element insertions.

The mobilization of the P family of elements is associated with a syndrome of germline abnormalities, collectively known as P-M hybrid dysgenesis (Kidwell et al. 1977; for more recent reviews see Bregliano and Kidwell 1983; Engels 1983). The characteristic features of the syndrome include temperature-dependent sterility, male recombination, chromosomal aberrations, transmission ratio distortion, and elevated rates of mutation, nondisjunction, and female recombination. Manifestations of P-M dysgenesis arise in the germline of progeny produced from matings between females of a maternally-contributing type (M strain) that do not carry autonomous elements and males of a paternally-contributing type (P strain) that carry one or more autonomous elements. Appreciable levels of hybrid dysgenesis do not usually occur in progeny from the reciprocal cross or from intrastrain matings. The nonreciprocal nature of this effect implicates specific cytoplasmic and chromosomal interactions, involving both transposable elements and their hosts, as being the underlying cause of the hybrid dysgenesis phenomenon.

Populations of *D. melanogaster* may be divided into several broad types based on the phenotypic characteristics produced by P element mobilization (Table 1). P activity refers to the potential capacity of a strain for mobilizing P elements when these elements are in an unregulated or susceptible state. The level of both P activity and the ability to regulate this activity are related to the number of autonomous and nonautonomous elements in the genome, as indicated in Table 1. However, in addition to the P element complement, a number of additional factors appear to influence the frequency of mobility of these elements. The purpose of this paper is to review and discuss our present knowledge of the nature and mode of action of those factors that affect P element mobility at the molecular level and the manifestations of that mobility at the level of the host organismal phenotype.

EXPRESSION AND CONTROL OF P ELEMENT MOBILITY

Variability of P Element Insertion and Excision of Different Genomic Locations

P element mutagenesis (Kidwell 1986) is now being widely used because of the high transposition of these elements, the ability to control their mobility by genetic manipulations, and their suitability for transposon tagging of specific genes (Bingham et al. 1981; Engels 1985). P elements can be highly effective as agents of both primary and secondary mutagenesis. Primary mutagenesis refers to the insertion of P elements in new genomic sites. Secondary mutagenesis involves the reactivation of a previously inserted element to produce precise or imprecise excisions, deletions, or chromosomal rearrangements. An example of secondary P element mutagenesis is provided

Table 1
Properties of Different Types of Strains Found in Natural Populations

Name	P activity	Regulation	No. autonomous elements	No. deficient elements
P	Low to high	Strong maternal (P cytotype)	Variable (2 to ~50+)	Many (up to 60)
Q	Very low	Strong maternal (P cytotype)	Variable	Many
M' (pseudo M)	None or very low	Chromosomal	Low or none	Many
M (true M)	None	None	None	None

by singed-weak (sn^w), an allele of the X-linked singed locus. This allele can mutate to wild-type or to a more extreme singed allele at inordinately high frequencies in excess of 40% per generation (Engels 1981).

A survey was recently carried out to compare the frequencies of P element mutagenesis that were observed in a large number of experiments. Target sites varied and included both large chromosomal regions and many different gene loci (Kidwell 1987). Based on the results of six independent studies, the frequency of P-induced X-chromosome lethal mutations varied between 1×10^{-2} and 9×10^{-4}. This is approximately an order of magnitude higher than the spontaneous lethal mutation rate. The results of the survey for specific loci are depicted graphically in Figure 2. Primary mutations were observed at frequencies between 10^{-4} and 10^{-5} in about 75% of the loci surveyed. This is at least two orders of magnitude higher than the normal spontaneous mutation rate. However, about 20% of the loci studied sustained no P element insertions, even though very large numbers of individuals were screened. The overall frequency of secondary mutations was about two orders of magnitude higher than that of primary mutations.

The survey results confirm the previously documented site specificity for P element insertion at the locus level (e.g., Golubovsky et al. 1977; Simmons and Lim 1980; Eeken et al. 1985). P elements are also known to insert rarely into centric heterochromatin relative to the euchromatic arms of chromosomes (Engels 1983). Preferential insertion of P elements has also been shown to occur at certain positions within affected genes. Kelley et al. (1987) observed thirteen out of fourteen insertions in the Notch locus to occur at or

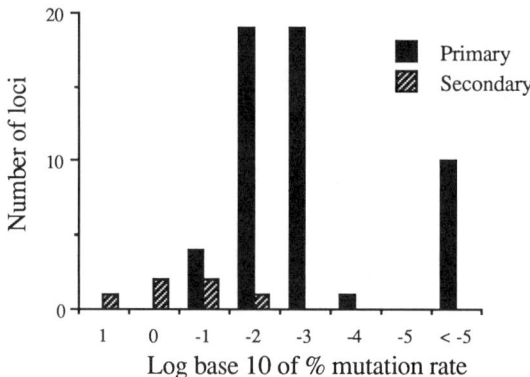

Figure 2
The distribution of P element-induced primary and secondary mutation rates at specific loci observed in more than 50 separate studies (data from Kidwell 1987).

near the transcription start site of the gene. Also, seven of nine P insertions at the *Rpll215* locus mapped within 1 kb of the transcription start site (Searles et al. 1986), and all of five insertions in the rudimentary locus were in the 5' end of the gene (Tsubota et al. 1985). Site specificity at this level may depend on variations in chromosome structure associated with activation or expression of target genes (Kelley et al. 1987). At the nucleotide level, derivatives of an 8-bp consensus sequence were observed at the insertion sites of P elements throughout the genome (O'Hare and Rubin 1983), suggesting preferential recognition of specific target sites.

Variables Affecting the Frequency of Genomic P Element Mobilization

A measure of the frequency of mobilization of P elements at the genomic level is provided by the trait of gonadal sterility (Engels and Preston 1979; Kidwell and Novy 1979). This type of sterility is highly dependent on temperature and is likely to be due to generalized chromosome breakage rather than insertion of P elements into specific loci.

Variables affecting the frequency of hybrid dysgenesis at the genomic level may conveniently be divided into those related to P element functions and those related to extrinsic factors. There is some evidence that the number of autonomous elements in the genome is directly proportional to the level of P factor activity (Todo et al. 1984; A.G. Good et al., pers. comm.). In other experiments, however, the relationship has not always been linear (e.g., Bingham et al. 1982). In addition, there is an important regulatory component to P element mobilization that is apparently mediated by the genomic complement of P elements and is referred to as P cytotype (Engels 1979). Nonautonomous rather than autonomous P elements appear to be involved in this regulatory role. The precise mechanisms involved in regulation are not understood and are the subject of intensive current research. However, it is known that there is an important regulatory component mediated by at least one class of truncated P elements (H. Robertson and W.R. Engels, pers. comm.) (see Fig. 1C). There is also some evidence for P element regulation by a second class of deleted elements, called KP elements (Black et al. 1987) whose structure is illustrated in Figure 1D. Unlike P cytotype, this type of regulation is not maternally inherited. KP elements are present in the genome in multiple copies in M' strains in Europe, Asia, Australia and other parts of the world (Black et al. 1987; I.A. Boussy et al., in prep.). It is formally possible that many different size classes of deleted elements may have a regulatory role but that there is wide variation in the regulatory efficiencies of different classes. Selection for the most efficient types would be expected to occur in order to mitigate the deleterious effects of transposition.

Simmons and Bucholz (1985) have proposed an additional regulatory mechanism by which transposase might act indirectly as its own inhibitor. This "transposase titration model" might explain P element regulation in strains in which the reciprocal cross effect is not observed (e.g., the M' strain, Sexi, studied by Kidwell 1985). Simmons and Bucholtz (1985) hypothesized that the transposase enzyme binds to sequences in the termini of both intact and degenerate elements and that the activity of functional elements leads to the generation of a large population of extrachromosomal P elements that are transpositionally inactive. This population would have many more P element copies than are in the genome and would subsequently act to bind, or titrate, the transposase, thus reducing the transposition of chromosomal elements. No evidence for the existence of extrachromosomal circles that would support this model has yet been obtained.

Several extrinsic factors that are not directly related to P element functions appear to play a role in P element regulation. The absence of mosaic flies in many experiments involving hybrid dysgenesis has strongly suggested that P element mobilization is normally restricted to the germline. McElwain (1986) has shown that dysgenesis-induced events rarely occur in somatic cells despite the induction of male recombination in mitotic divisions of germ line cells (Hiraizumi 1971). Laski et al. (1986) have shown that this tissue specificity is controlled at the level of mRNA splicing. Removal of the third intron of the intact P element, which is necessary for transposase production, occurs in germline but not in somatic cells.

In addition to tissue specificity there are other variables that appear to influence P element mobility. These include developmental stage (Bregliano and Kidwell 1983), sex, and environmental factors such as temperature (Kidwell and Novy 1979) and parental age (Kidwell et al. 1977). An important influence of unknown host factors on P element mobility is inferred from the results of experiments involving crosses between females of different true M strains and a single strong P strain (Daniels et al. 1987). As seen in Figure 3, the frequencies of F_1 hybrid sterility varied over a wide range according to the M strain used as the female parent.

The Evolution of P Factor Activity and P Element Regulation

Some insights into the control of P element mobility have been obtained in experiments designed to investigate the dynamics and evolution of laboratory lines into which P elements were introduced by controlled transposition (Kidwell 1986; M.G. Kidwell et al., in prep.). The molecular and phenotypic properties of P elements were monitored over a thirty generation period

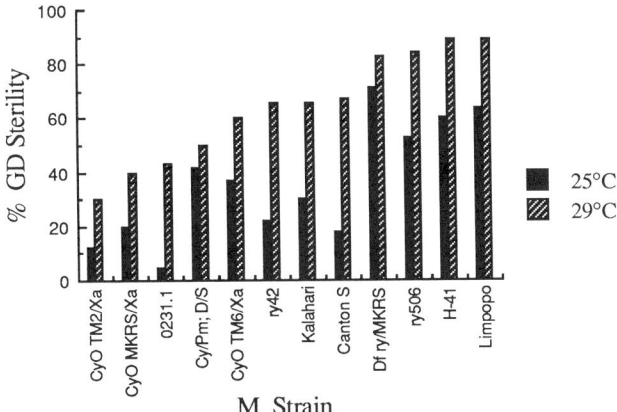

Figure 3
The distribution of female gonadal sterility in F_1 hybrids from crosses of females from 12 different true M strains and males from the same strong P strain, Agana (data from Daniels et al. 1987). The large variability suggests a role of maternal host functions in P element regulation.

following their introduction. A summary of the results is provided in Figures 4A,B, and C. Three general patterns of evolution were observed to occur, culminating in lines that were apparently stable and possessed many properties of the P, Q, and M' types that occur naturally (see Table 1). In each case, P factor activity arose rapidly in the initial generations (up to generation 5 or 6), usually in the absence of P element regulation. The subsequent pattern of genomic evolution of the family in any particular line appeared to be strongly influenced by whether or not P regulation evolved to suppress further transposition and associated element degeneration. Evolving P strains (Figure 4A) were associated with the rapid appearance of P cytotype at about the tenth generation, when P activity had apparently reached a maximum. P cytotype evolved somewhat earlier in an evolving Q strain (Figure 4B), and there was an early reduction of both potential and realized P activity to a stable low level. It should be noted that the type of regulation observed in the evolution of these P and Q strains (Fig. 4A and Fig. 4B, respectively) appeared to be similar to that found in the North American populations from which the original P (paternal stock) was derived. In some strains, P element regulation did not evolve at all (Fig. 4C), and the deletion process rapidly resulted in genomes with only nonautonomous elements and the consequent loss of P activity potential. This type of line possessed some of the properties of M' strains (Anxolabéhère et al. 1984; Kidwell 1985) but lacked the presence of KP elements and the intermediate level of regulation which is associated with them in M' natural populations (Black et al. 1987).

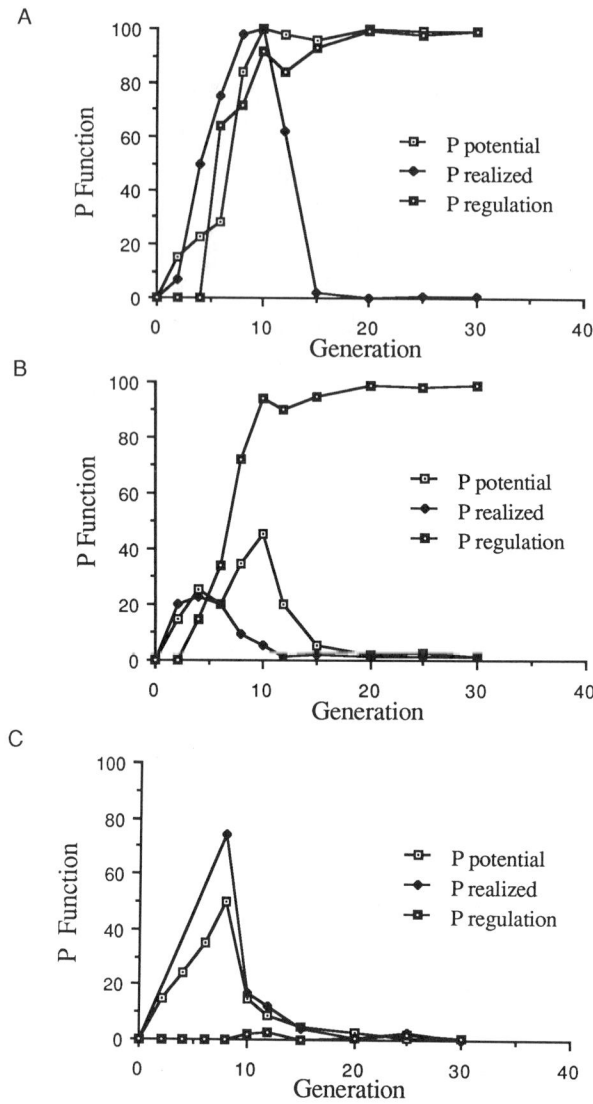

Figure 4
Reconstruction of three observed patterns of evolution in laboratory M lines into which P elements were introduced by controlled transposition. (*A*) Evolution to a P strain; (*B*) Evolution to a Q strain; (*C*) Evolution to an M' strain (data from M.G. Kidwell et al., in prep.). P potential is the frequency of gonadal sterility in the progeny of a cross between males of the evolving line and females of a standard M tester strain. P realized is the frequency of gonadal sterility in progeny of intrastrain matings. P regulation is the frequency of gonadal sterility in progeny of females of the evolving lines and males from a standard P tester strain.

DISCUSSION

A number of insights into various aspects of P element biology have been obtained from the broad perspective of population and evolutionary studies. Although P elements were first identified in *D. melanogaster*, there are several lines of evidence that suggest that they have only very recently invaded this species (Kidwell 1979, 1983). In contrast, a number of other, distantly related, species in the genus *Drosophila* appear to have harbored P elements for a long period of time (Daniels and Strausbaugh 1986). At least two species, in addition to *D. melanogaster*, have been successfully transformed by P elements (Brennan et al. 1984; Scavarda and Hartl 1984; Daniels et al. 1985a). The evolution of several transformed lines (Daniels et al. 1985a) has been studied in *Drosophila simulans*, a sibling species of *D. melanogaster*, at both the molecular and phenotypic levels (S.B. Daniels and M.G. Kidwell, in prep.). Many attributes of the P-M system seen in *D. melanogaster* were observed to evolve in these transformed lines of *D. simulans*, including correlated dysgenic traits, such as male recombination and gonadal sterility, and a type of P element-encoded regulation similar to P cytotype.

There are several implications of these results for our present interests. The apparent ability of P elements to retain their mobility and to evolve in a similar way when transferred between different *Drosophila* species strongly suggests, either that host functions are relatively unimportant for element survival and replication, or that the presence of essential host functions is evolutionarily strongly conserved within, and perhaps beyond, the genus *Drosophila*. The latter possibility seems more likely to be true because of the preliminary evidence for the strain variation in P element transposition rates which is independent of the elements themselves (Daniels et al. 1987).

In summary, our present limited knowledge suggests that the genomic complement of P elements can be viewed as an essentially self-regulatory system within a fairly broadly defined phylogenetic lineage. Transposition, excision, and transposition regulation all appear to be largely element-dependent processes. It can be reasonably argued that there has been strong natural selection to restrict transposition below the full level of which it is potentially capable. The products of insertional mutagenesis and other chromosomal changes associated with transposition are, on the average, highly deleterious to the carrier organism and their accumulation would be expected to lead to extinction of a lineage if transposition remained unregulated. Successful intraspecific transfer, integration, and survival in a "foreign" species may be critically dependent on the regulation of transposition being essentially an element-dependent rather than a host-dependent function.

ACKNOWLEDGMENTS

Thanks are extended to D.M. Black, I.A. Boussy, S.B. Daniels, W.R.Engels, K. Peterson, H. Robertson, and M.J. Simmons for discussions and willingness to share their unpublished results. Supported by National Institutes of Health grant GM-36715.

REFERENCES

Anxolabéhère, D., H. Kai, G. Périquet, and S. Ronsseray. 1984. The geographical distribution of P-M hybrid dysgenesis in *Drosophila melanogaster. Genet. Sel. Evol.* **16**: 15.

Bingham, P.M., M.G. Kidwell, and G.M. Rubin. 1982. The molecular basis of P-M hybrid dysgenesis: The role of the P element, a P strain-specific transposon family. *Cell* **29**: 995.

Bingham, P.M., R. Levis, and G.M. Rubin. 1981. Cloning of DNA sequence from the white locus of *Drosophila melanogaster* by a novel and general method. *Cell* **25**: 693.

Black, D.M., M.S. Jackson, M.G. Kidwell, and G.A. Dover. 1987. KP elements repress hybrid dysgenesis in *Drosophila melanogaster. EMBO J.* **6**: 4125.

Bregliano, J.C. and M.G. Kidwell. 1983. Hybrid dysgenesis determinants. In *Mobile genetic elements* (ed. J.A. Shapiro), p. 363. Academic Press, New York.

Brennan, M.D., R.G. Rowan, and W.J. Dickinson. 1984. Introduction of a functional P element into the germ line of *Drosophila hawaiiensis. Cell* **38**: 147.

Daniels, S.B. and L.D. Strausbaugh. 1986. The distribution of P-element sequences in *Drosophila*: The *willistoni* and *saltans* species groups. *J. Mol. Evol.* **23**: 138.

Daniels, S.B., L.D. Strausbaugh, and R.A. Armstrong. 1985a. Molecular analysis of P element behavior in *Drosophila simulans* transformants. *Mol. Gen. Genet.* **200**: 258.

Daniels, S.B., M. McCarron, C. Love, and A. Chovnick. 1985b. Dysgenesis induced instability of rosy locus transformation in *Drosophila melanogaster*: Analysis of excision events and the selective recovery of control element deletions. *Genetics* **109**: 95.

Daniels, S.B., I.A. Boussy, A. Tukey, M.C. Carillo, and M.G. Kidwell. 1987. Variability among true M lines in P-M gonadal dysgenesis potential. *Drosophila Inform. Serv.* (in press).

Eeken, J.C.J., F.H. Sobels, V. Hyland, and A.P. Schalet. 1985. Distribution of MR-induced sex-linked recessive lethal mutations in *Drosophila melanogaster. Mutat. Res.* **150**: 261.

Engels, W.R. 1979. Hybrid dysgenesis in *Drosophila melanogaster:* Rules of inheritance of female sterility. *Genet. Res.* **33**: 219.

———. 1981. Germline hypermutability in *Drosophila* and its relation to hybrid dysgenesis and cytotype. *Genetics* **98**: 565.

———. 1983. The P family of transposable elements in *Drosophila*. *Annu. Rev. Genet.* **17:** 315.
———. 1985. Guidelines for P-element transposon tagging. *Drosophila Inform. Serv.* **61:** 1.
Engels, W.R. and C.R. Preston. 1979. Hybrid dysgenesis in *Drosophila melanogaster:* The biology of male and female sterility. *Genetics* **92:** 161.
Golubovsky, M.D., Y.N. Ivanov, and M.M. Green. 1977. Genetic instability in *Drosophila melanogaster:* Putative multiple insertion mutants at the singed bristle locus. *Proc. Natl. Acad. Sci.* **74:** 2973.
Hiraizumi, Y. 1971. Spontaneous recombination in *Drosophila melanogaster* males. *Proc. Natl. Acad. Sci.* **68:** 268.
Karess, R.E. and G.M. Rubin. 1984. Analysis of P transposable element functions in *Drosophila*. *Cell* **38:** 135.
Kelley, M.R., S. Kidd, R.L. Berg, and M.W. Young. 1987. Restriction of P element insertions at the notch locus of *Drosophila melanogaster*. *Mol. Cell. Biol.* **7:** 1545.
Kidwell, M.G. 1979. Hybrid dysgenesis in *Drosophila melanogaster*: The relationship between the P-M and I-R interaction systems. *Genet. Res.* **33:** 205.
———. 1983. Evolution of hybrid dysgenesis determinants in *Drosophila melanogaster*. *Proc. Natl. Acad. Sci.* **80:** 1655.
———. 1985. Hybrid dysgenesis in *Drosophila melanogaster:* Nature and inheritance of P element regulation. *Genetics* **111:** 337.
———. 1986. The evolution of newly invasive P elements in M populations of *D. melanogaster*. *Genetics* **113:** s42.
———. 1987. A survey of P element mutagenesis in *Drosophila melanogaster*. *Drosophila Inform. Serv.* (in press).
Kidwell, M.G. and J.B. Novy. 1979. Hybrid dysgenesis in *Drosophila melanogaster:* sterility resulting from gonadal dysgenesis in the P-M system. *Genetics* **92:** 1127.
Kidwell, M.G., J.F. Kidwell, and J.A. Sved. 1977. Hybrid dysgenesis in *Drosophila melanogaster*: A syndrome of aberrant traits including mutation, sterility and male recombination. *Genetics* **36:** 813.
Laski, F.A., D.C. Rio, and G.M. Rubin. 1986. Tissue specificity of *Drosophila* P element transposition is regulated at the level of mRNA splicing. *Cell* **44:** 7.
McElwain, C.M. 1986. The absence of somatic effects of P-M hybrid dysgenesis in *Drosophila melanogaster*. *Genetics* **113:** 897.
O'Hare, K. and G.M. Rubin. 1983. Structures of P transposable elements and their sites of insertion and excision in the *Drosophila melanogaster* genome. *Cell* **34:** 25.
Rubin, G.M. 1983. Dispersed repetitive DNAs in *Drosophila*. In *Mobile genetic elements* (ed. J.A. Shapiro), p. 329. Academic Press, New York.
Scavarda, N.J. and D.L. Hartl. 1984. Interspecific DNA transformation in *Drosophila*. *Proc. Natl. Acad. Sci.* **81:** 7515.
Searles, L.L., A.L. Greenleaf, W.E. Kemp, and R.A. Voelker. 1986. Sites of P element insertion and structures of P element deletions in the 5' region of *Drosophila melanogaster RPII215*. *Mol. Cell. Biol.* **6:** 3312.
Simmons, M.J. and L.M. Bucholz. 1985. Transposase titration in *Drosophila*

melanogaster: A model of cytoype in the *P-M* system of hybrid dysgenesis. *Proc. Natl. Acad. Sci.* **82:** 8119.

Simmons, M.J. and J.K. Lim. 1980. Site specificity of mutations arising in dysgenic hybrids of *Drosophila melanogaster*. *Proc. Natl. Acad. Sci.* **77:** 6042.

Todo, T., Y. Sakoyama, S.I. Chigusa, A. Fukunaga, T. Honjo, and S. Kondo. 1984. Polymorphism in distribution and structure of P elements in natural populations of *Drosophila melanogaster* in and around Japan. *Jpn. J. Genet.* **59:** 441.

Tsubota, S., M. Ashburner, and P. Schedl. 1985. P-element induced control mutations at the *r* gene of *Drosophila melanogaster*. *Mol. Cell. Biol.* **5:** 2567.

Mutation Induction by *MR(P)* and Its Modification by Various Conditions

FRITS H. SOBELS* AND JAN C.J. EEKEN*[†]
*Department of Radiation Genetics and
Chemical Mutagenesis, University of Leiden
Sylvius Laboratories
Wassenaarseweg 72, Leiden, The Netherlands
[†]J.A. Cohen Institute
Interuniversity Institute for Radiopathology and
Radiation Protection
Leiden, The Netherlands

OVERVIEW

Findings obtained in our laboratory over the last eight years on the induction of mutations by transposition of P elements under the influence of male or meiotic recombination (*MR*) are reviewed: (1) Localization of 144 *MR-h12*-induced lethals showed differences with the distribution of X-ray-induced lethals and with those arising spontaneously in the absence of *MR*. In situ hybridization provided evidence that *MR*-induced mutation is associated with transposition of P elements. (2) The effect of mutagens was studied by employing reversion of an *MR*-induced singed (*sn*) mutation by excision of the P element. This process was not noticeably affected by exposure to high mutagenic concentrations of ethylnitrosourea (ENU), methylmethanesulfonate (MMS), formaldehyde, and X-irradiation. (3) Under conditions of defective DNA repair as with *mei-9* and *mei-41*, the frequency of forward mutations to *sn* and raspberry (*ras*) is enhanced, but reversion of *sn* is decreased. Forward mutation, involving transposition of P elements, thus may involve steps that differ from excision of such elements. (4) A synergistic effect between *MR* and X rays was observed for the induction of recessive lethals. Moreover, it was found that the presence of P elements predisposes to the induction of chromosome rearrangements by both X rays and *MR*, some of which have specific breakpoints at the sites of P elements. (5) From experiments in which the segregation of *MR* could be followed after having produced a mutation, it could be concluded that, after this event, *MR* retains its capacity to produce reverse mutations. (6) A molecular analysis of geographically different *MR* strains characterized by different potency to produce reverse mutations shows correlation of this property and the number of P elements present. (7) Following both heat and cold treatments, some enhancement of the frequency of reversion of different insertion element mutations was observed. A particular *sn* mutation, reverting autonomously and thus characterized by a complete P element, likewise reverted an *MR-h12*-induced *ras* mutation and a lethal associated with a tandem duplication.

Finally the relevance of some of the above findings, particularly those mentioned above (1, 2, and 4) for the validity of employing the "doubling dose method" for the assessment of genetic risks in man is discussed.

INTRODUCTION

In the assessment of genetic risks ensuing from exposure to irradiation or chemical mutagens, one of the problems confronting us concerns the translation of induced genetic damage into tangible human suffering. In this context, the doubling dose method is often used by, for example, the United Nation's Scientific Committee on the Effects of Atomic Radiation (UNSCEAR) 1982 or the Biological Effects of Ionizing Radiation (BEIR) Committee of the U.S. National Academy of Sciences (1980). The underlying principle is that the induced damage is expressed as an increase of the spontaneous load of genetic disease. The validity of employing the doubling dose method rests on two assumptions: (1) that there is a proportionality between spontaneous and induced mutations, and (2) that damage to the genetic material resulting from exposure to mutagenic agents is similar in nature to mutations arising spontaneously. It is these assumptions that prompted us to undertake a study into the causes underlying spontaneous mutation and to what extent this process can be affected by various mutagenic agents. Recent studies in a wide variety of species, ranging from bacteria to *Drosophila*, provide evidence that insertion of mobile DNA elements contributes in a significant manner to the process of spontaneous mutation (for review, see Shapiro 1983). Moreover, it was found that in bacteria, transposition of these mobile elements or transposons is not significantly modified by mutagenic agents. This paper reviews our findings on mutation by transposition of P elements in the *MR* system. *MR* stands for male or mitotic recombination which normally does not occur in *Drosophila* males. Hiraizumi (1971) observed that crossing males of particular natural populations with females of laboratory stocks, may result in male recombination. Apart from recombination, *MR* produces mutations, chromosome aberrations, and hybrid dysgenesis (Green 1977; Kidwell et al. 1977; Engels 1983).

RESULTS

Distribution and Nature of *MR*-induced Lethal Mutations

The purpose of this study (Eeken et al. 1985) was to examine whether (1) there is proportionality between the spontaneous mutation rate and the induced frequency and (2) the nature of the underlying lesions that lead to spontaneous and induced mutations is basically similar. Genetic techniques were employed to localize 144 X-chromosomal *MR-h12*-induced lethals. Fifteen duplications covering about 50% of the X chromosome were used for the localization. Lethals covered in a specific region were subsequently tested for allelism with deletions or complementation crosses. The analysis, first of all, showed that *MR*-induced lethals occur at many different sites along the X chromosome. A comparison of the localization of these *MR*-induced mutations with that of spontaneous (i.e., not induced by *MR*) (Schalet 1986) and X-ray-induced mutations (Lefevre 1981; Lefevre and Watkins 1986) revealed remarkable differences. Thus, at some hot spots for spontaneous or X-ray-induced lethals no obvious clustering of *MR*-induced lethals could be found, whereas in other cases, hot spots for *MR*-induced mutations did not coincide with those for X rays. The difference in the localization of hot spots for either X-ray- or *MR*-induced sex-linked recessive lethals, is shown in Table 1, based on data of Eeken et al. (1985).

To determine whether *MR*-induced lethals would be associated with transposition of P element, in situ hybridization was performed by employing the methodology of Bingham et al. (1982) and using a cloned P insertion element, p6.1 (Rubin et al. 1982). The analysis of 13 lethals localized at the tip of the X chromosome showed that the majority of these *MR*-induced lethals was indeed accompanied by the gain of a P element in the region where the lethal

Table 1
The Localization of Hot Spots for the Induction of Recessive Sex-linked Lethals by X-irradiation or *MR-h12*

Sites	X-ray-induced 1500 lethals	MR-induced 270 lethals	MR-induced 1500 lethals
N	16	8	30
HA18	3	3	15
ct	23	1	5
zw1	14	0	0/5
7A3	12	0/1	0/5
2C3	7	0/1	0/5
7D22	8	0	0/5
7A7	9	0/1	0/5
2F5	0/3	3	15
7B	0/2	3	15

The localization of hot spots for the induction of recessive sex-linked lethals by X-irradiation (Lefevre 1981; Lefevre and Watkins 1986) or *MR-h12* (Eeken et al. 1985). The two sites at the top show somewhat comparable frequencies; the next six exhibit high sensitivity to X rays but low to *MR;* the bottom two sites show low X ray but high *MR* sensitivity.

was mapped genetically. Obviously, the very nature of these lethals is different from X-ray-induced mutations. Particularly so, as the majority of X-ray-induced point mutations are characterized by deletions at the molecular level, as has been shown for the white (w) locus in a recent analysis of Pastink et al. (1987). In summary, it can be stated that in those cases where transposition of mobile DNA sequences is responsible for the occurrence of mutation (1) there is no complete proportionality between the frequencies with which MR (i.e., P-element-associated) mutations and X-ray mutations are induced, and (2) the nature of the underlying DNA lesions is clearly different for the two kinds of mutations. These differences thus raise questions about the validity of the assumptions involved in the application of the doubling dose method.

The Effect of Various Powerful Mutagens on Mutation Induction of MR

Another question which is likewise relevant for the applicability of the doubling dose concept is whether mutation taking place under the influence of MR can be affected by mutagenic agents. Forward mutations from wild-type to mutant phenotypes occur at too low frequencies to enable a comprehensive study of conditions possibly modifying this process. Consequently, we employed a reverse mutation system. Eeken (1982) found that an MR-induced sn mutation reverts to normal wild-type bristles with a frequency of 1.7% (255 reversions in a total of 15,163 chromosomes) in the presence of MR. In the absence of MR, however, no reversions were observed in a total of 20,676 progeny scored. Reverse mutation is associated with excision of the inserted P element. Four different mutagens, namely ENU, MMS, X rays, and formaldehyde were used for these studies. Since the mobility of P elements appears to occur predominantly in premeiotic stages of spermatogenesis, mutagen treatments were given early in development to first instar larvae hatching from eggs. To ensure that mutagenic concentrations were employed, recessive sex-linked lethal experiments were carried out simultaneously with the reversion studies.

The pooled data obtained in five experiments with ENU are summarized in Figure 1, left (Eeken and Sobels 1983b). It can be seen that 1 mM ENU does not increase the reversion frequency, despite a threefold increase of the recessive lethal frequencies. With 3 mM ENU the reversion frequency was doubled, but recessive lethals increased fivefold. The clear indication of a threshold of 1 mM ENU and the lack of proportionality with the recessive lethal frequencies at 3 mM both suggest that the high concentration of ENU has affected some indirect process rather than the reversion event itself.

The effects of MMS on the reversion of sn^{MR} are shown in Figure 1, right. It can be seen that treatment with 5 or 10 mM MMS did not result in an increase

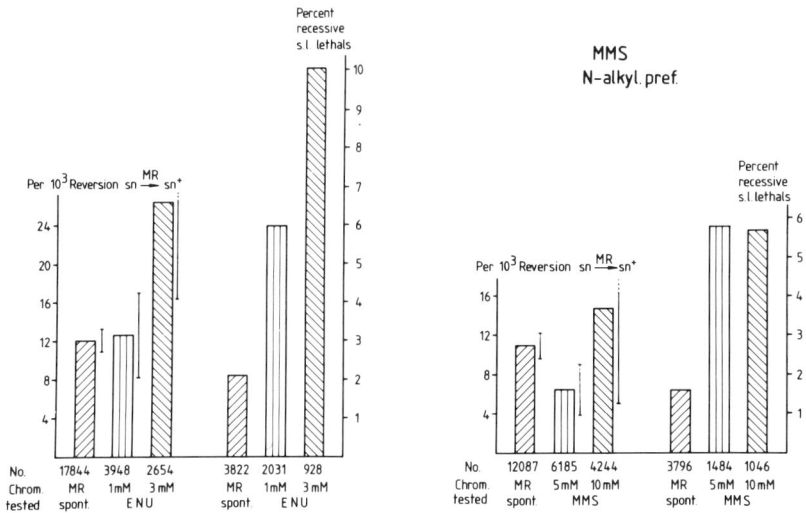

Figure 1
A comparison of the frequency of reversion from *sn* to wild-type with the induction of recessive sex-linked lethals after treatment of first instar larvae with ENU (*left*) or MMS (*right*) (from Eeken and Sobels 1983b).

of the reversion frequency, despite the fact that recessive lethals increase about fourfold. With 10 mM MMS the recessive lethal frequency is spuriously low as a consequence of germinal selection. Thus, in conclusion, neither ENU nor MMS affected reversion by P element excision of an *MR*-induced *sn* mutation in a manner proportional to the mutagenic effectiveness of these compounds (Eeken and Sobels 1983b).

A recent study with X rays and formaldehyde yielded somewhat surprising results (Eeken and Sobels 1986). With X rays no effect was observed in the presence of *MR* (1.6 ± 0.5% in 14,412 progeny from irradiated parents and 1.2 ± 0.2% in 15,012 progeny from unirradiated controls). However, in the absence of *MR*, i.e., in the controls with the homologous chromosome, the reversion frequency was consistently increased following X-ray exposure (0.22 ± 0.08% in 17,744 chromosomes tested) in comparison to that in unirradiated controls (0.07 ± 0.03% in 9,063 progeny). These data suggest that X-irradiation may exert a direct effect on the excision of the P element at sn^{MR}. Since the effect is small, it cannot be discerned in the presence of *MR*.

With formaldehyde in mutagenic concentrations producing 4–5% recessive lethals, the reversion of sn^{MR} is unaffected in the absence of *MR*, but in the presence of *MR* the frequency is significantly lowered ($P=0.001$); that is 0.5 ± 0.2% reversions were observed in 13,040 chromosomes with formaldehyde and 1.2 ± 0.2% in 15,012 progeny of the untreated control flies. A possible interpretation of this unexpected result could be that formaldehyde in some way inhibits *MR*-dependent transposase activity (Rio et al. 1986).

In summary, it can be stated that (1) some interactions of mutagenic agents with excision of mobile elements do occur, as in the case of ENU, X rays, and formaldehyde, (2) the types of interaction observed appear to be quite different, (3) these effects are small in comparison with the induction of recessive lethals, and (4) in no case a clear increase of the reversion frequency (as expected on the basis of the recessive lethal frequencies) was observed. These findings may, like the previous one described earlier, have consequences for the evaluation of induced genetic damage on the basis of the spontaneous load of genetic detriment in man.

The Possible Role of DNA Repair in Transposition and Excision of P elements

It is not yet known, at least in *Drosophila*, what causes the transposition of mobile DNA elements. To examine whether enzymes involved in the repair of damaged DNA possibly play a role in the transposition or excision of P elements, forward and reverse mutation under the influence of *MR* was studied in repair deficient strains. Stocks were used that were deficient for excision repair (*mei-9*), for postreplication repair (*mei-41*) or for both, with *mei-9* and *mei-41* combined on the same chromosome. In an earlier paper (Eeken and Sobels 1981), we were able to demonstrate that the frequency of forward mutation to *sn* and *ras* in the presence of *MR* is considerably enhanced in males deficient for both excision and postreplication repair (see Fig. 2, left). The low frequency of forward mutation does not lend itself to a more detailed analysis of the phenomena. Consequently, the reverse mutation system of an *MR*-induced *sn* mutation to wild-type, as described in the preceding section, was used. The data, illustrated in Figure 2, right, show that the presence of *mei-9*, *mei-41*, or both leads to a significant reduction of the reversion frequency (Eeken and Sobels 1983a). Assuming that reversion of *sn* to wild-type involves the excision of the insertion element followed by closing of the gap, it is conceivable that enzymes deficient in the meiotic mutants may possibly play a role in the completion of these processes. A selective elimination of cells in which the final steps of the excision process are less efficient may underly the reduction of the reversion frequencies observed.

The opposite effects of defective DNA repair on forward and reverse mutations suggest that the forward mutation process appears to involve mechanisms that are different from those in the reverse mutation process.

The Interaction of P-element Transposition and X-irradiation

A point of interest is whether *MR*, that is the presence of a P element causing transposition, would increase the probability of induction of genetic damage

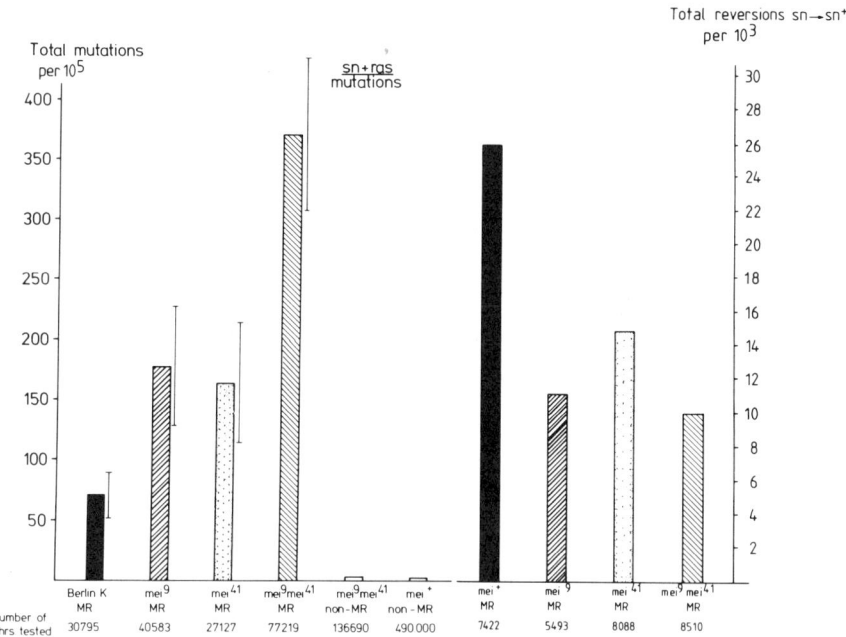

Figure 2
The effect of repair deficiencies *(mei-9ª, mei-41^{D5}*, or both) in the presence of *MR*, on the frequencies of forward mutation at the *sn* and *ras* loci (*left*), and the frequency of reversion of sn^{MR} to sn^+ (*right*) (from Eeken and Sobels 1981, 1983a).

by X-irradiation. For recessive lethals the presence of *MR* clearly resulted in an increase of the yield of X-ray-induced mutations. The data in Figure 3 show that in a number of replica experiments the mutation frequencies in the group with *MR* and X rays is greater than the sum of the mutation yields in the unirradiated *MR* and irradiated Curly flies (Curly is a balancer on the chromosome homologous of the second chromosome carrying *MR*). An intriguing aspect of these data is that mature sperm and late spermatids were used for these irradiation experiments, whereas the transposing activity of *MR* is thought to occur in dividing gonial stages. Experiments are now in progress to examine whether the observed interaction takes place in early zygotes derived from the irradiated spermatozoa.

A related question is whether the presence of P elements predisposes to the occurrence of other genetic changes. The problem was studied by inducing recessive lethals by *MR* or by the combination of *MR* and X-irradiation in a chromosome that carried an *MR*-induced *sn* and *ras* mutation (Eeken and Sobels 1984; Eeken et al. 1987a). Salivary-gland chromosome analysis of progeny of such flies showed that inversions and translocations had been induced. Of particular interest is the site specificity observed for the break

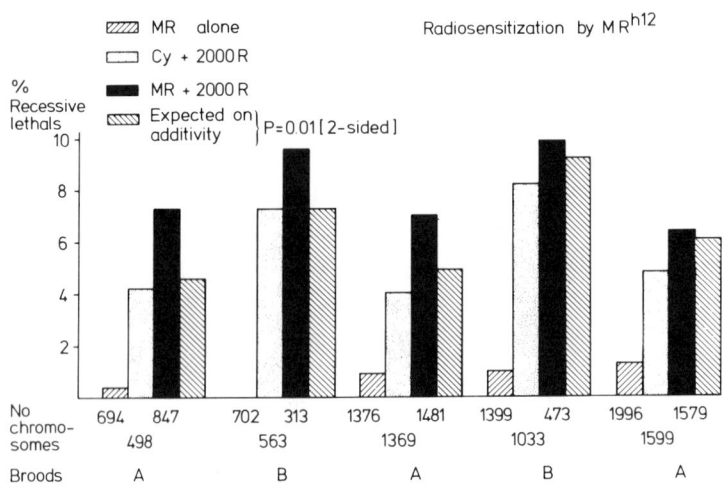

Figure 3
The synergistic effects of *MR* and X rays on the induction of recessive sex-linked lethal frequencies. The data show that *MR* and X rays induced more mutations than the additive effects of *MR* by itself or X rays by itself in the Curly homolog (data from Sobels and Eeken 1981).

points involved in these rearrangements. That is, sites where particular P elements are located, such as 7 C/D (*sn*) or 12 E, show a higher propensity to break under the influence of either X rays or *MR*. It is of interest that in chromosomes without a P element at *sn*, the specificity for breaks occurring at that site disappears. In this context, it may be noted that if mobile elements are part of the human genome (Evans 1984; Sankaranarayanan 1986; Sobels 1986) greater susceptibility to breakage at sites where such elements are located should be taken into consideration.

MR Retains Its Activity Following Mutation Induction by Insertion of P elements

Earlier, it was shown that the induction of recessive lethal mutations by *MR* is associated with the insertion of a P element. Little is known about the mechanism of this transposition process. Thus, we do not yet know whether *MR*-induced mutation involves transposition of *MR* itself, or whether alternatively *MR* produces P inserts by a replication process. In the first case, one would expect that *MR* activity is lost after production of an insertion mutation, whereas in the latter, *MR* activity should be retained. Experiments designed to distinguish between these two possibilities showed that in six independent cases where *MR* had induced a *sn* mutation, the *MR* chromosome segregating with the induced mutation had retained its capacity for inducing reversion of an unstable *sn* (Sobels and Eeken 1985). We conclude that the induction of mutation by insertion of P elements is, most probably, not associated with loss of *MR* activity and involves a replication process.

Correlation between the Strain-specific Potency to Produce Reverse Mutations and the Number of P elements

MR strains of different geographical origin exhibit pronounced differences in their capacity to produce reversions of *MR*-induced mutations, such as *sn* and *ras*. Functional P elements are 2.9-kb long, including a 2.4-kb *Acc*I fragment. Eeken et al. (1987b) determined the number of copies of P elements carrying an intact 2.4-kb *Acc*I fragment in the various *MR* strains. The data illustrated in Figure 4 show that different *MR* strains have different numbers of copies of the 2.4-kb *Acc*I fragment.

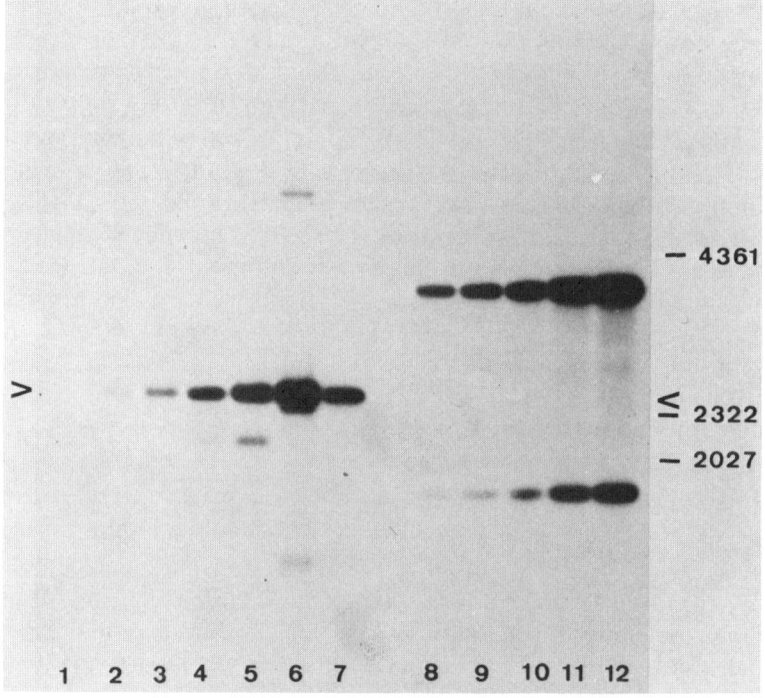

Figure 4
Autoradiogram of *Acc*I-digested genomic DNA of various *MR* strains (lanes *2–7*). In lane *1*, a control strain free of P elements is shown. Lanes *8–9* show different genomic equivalents of *Acc*I-digested pHS used as a reference to calculate the number of *Acc*I 2.4-kb P fragment copies in the tested *MR*-strains (lanes *2–7*). Hybridization was performed with pHS as probe.

It is of interest to note that the numer of P elements present in the different strains appears to correspond well with the capacity to revert *MR*-induced mutations. The reversion potency exerted by *MR-T007* is about one order of magnitude higher than that of *MR-h12*. Recent cloning experiments by Eeken have shown that the *MR-T007* chromosome contains eight complete P elements, whereas the *MR-h12* one has only two. The capacity to produce forward mutations, either recessive lethals or visibles, does not greatly differ between the various *MR* strains carrying different numbers of P elements. The reason for this discrepancy in response between forward and reverse mutation is not known, but it may be recalled that, under conditions of defective DNA repair, likewise contrasting results were obtained for *MR*-induced forward and reverse mutations.

Other Experiments Designed to Study the Mechanism of Reversion of Insertion Mutations

In experiments by Eeken, Sobels, and M.M. Green to explore the effects of genomic stress (McClintock 1984), first instar larvae were exposed to 38°C for 1 hour or to 4°C for 2 days. Subsequent matings were designed to study reversion of (yellow) y^2 and (scute) sc^1 (resulting from insertion of gypsy), (white) w^a (from insertion of copia), $sn^{83i7(1)}$ (from insertion of a complete P) and v^{36f} (from insertion of roo). The results showed a slight enhancement after both heat and cold treatments of male larvae (Sobels 1986). In addition the data suggest that the presence of a complete P element, as in $sn^{83i7(1)}$, resulted not only in reversion of this singed mutation but also caused reversion of mutants with different insertion elements (see also Gerasimova et al. 1984). Characteristic for $sn^{83i7(1)}$ that had been induced by $MR + (3)KMCG$ is that it does revert autonomously.

Possible Implications for the Assessment of Genetic Risk in Man

The studies reviewed above are of interest, we believe, for elucidating some of the processes that may possibly play a role in mutation induction by transposition of P elements in *Drosophila*. Another question concerns their possible significance for the assessment of genetic risks in man, using the doubling dose method. The following data appear relevant in this respect. For the induction of recessive lethal mutations caused by transposition of P elements, we were able to show (1) that their distribution over the chromosome differs from that of similar mutations induced by radiation or arising spontaneously in the absence of P elements; (2) that the nature of the underlying DNA lesions is different for P-element-associated mutations and X-ray-induced mutants, in that the first are characterized by insertions, the latter mainly by deletions. Studying the effects of powerful mutagens on reverse mutation of an *MR*-induced *sn* mutation, we found that reverse mutation by excision of P elements in the presence of *MR* is not noticeably affected by ENU, MMS, formaldehyde, or X rays. If, indeed, a significant fraction of the spontaneous mutations in man would arise as a result of transposition of moveable genetic elements, as is the case in *Drosophila*, the above findings would cast doubt on the validity of the doubling dose concept that is based on a comparison of induced mutation frequencies in the mouse and spontaneous mutation frequencies in man.

The finding that the presence of P elements predisposes to the formation of chromosome aberrations with specific breakpoints appears likewise of relevance for the assessment of genetic radiation hazards.

ACKNOWLEDGMENTS

The authors would like to thank their colleague, Prof. K. Sankaranarayanan, for critically reading the manuscript, and they want to acknowledge the many stimulating discussions with Prof. M.M. Green and Dr. A. Schalet. This work was supported in part by Euratom Contract No. BIO E406-81-NL with the University of Leiden.

REFERENCES

BEIR Report. 1980. *The effects on populations of exposure to low levels of ionizing radiation*, National Academy of Science, National Research Council, Washington, D.C.

Bingham, P.M., M.G. Kidwell, and G.M. Rubin. 1982. The molecular basis of P-M hybrid dysgenesis: The role of the P-element, a P strain-specific transposon family. *Cell* **29:** 995.

Eeken, J.C.J. 1982. The stability of mutator (*MR*)-induced X-chromosomal recessive visible mutations in *Drosophila melanogaster*. *Mutat. Res.* **96:** 213.

Eeken, J.C.J. and F.H. Sobels. 1981. Modification of *MR* mutator activity in repair-deficient strains of *Drosophila melanogaster*. *Mutat. Res.* **83:** 191.

———. 1983a. The influence of deficiencies in DNA-repair on *MR*-mediated reversion of an insertion-sequence mutation in *Drosophila melanogaster*. *Mutat. Res.* **110:** 287.

———. 1983b. The effect of two chemical mutagens, ENU and MMS on *MR*-mediated reversion of an insertion-sequence mutation in *Drosophila melanogaster*. *Mutat. Res.* **110:** 297.

———. 1984. P-element transposition-related events in *Drosophila melanogaster* carrying *MR-h12*. *J. Cell. Biochem.* (Suppl.) **8B:** 152. (Abstr.).

———. 1986. The effect of X-irradiation and formaldehyde treatment of spermatogonia on the reversion of an unstable P-element insertion mutation in *Drosophila melanogaster*. *Mutat. Res.* **175:** 61.

Eeken, J.C.J., R. Romeyn, and A.W.M. de Jong. 1987a. An analysis of X-linked recessive lethal mutations induced by *MR* and X-rays in *Drosophila melanogaster* carrying P-elements. *Mutat. Res.* (in press).

Eeken, J.C.J., F.H. Sobels, V. Hyland, and A.P. Schalet. 1985. Distribution of *MR*-induced sex-linked recessive lethal mutations in *Drosophila melanogaster*. *Mutat. Res.* **150:** 261.

Eeken, J.C.J., R. Romeyn, A.W.M. de Jong, G. Yannopoulos, and N. Stamatis. 1987b. A correlation between the genetic effects and intact P-elements in *MR*-strains of *Drosophila melanogaster*. *Mutat. Res.* (in press).

Engels, W.R. 1983. The P family of transposable elements in *Drosophila*. *Annu. Rev. Genet.* **17:** 315.

Evans, H.J. 1984. Structure and organization of the human genome. In *Mutation in man* (ed. G. Obe), p. 58. Springer-Verlag, Berlin.

Gerasimova, T.I., L.V. Matyunina, Y.V. Ilyin, and G.P. Georgiev. 1984. Simultaneous transposition of different mobile elements: Relation to multiple mutagnesis in *Drosophila melanogaster*. *Mol. Gen. Genet.* **194:** 517.

Green, M.M. 1977. Genetic instability in *Drosophila melanogaster*: De novo induction of putative insertion mutations. *Proc. Natl. Acad. Sci.* **74:** 3490.

Hiraizumi, Y. 1971. Spontaneous recombination in *Drosophila melanogaster* males. *Proc. Natl. Acad. Sci.* **68:** 268

Kidwell, M.G., J.F. Kidwell, and J.A. Sved. 1977. Hybrid dysgenesis in *Drosophila melanogaster*: A syndrome of aberrant traits including mutation, sterility and male recombination. *Genetics* **86:** 813.

Lefevre, G. 1981. The distribution of randomly recovered X-ray-induced sex-linked genetic effects in *Drosophila melanogaster*. *Genetics* **99:** 461.

Lefevre, G. and W. Watkins. 1986. The question of the total gene number in *Drosophila melanogaster*. *Genetics* **113:** 869.

McClintock, B. 1984. The significance of responses of the genome to challenge. *Science* **226:** 792.

Pastink, A., A.P. Schalet, C. Vreeken, and J.C.J. Eeken. 1987. The nature of radiation induced mutations at the white locus of *Drosophila melanogaster*. *Mutat. Res.* **177:** 101.

Rio, D.C., F.A. Laski, and G.M. Rubin. 1986. Identification and immunochemical analysis of biologically active *Drosophila* P-element *transposase*. *Cell* **44:** 21.

Rubin, G.M., M.G. Kidwell, and P.M. Bingham. 1982. The molecular basis of P-M hybrid dysgenesis: The nature of induced mutations. *Cell* **29:** 987.

Sankaranarayanan, K. 1986. Transposable genetic elements, spontaneous mutations and the doubling dose method of radiation genetic risk evaluation in man. *Mutat. Res.* **160:** 73.

Schalet, A.P. 1986. The distribution of and complementation relationships between spontaneous X-linked recessive lethal mutations recovered from crossing long-term laboratory stocks of *Drosophila melanogaster*. *Mutat. Res.* **163:** 115.

Shapiro, J.A., ed. 1983. *Mobile genetic elements*. Academic Press, New York.

Sobels, F.H. 1986. Changing concepts in mutation research. In *Genetic toxicology of environmental chemicals, part A: Basic principles and mechanisms of action*, p. 129. Alan R. Liss, New York.

Sobels, F.H. and J.C.J. Eeken. 1981. Influence of the *MR* (mutator) factor on X-ray-induced genetic damage. *Mutat. Res.* **83:** 201.

———. 1985. *MR* retains its activity following mutation induction by insertion of P-elements in *Drosophila. Mutat. Res.* **144:** 73.

UNSCEAR. 1982. *United Nation Scientific Committee on the Effects of Atomic Radiation: Sources and Biological Effects*, p. 513. United Nations, New York.

The Structure of Mutations Induced by I-R Hybrid Dysgenesis in *Drosophila melanogaster*

DAVID J. FINNEGAN,* ISABELLE BUSSEAU,† MICHAEL LYNCH,*
DIANA H. FAWCETT,* CLAIRE K. LISTER,* ALAIN PÉLISSON,†
HELEN M. SANG,* AND ALAIN BUCHETON†

*Department of Molecular Biology
King's Buildings
Edinburgh EH9 3JR, United Kingdom
†Laboratoire de Génétique
Université de Clermont-Ferrand
Aubière, France

INTRODUCTION

Hybrid dysgenesis is the production of a set of abnormal characteristics in the progeny produced when particular strains of *Drosophila melanogaster* are crossed in an appropriate fashion (Bregliano and Kidwell 1983). These traits include reduced fertility and increased frequencies of apparent point mutations and cytologically visible chromosome rearrangements. Most strains of *D. melanogaster* can be classified as being one or the other of two types: inducer (I) or reactive (R), with respect to the I-R system of hybrid dysgenesis, and dysgenesis is seen in the progeny of crosses between R strain females and I strain males. The progeny of the reciprocal cross are apparently normal as are the progenies of crosses between any two R or any two I strains.

The characteristics of I strains are controlled by transposable elements called I factors. These are quiescent in I strains but are activated when introduced into the cytoplasmic background of an R strain. This is believed to reflect the presence of regulatory molecules that are produced by I factors in I strains but are not present in R strains.

We have identified the DNA of the I factor by analyzing molecular lesions associated with eight white gene mutations, w^{IR1-8}, that were produced in I-R dysgenic females (Bucheton et al. 1984; Sang et al. 1984). Six of these mutations determine colored eye phenotypes and have insertions of indistinguishable 5.4-kb elements within the white gene. We know that these insertions are of I factor DNA as several of these mutations are linked to I factor activity (Pélisson 1981 and unpubl.), and we have shown that the insertion associated with the w^{IR3} mutation can confer the inducer phenotype on an R strain into which it is introduced by transformation (M. Pritchard et al., unpubl.).

Banbury Report 30: Eukaryotic Transposable Elements as Mutagenic Agents
© Cold Spring Harbor Laboratory. 0-87969-230-8/88. $1.00 + .00

Structure of the I Factor

We have determined the complete base sequence of the I factor associated with the w^{IR1} mutation (Fawcett et al. 1986). This element is 5371-bp long, and is flanked by direct repeats of a 12-bp target sequence. It has no terminal repeats, but it has four copies of the sequence TAA at the 3' (right-hand) end of one strand (Fig. 1). We have confirmed that this sequence organization is a general property of I factors by sequencing the ends of five other insertions within the white gene, those associated with mutations w^{IR2-6} and one associated with a mutation of the bithorax complex, bx^{F31} (Peifer and Bender 1986). Each element is flanked by a target site duplication, although these can vary slightly in length from one element to another. Three elements, those associated with the mutations w^{IR1}, w^{IR3}, and w^{IR4}, have inserted at exactly the same site. This suggests that I elements have preferred targets, although we have been unable to derive a convincing consensus sequence from the mutations that we have studied so far. The ends of the elements themselves are highly conserved (Fig. 1), and the only differences between them are in the third base, which can be either G or T, and in the number of TAA repeats at the right-hand end. All the coding capacity of the I factor appears to be in one strand which contains two long open reading frames (ORFs) (Fig. 1).

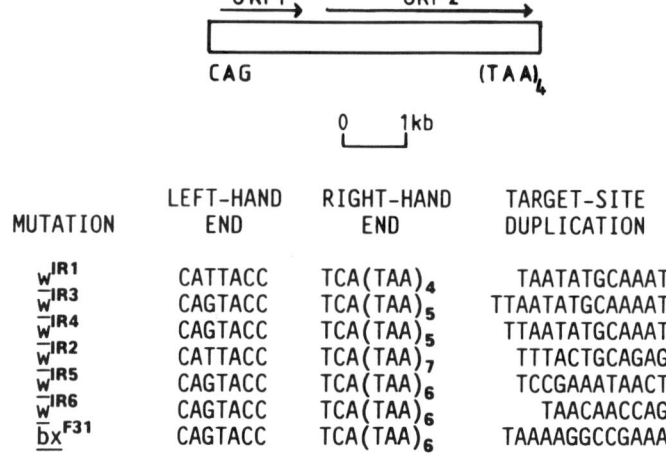

MUTATION	LEFT-HAND END	RIGHT-HAND END	TARGET-SITE DUPLICATION
w^{IR1}	CATTACC	TCA(TAA)$_4$	TAATATGCAAAT
w^{IR3}	CAGTACC	TCA(TAA)$_5$	TTAATATGCAAAAT
w^{IR4}	CAGTACC	TCA(TAA)$_5$	TTAATATGCAAAT
w^{IR2}	CATTACC	TCA(TAA)$_7$	TTTACTGCAGAG
w^{IR5}	CAGTACC	TCA(TAA)$_6$	TCCGAAATAACT
w^{IR6}	CAGTACC	TCA(TAA)$_6$	TAACAACCAG
bx^{F31}	CAGTACC	TCA(TAA)$_6$	TAAAAGGCCGAAA

Figure 1
Map of the I factor from w^{IR1} mutation and sequences from the ends of seven I factors. (□) indicates the I factor; (→) indicates the positions of the two long ORFs within it. The table below the map lists sequences at the ends of I factors associated with different mutations. The last column gives the target site duplications flanking each insertion. We cannot be sure of the precise length of the duplications next to the w^{IR1}, w^{IR6}, and bx^{F31} insertions. In each case, one copy of the TAA repeat could be part of either the target site duplication or the I factor.

These are translated from left to right, and both have methionine codons near their amino-termini that could be used to initiate translation.

The sequence organization of the I factor is unlike that of most transposable elements, but is similar to that of mammalian LINE or L1 elements (Singer 1982), processed pseudogenes, and other retroposons (Rogers 1985). Several authors have suggested that these elements transpose by reverse transcription of an RNA intermediate, so we have compared the amino acid sequences of ORF1 and ORF2 with those of viral reverse transcriptases. ORF2 contains regions that correspond to amino acid sequence motifs that are highly conserved between these viral enzymes (Fawcett et al. 1986), and we think that it may encode a reverse transcriptase that is responsible for I factor transposition. The amino acid sequence of ORF1 has no extensive homology to any polypeptide in the data bases with which it has been compared but does contain a region that matches a conserved motif, $CX_2CX_4HX_4C$, found in viral nucleic acid binding proteins, including *gag* polypeptides of retroviruses and the coat protein of cauliflower mosaic virus (Covey 1986). This would be consistent with ORF1 encoding a regulatory function. Open reading frames with coding capacities similar to those of ORF1 and ORF2 have been found in several mammalian L1 elements (D'Ambrosio et al. 1986; Hattori et al. 1986; Loeb et al. 1986; Fanning and Singer 1987).

Other Types of Mutations Associated with I Factor Activity

Not all mutations induced by I-R hybrid dysgenesis are due to insertions of 5.4-kb I factors. We have recently studied five dysgenesis-induced mutations of the yellow (*y*) gene (I. Busseau and A. Pélisson, unpubl.). Three of them are associated with insertions, but only one, y^{IR8}, contains a 5.4-kb insertion like those described above. The other two mutations, y^{IR3} and y^{IR4}, have insertions of incomplete I elements. These are truncated at their left-hand, or 5', ends and differ in length, one being 3891-bp long and the other 3811-bp long. The right-hand, or 3', ends of these elements are like those of complete I factors, except that one has the terminal sequence $TAA(TAAA)_3$ rather than simply a run of TAA repeats. Both of these elements are flanked by target site duplications suggesting that they inserted as truncated elements rather than having been deleted after insertion. We cannot say whether they are exact copies of deleted donor elements, or whether they were truncated during transposition. In either case, they indicate that the sequence at the left-hand end of complete I factors is not required absolutely for integration.

The remaining two I-R induced yellow mutations that we have studied are associated with large DNA rearrangements, both of which have I factor sequences at at least one breakpoint. Chromosomes carrying the y^{IR5} mutation have a cytologically visible inversion. This runs proximally from the yellow locus, and has an apparently complete I factor at the breakpoint within yellow (Fig. 2a). We have not yet analyzed the other end of the inversion. There is no cytologically visible rearrangement on chromosomes carrying the y^{IR6} mutation, but it is also associated with an inversion. This starts within yellow and extends proximally for an unknown distance. There is a complete I factor at the yellow breakpoint, and a 5' truncated I element at the other (Fig. 2b). The organization of the sequences surrounding these elements suggests that this rearrangement is the result of two events, insertion of a truncated I element at yellow followed by recombination between this element and a complete I factor proximal to it and in the opposite orientation. This would have inverted the DNA between them (Fig. 2c). The recombination step could have been stimulated by I-R dysgenesis directly or could have taken place after transposition and independently of it.

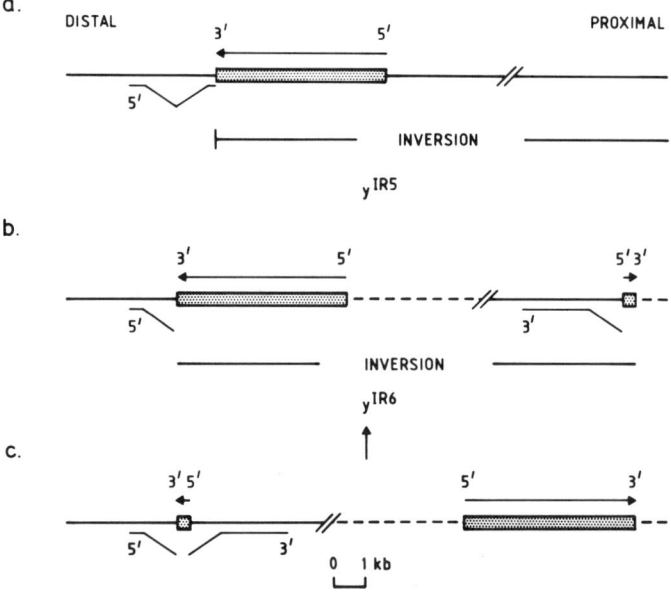

Figure 2
Inversions associated with I-R induced mutations. (*a*) Map of the inversion associated with the y^{IR5} mutation. The horizontal line immediately below the map indicates the two exons of the yellow gene (Chia et al. 1986). The I element at the distal breakpoint appears to be complete. The proximal breakpoint has not been analyzed. (*b*) Map of the inversion associated with the y^{IR6} mutation. The I element at the distal breakpoint appears to be complete, whereas that at the proximal breakpoint is truncated at the 5'end. The length of the inversion is not known. (*c*) Hypothetical intermediate in the formation of the y^{IR6} inversion, showing the truncated I element inserted into the yellow gene prior to recombination with a complete I factor lying proximal to it. A more detailed explanation is given in the text.

Inversions are not the only events that can take place next to I sequences in the genome. Pélisson (1981) tested the w^{IR1} mutation to find out whether it could revert during I-R hybrid dysgenesis, as would occur if the I factor associated with it excised precisely. He did not find any wild-type revertants, but he did generate other mutant white alleles. These determine a bleached white eye color or an eye-color that is less pigmented than that associated with w^{IR1}. Presumably, these alleles have DNA rearrangements at or near the original w^{IR1} I factor. We have studied in detail A30, one of the derivatives that determines a bleached white eye color. Restriction analysis of A30 genomic DNA indicates that there are no changes to the left of the original I factor, but that there has been a deletion of 2.4 kb at the right-hand end (Fig. 3a). We have cloned and sequenced this region, and we have found that the right-hand end of the I factor is intact and that the deletion has removed white sequences adjacent to it. This has increased the number of TAA repeats at the right-hand end of the I factor from five to seven (Fig. 3b). We cannot say exactly how this structure was generated. It is not the result of a simple deletion, as the sequence ATA, at least, must have been generated by whatever event took place. Since variation in the numer of TAA repeats is a characteristic of I elements, we wonder whether this deletion was generated by an enzyme that is normally involved in transposition, possibly at the integration step.

Two of the I-R induced white mutations that we analyzed initially, w^{IR7} and w^{IR8}, determine a bleached white eye color and are due to deletions of part of the white gene (Sang et al. 1984). It is not clear whether I-R dysgenesis was

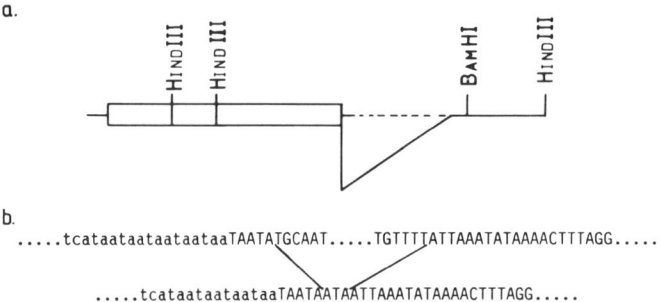

Figure 3
Deletion associated with the A30 mutation. (*a*) Restriction map of A30. (□) indicates the I factor; (———) indicates white DNA; (----) indicates the region that has been deleted. (*b*) The upper sequence is that of the w^{IR1} chromosome. The I factor sequence is given in lower case and the white sequence in upper case. The lower sequence is that of the A30 deletion. The breakpoint is indicated by lines and shows the sequence ATA that cannot be derived from the w^{IR1} sequence.

directly responsible for generating these deletions, as neither of them has I factor sequences at its breakpoint (Fawcett et al. 1986).

DISCUSSION

Transposable elements with a sequence organization similar to that of I elements are found in a wide range of organisms. These include L1 elements in mammals, F and G elements in *D. melanogaster* (Di Nocera et al. 1983; Di Nocera and Dawid 1983; Burton et al. 1986; Di Nocera and Casari 1987), and an element called ingi found in the genome of *Trypanosoma brucei* (Kimmel et al. 1987). The Doc and jockey sequences of *D. melanogaster* may also be elements of this type (Schneuwly et al. 1987; Y. Ilyin, pers. comm.). We have shown that I elements can cause mutations by inserting into genes and can be responsible for gross rearrangements of adjacent sequences. The frequency with which these events take place is low in I strains but is greatly increased when complete I factors are expressed at high levels under conditions of I-R hybrid dysgenesis. This raises the possibility that other elements of this type may produce similar genomic changes and that these might make a significant contribution to mutation frequencies if expression of functional elements were to be stimulated by outcrossing or other forms of genomic shock.

ACKNOWLEDGMENTS

M.L. is supported by a studentship from the Science and Engineering Research Council. This research was supported by grants from the Medical Research Council, grants UA360 and ATP8304 from the Centre National de la Recherches Scientifique, the Université de Clermont-Ferrand II, the Association pour la Recherche contre le Cancer, and Foundation pour Recherche Medicale Française. We are grateful to E. Kellett and S. Walsh for technical assistance, to A. Wilson for help in preparing figures, and to G. Brown for photography.

REFERENCES

Bregliano, J.C. and M.G. Kidwell. 1983. Hybrid dysgensis determinants. In *Mobile genetic elements* (ed. J. Shapiro), p. 363. Academic Press, New York.

Bucheton, A., R. Paro, H.M. Sang, A. Pélisson, and D.J. Finnegan. 1984. The molecular basis of I-R hybrid dysgenesis: Identification, cloning and properties of the I factor. *Cell* **38:** 153.

Burton, F.H., D.D. Loeb, C.F. Voliva, S.L. Martin, M.H. Edgell, and C.A. Hutchison. 1986. Conservation throughout mammalia and extensive protein encoding capacity of the highly repeated DNA long interspersed sequence one. *J. Mol. Biol.* **187:** 291.

Chia, W., G. Howe, M. Martin, Y.B. Meng, K. Moses, and S. Tsubota. 1986. Molecular analysis of the *yellow* locus of *Drosophila*. *EMBO. J.* **5:** 3597.

Covey, S.N. 1986. Amino acid sequence homology in *gag* region of reverse transcribing elements and the coat protein gene of cauliflower mosaic virus. *Nucleic Acids Res.* **14:** 623.

D'Ambrosio, E., S.D. Waitzkin, F.R. Whitney, A. Salemme, and A.V. Furano. 1986. Structure of highly repeated, long interspersed DNA family (LINE or L1Rn) of the rat. *Mol. Cell. Biol.* **6:** 411.

Di Nocera, P.P. and G. Casari. 1987. Related polypeptides are encoded by *Drosophila* F elements, I factors and mammalian L1 sequences. *Proc. Natl. Acad. Sci.* **84:** 5843.

Di Nocera, P.P. and I.B. Dawid. 1983. Interdigitated arrangement of two oligo(A)-terminated DNA sequences in *Drosophila*. *Nucleic Acids Res.* **11:** 5475.

Di Nocera, P.P., M.E. Digan, and I. Dawid. 1983. A family of oligo-adenylated transposable sequences in *Drosophila melanogaster*. *J. Mol. Biol.* **168:** 715.

Fanning, T. and M. Singer. 1987. The LINE-1 DNA sequences in four mammalian orders predict proteins that conserve homologies to retrovirus proteins. *Nucleic Acids Res.* **15:** 2251.

Fawcett, D.H., C.K. Lister, E. Kellett, and D.J. Finnegan. 1986. Transposable elements controlling I-R hybrid dysgenesis in D. melanogaster are similar to mammalian LINEs. *Cell* **47:** 1007.

Hattori, M., S. Kuhara, O. Takenaka, and Y. Sakaki. 1986. L1 family of repetitive DNA sequences in primates may be derived from a sequence encoding a reverse transcriptase-related protein. *Nature* **321:** 625.

Kimmel, B.E., O.K. Moiyoi, and J.R. Young. 1987. Ingi, a 5.2kb dispersed sequence element from *Trypanosoma brucei* that carries half of a smaller mobile element at either end and has homology with mammalian LINEs. *Mol. Cell. Biol.* **7:** 1465.

Loeb, D.D., R.W. Padgett, S.C. Hardies, W.R. Shehee, M.B. Comer, M.H. Edgell, and C.A. Hutchison. 1986. The sequence of a large L1Md element reveals a tandemly repeated 5' end and several features found in retrotransposons. *Mol. Cell. Biol.* **6:** 168.

Peifer, M. and W. Bender. 1986. The anterobithorax and bithorax mutations of the bithorax complex. *EMBO J.* **5:** 2293.

Pélisson, A. 1981. The I-R system of hybrid dysgenesis in *Drosophila melanogaster*: Are I factor insertions responsible for the mutator effect of the I-R interaction? *Mol. Gen. Genet.* **183:** 123.

Rogers, J.E. 1985. The origin and evolution of retrotransposons. *Int. Rev. Cytol.* **93:** 188.

Sang, H.M., A. Pélisson, A.A. Bucheton, and D.J. Finnegan. 1984. Molecular lesions associated with *white* gene mutations induced by I-R hybrid dysgenesis in *Drosophila melanogaster*. *EMBO J.* **3:** 3079.

Schneuwly, S., A. Kuroiwa, and W.J. Gehring. 1987. Molecular analysis of the dominant homeotic *Antennapedia* phenotype. *EMBO J.* **6:** 201.

Singer, M.F. 1982. SINEs and LINEs: Highly repeated short and long interspersed sequences in mammalian genomes. *Cell* **28:** 433.

Inducers/Regulators of Transposable Element Expression and Transposition: Genomic Stress and Environmental Effects

Evidence of Host-mediated Regulation of Retroviral Element Expression at the Posttranscriptional Level

JOHN F. MCDONALD,* DENNIS J. STRAND,* MARK R. BROWN,[†]
SUSAN M. PASKEWITZ,[†] AMY K. CSINK,* AND SUSAN H. VOSS[†]
Departments of Genetics* and Entomology[†]
University of Georgia
Athens, Georgia 30602

OVERVIEW

The transposition of mobile genetic elements is today recognized as a major cause of functionally significant mutations (Shapiro 1983). Perhaps the largest class of mobile genetic elements consists of those whose structure and mode of replication parallels that of mammalian retroviruses (Baltimore 1985). Thus far, the existence of retroviral elements[1] has been documented within the genomes of baculovirus (Miller and Miller 1982), *Dictyostelium* (Rosen et al. 1983), yeast (Roeder and Fink 1983), insects (Finnegan and Fawcett 1986), mollusks (Oprandy et al. 1981), fish (Papas et al. 1976), reptiles (Andersen et al. 1979), birds (Vogt and Friis 1971), mammals (Benveniste 1985), and plants (Johns et al. 1985), and there is reason to believe that they are constituents of all eukaryotic genomes.

Retroviral elements are characterized by their dependence on a reverse transcriptase for replication and by the possession of similarly organized and sequentially homologous genomes (Coffin 1985). The first gene in the retroviral-element genome, *gag*, encodes proteins that interact with the genomic RNA to form a nucleocapsid. The nucleocapsid is believed to play an essential role in regulating the use of genomic RNA in reverse transcription (Hull and Covey 1986; also see chapter by Boeke et al., this volume). The second gene shared by all retroviral elements, *pol*, encodes a polypeptide that has been associated with several enzymatic functions, including reverse transcription and integration activity required for incorporation of the DNA form of the retroviral element into the host genome. In addition to *gag* and

---

[1]The extensive array of viruses and viral-like elements that rely on reverse transcriptase for replication are collectively referred to as retroid elements (Hull and Covey 1986). However, reverse transcription is not limited to true retroviruses and retrotransposons but also plays a role in the replicative cycle of some DNA viruses such as hepadnavirus and caulimovirus (Hohn et al. 1985; Tiollais et al. 1985). Because many of the generalizations made in this paper do not necessarily apply to retroid DNA viruses, we have chosen to employ the less broadly encompassing term "retroviral element" to refer to true infectious retroviruses, endogenous retroviruses, and retrotransposons.

Banbury Report 30: Eukaryotic Transposable Elements as Mutagenic Agents
© Cold Spring Harbor Laboratory. 0-87969-230-8/88. $1.00 + .00

pol, the genomes of retroviruses may contain sequences that encode peptides required for cell-to-cell transmission (e.g., *env*) and other more specialized functions associated with the life cycle of particular classes of retroviral elements.

Although, in general, retroviral elements contain most of the genetic information necessary for their own replication, the fact that the retroviral and retrotransposon life cycles can be mediated by host-encoded functions is well documented. Host-mediated regulation is believed to play an important role in modulating the mutagenic potential of retrotransposons (Fink et al. 1986), as well as in controlling the duration of proviral latency periods (Curran et al. 1985). Thus, an understanding of the mechanisms that underlie host-mediated control of retroviral-element expression is likely to have relevancy to a broad spectrum of biological issues ranging from evolution (McDonald 1983; McDonald et al. 1987) to the etiology of retroviral-associated disease (Curran et al. 1985).

Specific host genes that can influence retrotransposon and retroviral expression have been thus far unequivocably identified in yeast (Winston et al. 1984), *Drosophila* (Parkhurst and Corces 1986a), and mice (Varmus and Swanstrom 1985), and their existence has been inferred in humans (Curran et al. 1985) and other primates (Benveniste 1985). Studies of retrotransposon expression in yeast (Winston et al. 1987 and this volume) and *Drosophila* (Parkhurst and Corces 1986b), have demonstrated that host genes may exert regulatory effects on a transcriptional level. For example, in yeast three loci have been identified thus far that can significantly modulate steady-state levels of Ty transcripts (Winston et al. 1987). A recent analysis of copia regulation in *Drosophila melanogaster* suggests that at least one host gene, suppressor-of-white-apricot, may mediate copia expression at the level of RNA processing (Zachar et al. 1985; see also chapter by Bingham et al., this volume). In this paper, we present evidence that regulation on a post-transcriptional level may also be an important and general component of host-mediated control of retroviral element expression.

COMPARATIVE ANALYSIS OF RETROVIRAL ELEMENT EXPRESSION IN CULTURED CELLS

Our laboratory has recently initiated a comparative analysis of the ability of a phylogenetically diverse array of retroviral elements to be expressed in a variety of insect and mammalian cell lines (J.F. McDonald and D. Strand, in prep.). The three retroviral elements analyzed in the studies reported below are representative of the broad spectrum of structural and functional variability that exists among eukaryotic retroviruses and retroviral transposable elements.

Copia is a moderately repeated retrotransposon that is primarily, if not exclusively, inherited through germ-line transmission from parent to offspring. We have recently demonstrated that copia exists within the genomes of a number of *Drosophila* species (A. Csink and J. McDonald, in prep.) and is expressed within the nuclei of a variety of cell types over *Drosophila* development (see below). Other retroviral elements that display structural and functional properties similar to copia are the *Dictyostelium* DIRS-1 element (Rosen et al. 1983), the yeast Ty element (Roeder and Fink 1983), the maize *Bs1* element (Johns et al. 1985), and the murine intracisternal-A-particle-encoding element (Kuff et al. 1981).

The Rous sarcoma virus (RSV) and the human immunodeficiency virus (HIV) are examples of vertebrate retroviruses (Teich 1985). In contrast to copia-like transposable elements, true retroviruses are typically chromosomally integrated only in infected somatic cells; thus, retroviruses are not usually inherited in a Mendelian fashion. The RSV genome, in addition to the *gag*, *pol*, and *env* genes characteristic of most infectious retroviruses, contains an additional sequence that imparts oncogenic potential (Purchio et al. 1978). A number of similar RNA tumor viruses have been isolated from various vertebrate species and have been extensively characterized (Bishop and Varmus 1985). HIV represents a more recently identified class of mammalian retroviruses whose genomes contain, in addition to the standard *gag*, *pol*, and *env* genes, additional sequences which code for proteins that *trans*-regulate viral expression. In HIV, at least one of these regulatory genes, designated *tat*III (*trans*-acting transcriptional activator) interacts with *cis*-acting sequences within the viral promoter region to stimulate HIV expression (Arya et al. 1985; Wong-Staal and Gallo 1985).

Despite the significant structural and functional variability that exists among retroviral elements, there are, nevertheless, a number of shared properties. For example, as mentioned above, sequence similarity exists within those regions of the genome (*gag* and *pol*) encoding functions essential to replication by reverse transcriptase. In addition, the coding regions of integrated retroviral elements are bordered on their 5' and 3' ends by direct long terminal repeat sequences (LTRs) that contain promoter, enhancer, and other sequences critical to the regulation of retroviral gene expression and replication.

To explore the effect of host-encoded factors on retroviral element expression, we assayed the ability of the copia, RSV, and HIV LTRs to drive expression of the bacterial chloramphenicol acyltransferase (*cat*) reporter gene in various cell lines. We monitored the ability of three LTR-CAT fusion plasmids, copia-CAT (DiNocera and Dawid 1983), RSV-CAT (Gorman et al. 1982) and HIV-CAT (Feinberg et al. 1986), to be expressed in a *Drosophila* cell line, S-2 (Schneider 1972) (embryo derived); a murine cell line, NIH-3T3

(Todaro and Green 1963) (embryo derived); and a human cell line, Y-79 (Lee et al. 1984) (retinoblastoma derived). The *tat*III inducibility of the HIV LTR in these cells was determined by cotransfecting HIV-CAT with an HIV-TAT-III plasmid (Feinberg et al. 1986) that contains the *tat*III coding sequence controlled by the HIV-LTR. Transient expression of the LTR-CAT plasmids was determined on both the protein and RNA levels. Levels of *cat*-encoded protein were measured by determining the levels of CAT enzymatic activity present in total protein extracts isolated from transfected cells (Gorman et al. 1982). Relative levels of *cat* mRNA isolated from transfected cells were determined by quantitative slot-blot and S1 protection analyses (J.F. McDonald and D. Strand, in prep.).

Levels of LTR-initiated *cat* mRNAs Are Not Always Correlated with Levels of CAT Activity

The results, summarized in Table 1, indicate that the relative efficiencies of the HIV, RSV, and copia LTRs in initiating transcription are not necessarily correlated with the degree to which the mRNAs are translated. For example, relative levels of *cat* mRNAS were nearly identical in *Drosophila* cells trans-

Table 1

Relative Levels of CAT Activity and *cat* Transcripts in LTR-CAT Transfected Cell Lines

	(a) CAT activity		
	Drosophila S-2	Murine NIH-3T3	Human Y-79
Copia	95	0	0
RSV	20	0.3	15
HIV	0.3	1.0	0.7
HIV/ HIV-TAT	1.0	20	7
	(b) cat RNA		
Copia	1.8	1.6	1.4
RSV	1.6	2.4	1.7
HIV	1.0	1.0	1.0
HIV/ HIV-TAT	2.3	10.8	6.3

CAT assays were conducted as described by Gorman et al. (1982). Relative CAT activities are average percent conversion values for each cell line. Percent conversion was determined by cutting out spots containing unreacted 14C-chloramphenicol and acetylated forms, and quantitating the amount of radioactivity in a liquid scintillation counter. Relative levels of *cat* RNA were determined by a slot blot procedure using nick-translated *cat* probe. Autoradiographs were scanned on a Beckman DU-8 scanner and values normalized to the level present in HIV-CAT transfected cells. Values presented are means of at least 4 replicate assays. Percentage error (SE/Mean) for all values <20%. (After J.F. McDonald and D. Strand, in prep.)

fected with HIV-CAT (and HIV-TAT), RSV-CAT, and copia-CAT and yet the level of CAT activity in copia-CAT transfected cells was >5-fold higher than in RSV-CAT transfected cells and >100-fold higher than in HIV-CAT/ HIV-TAT-III transfected *Drosophila* cells. These results indicate that RSV and HIV initiated *cat* transcripts are posttranscriptionally repressed in *Drosophila* cells. In contrast, essentially no CAT activity could be detected in copia-CAT transfected murine NIH-3T3 or human Y-79 cells, despite the fact that copia-LTR-initiated *cat* transcripts were relatively abundant. This indicates that copia-LTR-initiated transcripts are posttranscriptionally repressed in these mammalian cells.

Although *tat*III mediated effects on the HIV LTR are known to involve interaction between the *tat*III-encoded protein (TAT-III) and sequences within the HIV LTR (Rosen et al. 1985), whether the TAT-III response occurs at the transcriptional or posttranscriptional levels, or both, is unresolved (Chen 1986; Rosen et al. 1986; Marx 1987; Muesing et al. 1987). In our studies, cotransfection of the HIV-CAT plasmid with HIV-TAT-III resulted in a 2–3-fold increase in levels of *cat* mRNA within the *Drosophila* cells but a 6–12-fold increase in the murine and human cell lines tested. In the mammalian cell lines, the magnitude of the TAT-III response on the level of CAT activity was equivalent to or slightly higher than that observed on the level of *cat* mRNA. By contrast, only very low levels of CAT activity could be detected in the HIV-CAT/HIV-TAT-III transfected *Drosophila* cells despite the fact that *cat* mRNA was present.

It has recently been reported that HIV can infect insect cells, including *Drosophila*, and become integrated into the host cell's genome (Becker et al. 1986). However, no evidence of HIV replication was detected in any of the insect cells tested. Our data suggest that the apparent block to HIV replication in insect cells may, at least in part, be due to a highly effective mechanism of gene represssion operating at a posttranscriptional level (J.F. McDonald and D. Strand, in prep.).

The Copia LTR Is Posttranscriptionally Induced by Radiation Stress

The expression of a number of retroviral elements has been shown to be responsive to environmental stress. For example, we have previously demonstrated that heat-shock and heat-shock mimetic treatments transcriptionally activate copia expression in whole flies and in mammalian cells transfected with a copia LTR-CAT fusion plasmid (Strand and McDonald et al. 1985; McDonald et al. 1987). The Ty element has recently been shown to be transcriptionally induced when yeast cells are exposed to ultraviolet light (Rolfe et al. 1986; see also McEntee and Bradshaw, this volume). X-ray and gamma radiation, long known as effective activators of latent vertebrate retroviruses (Weiss et al. 1971; Tennant and Rascati 1980), also induce formation of retroviral particles in Drosophila (Philpott et al. 1969; Gartner 1971, 1972).

To explore the effect of gamma radiation on copia LTR expression, we compared the relative efficiency of copia-CAT expression in irradiated versus nonirradiated Drosophila S-2 cells. The results, presented in Table 2, show that radiation has little effect on levels of copia LTR initiated *cat* mRNA. By contrast, levels of CAT enzymatic activity were significantly higher in irradiated cells than in control cells. The radiation response was transient, i.e., it was most pronounced in cells monitored immediately after the radiation treatment. Levels of CAT activity in irradiated cells that were allowed to recover for several hours following treatment were not significantly different from control levels.

IN SITU ANALYSIS OF COPIA EXPRESSION OVER DROSOPHILA DEVELOPMENT

Retroviral-like particles (RLPs) have been reported by many investigators to be present within the nuclei of a variety of Drosophila cells. Most, if not all, of these RLPs are believed to contain RNA homologous to the copia retrotransposon (Shiba and Saigo 1983; Ryo et al. 1984). The presence and abundance of RLPs in Drosophila appears to follow precise developmental and tissue-specific patterns of expression. RLPs do not usually appear in Drosophila cells until late in the fly's life-span; therefore, the copia retroviral elements may be considered to be in a latent state in developing embryos and young flies. For example, no RLPs have been detected in wild-type flies younger than 15 days posteclosion (Gateff 1978). By 15–20 days posteclosion, however, isolated particles can be observed in the nuclei of salivary gland and

Table 2
Effect of Environmental Stress on LTR-CAT expression in Drosophila Cells

LTR-CAT	Cell line	Treatment	CAT activity	*cat* RNA
Copia	2	control	1.00	1.00
Copia	2	gamma radiation (2000 rads)	3.50 (0.5 hr recovery)	1.10
Copia	2	gamma radiation (2000 rads)	1.70 (1.0 hr recovery)	1.10
Copia	2	gamma radiation (2000 rads)	0.98 (4.0 hr recovery)	1.07

Methods of quantitation were as described in legend to Table 1. Values normalized to unstressed controls. Cells were allowed to equilibrate 5 hr after transfection prior to radiation treatment.

fat body cells (Philpott et al. 1969). In flies older than 30 days, the number of RLPs within cells has been reported to increase markedly, and by 50–70 days posteclosion, more than 1000 RLPs per nucleus can be detected in a variety of cell types (Filshie et al. 1967; Gartner 1971, 1972; Herman et al. 1971).

We have recently completed a survey of the tissue-specific expression of copia over *Drosophila* development (D. Strand et al., in prep.). Table 3 summarizes the results of this survey and provides a comparison of the tissue-specific pattern of copia expression with the previously gathered data regarding the presence and absence of RLPs in *Drosophila* tissues over development. In general, the correlation between the presence of RLPs and the occurrence of copia transcripts is good. Tissues in which copia transcripts are clearly not present (e.g., larval brain and oocytes of adult females) have never been reported to harbor RLPs at any developmental stage. Moreover, tissues in which RLPs have been observed are the same ones in which we detect the presence of copia transcripts (Fig. 1, Table 3). Copia transcripts are generally most abundant in cell nuclei. In larval fat body and salivary gland cells, copia transcripts appear to be strictly limited to the nuclei but in other tissues such as larval and adult midgut, transcripts can be detected in the cytoplasm as well (Fig. 1).

In addition to these correlations, a most interesting finding that has emerged from our study is the fact that copia transcripts are abundant within cells and at developmental periods where RLPs are not present. For example, RLPs have not been detected within the cells of *Drosophila* flies < 10 days

Table 3
RLPs and copia Transcripts in Tissues of Wild-type *Drosophila melanogaster*

Life stage	Tissue	RLPs[a]	copia Transcripts
Larvae			
3rd instar	eye disk[1]	+	+
	midgut[6]	+++	+++
	muscle[2]	++	++
	fat body[3]	—	+++
	salivary[2] gland	—	++
	brain[1]	—	—
Adult	midgut[5]	+++	+++
	muscle[2]	++	++
	fat body[4]	+	+++
	oocytes[2]	—	—
	brain[4]	++	++

[a]Data on the presence/absence of RLPs were taken from the following references: (1) Akai et al. 1967; (2) Gateff 1978; (3) Felluga et al. 1971; (4) Philpott et al. 1969; (5) Gartner 1971, 1972; (6) Filshire et al. 1967. (After D. Strand et al., in prep.)

posteclosion, yet we find copia transcripts to be abundant in a variety of tissues in five day posteclosion adults (Fig. 1). Northern analysis of copia expression over *Drosophila* development has confirmed the presence of copia transcripts in adult flies as early as one day posteclosion (Parkhurst and Corces 1987; D. Strand et al., in prep.). Moreover, we have detected an abundance of copia transcripts within the nuclei of larval fat body and salivary gland cells (Fig. 1), and yet the presence of RLPs has not been reported within these same cells in wild-type *Drosophila*.

Clearly, additional studies will be required before conclusions can be drawn from these comparisons. For example, immunocytochemical analyses utilizing antibody against *trpE*-copia-*gag* fusion protein are currently in progress in our laboratory. These studies will allow us to determine whether or not the presence or absence of RLPs in a particular cell type is a true reflection of the tissue-specific pattern of copia translational level expression. However, if, as sequence analysis suggests (Mount and Rubin 1985), copia does encode its own capsid protein, our finding that copia transcripts are abundant within tissues and at developmental stages where RLPs are not present would indicate posttranscriptional regulatory control.

DISCUSSION

Evidence for the existence of host-mediated regulation of retroviral element expression derives from a number of sources. Variation in the expression of the same element in different cell lines is a well-documented phenomenon that is attributed to host-encoded factors (Gorman et al. 1982). Tissue-specific patterns of retroviral-element expression that change over development is also evidence of host-encoded *trans*-regulatory control (Overbeek et al. 1986; Soriano et al. 1986). Finally, both retroviruses and retrotransposons have been found to be subject to regulation by the two major host-encoded stress response systems in eukaryotes: the DNA damage response system (Rolfe et al. 1986; see also chapter by McEntee, this volume) and heat shock response (Strand and McDonald 1985; McDonald et al. 1987).

In an effort to understand the molecular mechanisms that underlie host-mediated regulation of retroviral element expression, we have initiated a comparative analysis of the expression of copia, RSV, and HIV LTRs in *Drosophila* and mammalian cell lines. The initial results demonstrate the existence of significant cell-line-dependent effects on the expression of each of the LTRs studied. We have found that host-cell-mediated controls significantly influence steady-state levels of LTR-initiated mRNAS, but that the abundance of these transcripts is not always correlated with the presence or abundance of their encoded protein products. Our findings are consistent with previous observations (Wright et al. 1986) and indicate that a significant

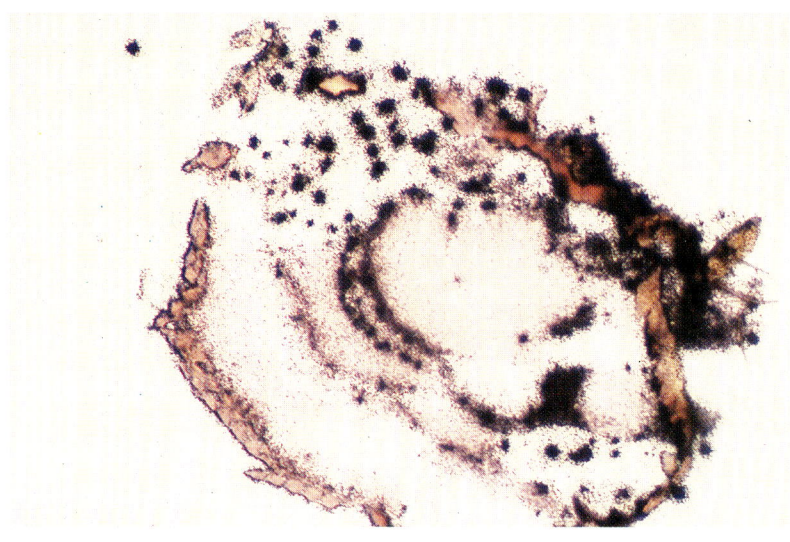

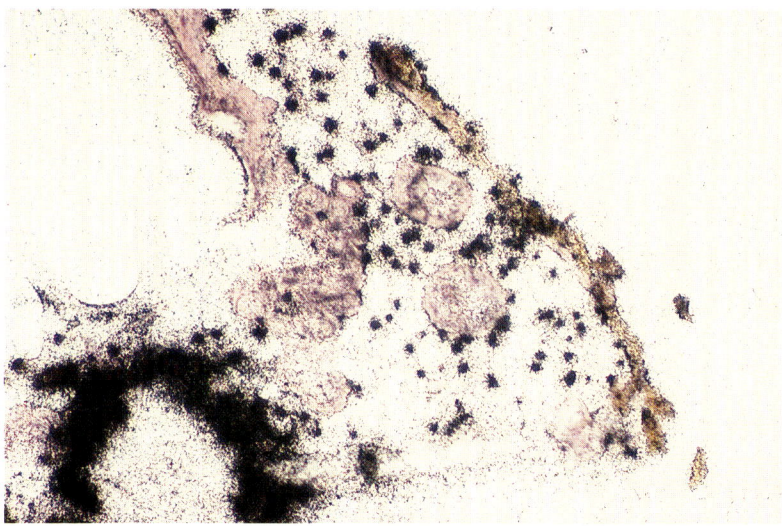

Figure 1
Expression of full-length (5 kb) copia transcripts in *Drosophila* tissues by in situ hybridization using a radiolabeled synthetic RNA probe. (*Upper*) Section through adult brain showing copia transcripts present in nuclei of cells at periphery of the lobes (expression in fat body cell nuclei also visible). (*Lower*) Section through adult midgut showing copia transcripts abundant in both nuclei and cytoplasm (expression in fat body cell nuclei also visible) (D. Strand et al., in prep.).

component of retroviral element expression may involve regulation at the posttranscriptional level.

Posttranscriptional regulatory mechanisms also appear to be involved in the induction of the copia LTR by radiation stress. This is consistent with studies that indicate that the response of eukaryotic genes to stress includes regulatory changes at the posttranscriptional level (Klemenz et al. 1985; McGarry and Lindquist 1985; Klemenz and Gehring 1986; see also the chapter by Lambert et al., this volume).

Eukaryotic cells are known to be capable of regulating expression of both viral (Vaidya et al. 1983; Katz et al. 1986; Peterson and Nuss 1986) and host genes (Lodish 1976; Revel and Groner 1978) by a variety of posttranscriptional mechanisms. For example, it has recently been demonstrated that inhibition of muscle-specific protein synthesis in undifferentiated rat skeletal muscle cells involves the exclusion of the muscle-specific mRNAs from polysomes (Endo and Nadal-Ginard 1987). In another study, it was found that cessation of hemoglobin synthesis in transformed chicken erythroblasts is due to the sequestering of the globin transcripts within cell nuclei, which effectively prevents the association of hemoglobin message and polysomes (Therwath and Scherrer 1978). Whether or not analogous mechanisms may contribute to the repression of copia LTR expression in mammalian cells and HIV LTR expression in *Drosophila* cells is currently under investigation. It is relevant to note that our in situ analysis of copia expression over *Drosophila* development does indicate that copia transcripts are sequestered in the nuclei of cells in which the formation of copia-encoded RLPs is blocked. It remains to be determined whether this mechanism of nuclear localization is the primary method by which *Drosophila* represses translation of copia transcripts over development.

CONCLUSION

In summary, our studies of copia, RSV, and HIV LTR expression in *Drosophila*, murine, and human cells, as well as our survey of copia tissue-specific expression over *Drosophila* development, suggest that a significant component of the host-mediated *trans*-regulatory control of retroviral element expression occurs on the posttranscriptional level. Interestingly, posttranscriptional regulation is emerging as an important component of eukaryotic gene expression in general. The developing consensus is that eukaryotic gene expression is a coordinated process, involving a complex of cell- and stage-specific factors that exert their regulatory effects at multiple levels. Since retroviral element expression appears to be mediated by pre-established, host-encoded regulatory networks, it is perhaps not surprising that the underlying molecular mechanisms also involve multiple levels of gene control.

ACKNOWLEDGMENTS

We are grateful to the following individuals for providing constructs or cell lines used in our studies: F. Wong-Staal and S. Josephs (HIV-CAT, HIV-TAT-III); I. Dawid (copia-CAT); C. Gorman (RSV-CAT); R. Franza (NIH-3T3); and W. Bennedict (Y-79).

REFERENCES

Akai, H., E. Gateff, L. Davis, and H. Schneiderman. 1987. Virus-like particles in cultured and mutant tissues of *Drosophila*. *Science* **157**: 810.

Andersen, P., M. Barbacid, S. Tronick, H. Clark, and S. Aaronson. 1979. Evolutionary relatedness of viper and primate endogenous retroviruses. *Science* **204**: 318.

Arya, S., C. Guo, S. Josephs, and F. Wong-Staal. 1985. *Trans*-activator gene of human T-lymphotropic virus type III. *Science* **229**: 69.

Baltimore, D. 1985. Retrovirus and retrotransposons: The role of reverse transcription in shaping the eukaryotic genome. *Cell* **40**: 481.

Becker, J., U. Hazan, M. Nugeyre, F. Rey, B. Spine, F. Barre-Sinooussi, A. Georges, L. Teulieres, and J. Cherman. 1986. Infection of insect cell cultures by the HIV virus and deletion of this virus' presence in insects of African origin. *C.R. Acad. Sci. (Ser. III)* **3**: 303.

Benveniste, R. 1985. The contributions of retroviruses to the study of mammalian evolution. In *Molecular evolutionary genetics* (ed. R. MacIntyre), p. 359. Plenum Press, New York.

Bishop, J.M. and H. Varmus. 1985. Functions and origins of retroviral transforming genes. In *Molecular biology of tumor viruses,* 2nd edition: *RNA tumor viruses* (ed. R. Weiss et al.), vol. 2, p. 999. Cold Spring Harbor Laboratory, Cold Spring Harbor, New York.

Chen, I. 1986. Regulation of AIDS virus expression. *Cell* **47**: 1.

Coffin, J. 1985. Structure of the retroviral genome. In *Molecular biology of tumor viruses,* 2nd edition: *RNA tumor viruses* (ed. R. Weiss et al.), vol. 2, p. 261. Cold Spring Harbor Laboratory, Cold Spring Harbor, New York.

Curran, J., W. Morgan, A. Hardy, H. Jaffe, W. Darrow, and W. Dowdle. 1985. The epidemiology of AIDS: Current status and future prospects. *Science* **229**: 1352.

DiNocera, P. and I. Dawid. 1983. Transient expression of genes introduced into cultured cells of *Drosophila*. *Proc. Natl. Acad. Sci.* **80**: 7095.

Endo, T. and B. Nadal-Ginard. 1987. Three types of muscle-specific gene expression in fusion-blocked rat skeletal muscle cells: Translational control in EGTA-treated cells. *Cell* **49**: 515.

Feinberg, M., R. Jarrett, A. Aldovini, R. Gallo, and F. Wong-Staal. 1986. HTLV-III expression and production involve complex regulation at the levels of splicing and translation of viral RNA. *Cell* **46**: 807.

Felluga, B., V. Jonsson, and M. Liljeros. 1971. Ultrastructure of new viruslike particles in *Drosophila*. *J. Invertebr. Pathol.* **17**: 339.

Filshie, B., T. Grace, D. Poulson, and J. Rehacek. 1967. Viruslike particles in insect cells of three types. *J. Invertebr. Pathol.* **9:** 271.

Fink, G., J. Boeke, and D. Garfinkel. 1986. The mechanisms and consequences of retrotransposition. *Trends Genet.* **2:** 119.

Finnegan, D. and D. Fawcett. 1986. Transposable element in *Drosophila melanogaster*. *Oxf. Surv. Eukaryotic Genes* **3:** 1.

Gartner, L. 1971. Two types of viruslike particles in *Drosophila* midgut. *Experientia* **27:** 562.

———. 1972. Viruslike particles in the adult *Drosophila* midgut. *J. Invertebr. Pathol.* **20:** 364.

Gateff, E. 1978. Malignant neoplasms of genetic origin in *Drosophila melanogaster*. *Science* **200:** 1448.

Gorman, C., G. Merlino, M. Willingham, I. Pastan, and B. Howard. 1982.The Rous sarcoma virus long terminal repeat is a strong promoter when introduced into a variety of eukaryotic cells by DNA-mediated transfection. *Proc. Natl. Acad. Sci.* **79:** 6777.

Herman, M., M. Johnson, and J. Miguel. 1971. Viruslike particles and related filaments in neurons and glia of *Drosophila melanogaster* brain. *J. Invertebr. Pathol.* **17:** 442.

Hohn, T., B. Hohn, and P. Pfeiffer. 1985. Reverse transcription in CaMV. *Trends Biochem. Sci.* **4:** 205.

Hull, R. and S. Covey. 1986. Genome organization and expression of reverse transcribing elements: Variations and a theme. *J. Gen. Virol.* **67:** 175.

Johns, M., J. Mottinger, and M. Freeling. 1985. A low copy number *copia*-like transposon in maize. *EMBO J.* **4:** 1093.

Katz, R., B. Cullen, R. Malavarca, and A. Skalka. 1986. Role of the avian retrovirus mRNA leader in expression: Evidence for novel translational control. *Mol. Cell. Biol.* **6:** 372.

Klemenz, R. and W. Gehring. 1986. Sequence requirement for expression of the *Drosophila melanogaster* heat shock protein hsp22 gene during heat shock and normal development. *Mol. Cell. Biol.* **6:** 2011.

Klemenz, R., D. Hultmark, and W. Gehring. 1985. Selective translation of heat shock mRNA in *Drosophila melanogaster* depends on sequence information in the leader. *EMBO J.* **4:** 2053.

Kuff, E., L. Smith, and K. Lueders. 1981. Intracisternal A-particle genes in *Mus musculus:* A conserved family of retrovirus-like elements. *Mol. Cell. Biol.* **1:** 216.

Lee, W., A. Murphee, and W. Benedict. 1984. Expression and amplification of the *N-myc* gene in primary retinoblastomas. *Nature* **309:** 458.

Lodish, H. 1976. Translational control of protein synthesis. *Annu. Rev. Biochem.* **45:** 39.

Marx, J. 1987. The AIDS virus: Well known but a mystery. *Science* **236:** 391.

McDonald, J. 1983. Molecular basis of adaptation: A critical review of relevant ideas and observations. *Annu. Rev. Ecol. Syst.* **14:** 77.

McDonald, J., D. Strand, M. Lambert, and I.B. Weinstein. 1987. Responsive genome: Evidence and evolutionary implications. In *Development as an evolutionary process* (ed. R. Rauff and E. Rauff), p. 239. Liss Press, New York.

McGarry, T. and S. Lindquist. 1985. The preferential translation of *Drosophila hsp-70* mRNA requires sequences in the untranslated leader. *Cell* **42:** 903.

Miller, D. and L. Miller. 1982. A virus mutant with an insertion of a *copia*-like transposable element. *Nature* **299:** 562.

Mount, S. and G. Rubin. 1985. Complete nucleotide sequence of the *Drosophila* transposable element *copia:* Homology between *copia* and retroviral proteins. *Mol. Cell. Biol.* **5:** 1630.

Muesing, M., D. Smith, and D. Capon. 1987. Regulation of mRNA accumulation by a human immunodeficiency virus trans-activator protein. *Cell* **48:** 691.

Oprandy, J., P. Chang, A. Pronovost, K. Cooper, R. Brown, and V. Yates. 1981. Isolation of a viral agent causing hematopoietic neoplasia in the soft-shell clam. *J. Invertebr. Pathol.* **38:** 45.

Overbeek, P., S.-P. Lai, K. Van Quill, and H. Westphall. 1986. Tissue-specific expression in transgenic mice of a fused gene containing RSV terminal sequences. *Science* **231:** 1573.

Papas, T., J. Dahlberg, and R. Sonstegard. 1976. Type C virus in lymphosarcoma in northern pike. *Nature* **261:** 506.

Parkhurst, S. and V. Corces. 1986a. Interactions among the Gypsy transposable element and the yellow and the suppressor of Hairy-wing loci in *D. melanogaster*. *Mol. Cell. Biol.* **6:** 47.

———. 1986b. Retroviral elements and suppressor genes in *Drosophila*. *Bioessays* **5:** 52.

———. 1987. Developmental expression of *D. melanogaster* retrovirus-like transposable elements. *EMBO J.* **6:** 419.

Peterson, A. and D. Nuss. 1986. Regulation of expression of the wound tumor virus genome in persistently infected vector cells is related to change in translational activity of viral transcripts. *J. Virol.* **59:** 195.

Philpott, D., J. Weibel, H. Atlan, and J. Miguel. 1969. Virus-like particles in fat body, oenocytes and central nervous tissue of *D. melanogaster* imagoes. *J. Invertbr. Pathol.* **14:** 31.

Purchio, A., E. Erikson, J. Brugge, and R. Erikson. 1978. Identification of a polypeptide encoded by the avian sarcoma virus *src* gene. *Proc. Natl. Acad. Sci.* **75:** 1567.

Revel, M. and Y. Groner. 1978. Post-translational and translational controls of gene expression in eukaryotes. *Annu. Rev. Biochem.* **47:** 1079.

Roeder, G. and G. Fink. 1983. Transposable elements in yeast. In *Mobile genetic elements* (ed. J. Shapiro), p. 300. Academic Press, New York.

Rolfe, M., A. Spanos, and G. Banks. 1986. Induction of yeast Ty element transcription by ultraviolet light. *Nature* **319:** 339.

Rosen, E., A. Siverten, and R. Firtel. 1983. An unusual transposon encoding heat shock inducible and developmentally regulated transcripts in *Dictyostelium*. *Cell* **35:** 243.

Rosen, C., J. Sodroski, and W. Haseltine. 1985. Location of cis-acting regulatory sequences in the human T cell lymphotropic virus type III long terminal repeat. *Cell* **41:** 813.

Rosen, C., J. Sodroski, W. Goh, A. Dayton, J. Lippke, and W. Haseltine. 1986. Post-transcriptional regulation accounts for trans-activation of human T-lymphotropic virus type III. *Nature* **319:** 555.

Ryo, H., T. Shiba, S. Kondo, and E. Gateff. 1984. Chromosomal aberrations and retrovirus-like particles by *in vitro* transplantation in neoplastic brain cells of a *Drosophila* mutant strain. *Gann* **75:** 22.

Schneider, I. 1972. Cell lines derived from late embryonic stages of *Drosophila melanogaster*. *J. Embryol. Exp. Morphol.* **27:** 353.

Shapiro, J., ed. 1983. *Mobile genetic elements*. Academic Press, New York.

Shiba, T. and K. Saigo. 1983. Retrovirus-like particles containing RNA homologous to the transposable element *copia* in *D. melanogaster*. *Nature* **302:** 119.

Soriano, P., R. Cone, R. Mulligan, and R. Jaenisch. 1986. Tissue-specific and ectopic expression of genes introduced into transgenic mice by retroviruses. *Science* **234:** 1409.

Strand, D. and J. McDonald. 1985. *Copia* is transcriptionally responsive to environmental stress. *Nucleic Acids Res.* **13:** 4401.

Teich, N. 1985. Taxonomy of retroviruses. In *Molecular biology of tumor viruses*, 2nd edition: *RNA tumor viruses* (ed. R. Weiss et al.), vol. 2, p. 1. Cold Spring Harbor Laboratory, Cold Spring Harbor, New York.

Tennant, R. and R. Rascati. 1980. Mechanisms of cocarcinogenesis involving endogenous retroviruses. In *Modifiers of chemical carcinogenesis* V. (ed. T. Slaga), p. 185. Raven Press, New York.

Therwath, A. and K. Scherrer. 1978. Post-transcriptional suppression of globin gene expression in cells transformed by avian erythroblastosis virus. *Proc. Natl. Acad. Sci.* **75:** 3776.

Tiollais, P., C. Pourcell, and A. De Jean. 1985. The hepatitis B virus. *Nature* **317:** 489.

Todaro, G. and H. Green. 1963. Quantitative studies of the growth of mouse cells in culture and their development into established cell lines. *J. Cell. Biol.* **17:** 299.

Vaidya, A., N. Taraschi, S. Tancin, and C. Long. 1983. Regulation of endogenous murine mammary tumor virus expression in C57BL mouse lactating mammary glands: Transcription of functional mRNA with a block at the translational level. *J. Virol.* **46:** 818.

Varmus, H. and R. Swanstrom. 1985. Replication of retroviruses. In *Molecular biology of tumor viruses*, 2nd edition: *RNA tumor viruses* (ed. R. Weiss et al.), vol. 2, p. 369. Cold Spring Harbor Laboratory, Cold Spring Harbor, New York.

Vogt, P. and R. Friis. 1971. An avian leukosis virus related to RSV (O): Properties and evidence of helper activity. *Virology* **43:** 223.

Weiss, R., R. Friis, E. Katz, and P. Vogt. 1971. Induction of avian tumor viruses in normal cells by physical and chemical carcinogens. *Virology* **49:** 920.

Winston, F., K. Durbin, and G. Fink. 1984. The *SPI3* gene is required for normal transcription of Ty elements in *S. cerevisiae*. *Cell* **39:** 675.

Winston, F., C. Dollard, B. Malone, J. Clare, J. Kapakos, P. Farabaugh, and P. Minehart. 1987. Three genes are required for trans-activation of Ty transcription in yeast. *Genetics* **115:** 649.

Wong-Staal, P. and R. Gallo. 1985. Human T-lymphotropic retroviruses. *Nature* **317:** 395.

Wright, C., B. Felber, H. Paskalis, and G. Pavlakis. 1986. Expression and characterization of the trans-activator of HTLV-III/LAV virus. *Science* **234:** 988.

Zachar, Z., D. Davison, D. Garza, and P. Bingham. 1985. A detailed developmental and structural study of the transcriptional effects of insertion of the *copia* transposon into the *white* locus of D. melanogaster. *Genetics* **111:** 495.

Effect of Temperature on Ty Transposition

CHARLOTTE ELDER PAQUIN* AND VALERIE MOROZ WILLIAMSON[†]
Department of Biological Sciences
University of Cincinnati
Cincinnati, Ohio 45221
[†]Department of Nematology
University of California
Davis, California 95616

OVERVIEW

A method of estimating the transposition rate of the transposable element Ty (transposon yeast) in the yeast *Saccharomyces cerevisiae* has been developed. The transposition assay is based on the altered expression of two alcohol dehydrogenase (ADH) structural genes, *ADH2* and *ADH4*, upon insertion of a Ty adjacent to these genes. Ty transposition to both *ADH2* and *ADH4* is 100-fold more frequent at 15°C or 20°C than at 30°C, the optimum growth temperature for *S. cerevisiae*. These results demonstrate that transposition rates can be dramatically altered by changes in environmental conditions.

INTRODUCTION

Ty elements are a homologous family of transposable elements found in the yeast *S. cerevisiae* (Cameron et al. 1979; Williamson 1983). These elements are 5.9-kb DNA sequences with 330 base-pair direct repeats, called σs, on each end and a unique internal ϵ region (Fig. 1). Ty elements show considerable homology to retroviruses (Clare and Farabaugh 1985) and transpose through an RNA intermediate (Boeke et al. 1985). Laboratory strains of *S. cerevisiae* carry approximately 30 copies of Ty and more than a hundred copies of σ. Mutations due to Ty insertions have been reported at many loci (Williamson 1983). However, Ty transposition rates have been difficult to measure using marked elements because transposition occurs at a much lower rate than recombination events between Ty elements or between Ty elements and solo σs (Roeder and Fink 1980). Ty carries enhancer-like sequences that have been shown to increase expression of adjacent genes (Errede et al. 1985). We developed an assay for Ty transposition, using the ability of Ty elements to increase the expression of adjacent genes when inserted in the correct location and orientation.

Ty insertions were detected by their ability to increase the expression of ADH genes. Four isozymes of ADH (EC 1.1.1.1, ADH) from the yeast, *S.*

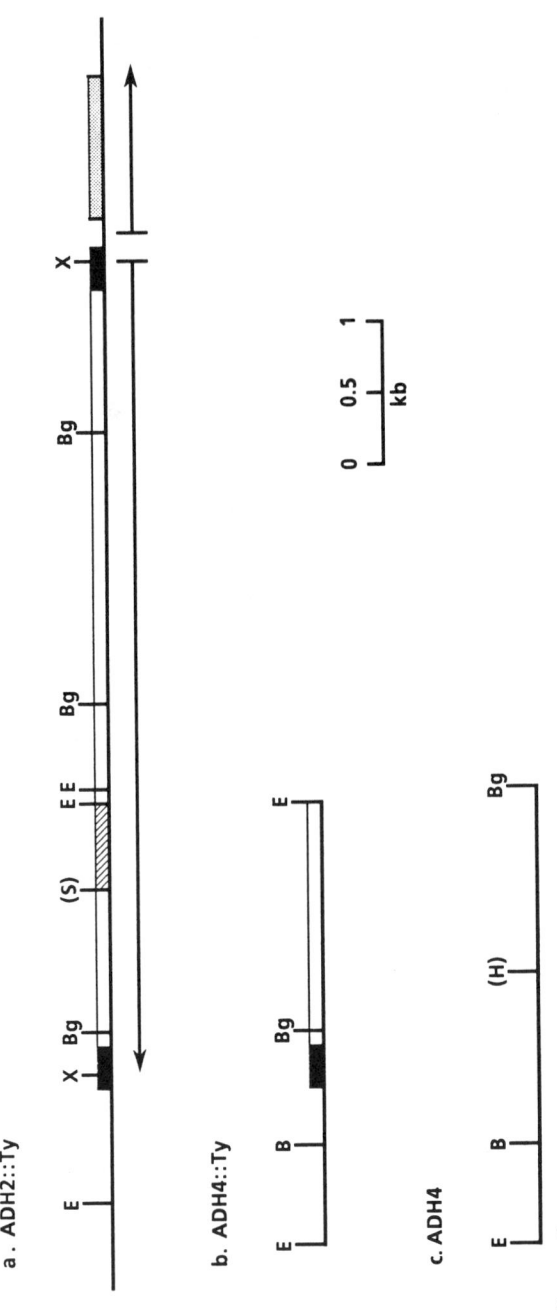

Figure 1

Ty insertions at *ADH2* and *ADH4*. (*a*) Retriction map of a Ty insertion adjacent to *ADH2* which results in expression of *ADH2* during growth on glucose (Williamson et al. 1981). (□) the ε region of a Ty element and (■) the σ regions. The internal fragment of Ty used as a probe in Fig. 2 and Fig. 3 is indicated by a crosshatched box within the ε region. The stippled box indicates the location of the *ADH2* structural gene. Direction and extent of transcription from *ADH2* and the Ty element are indicated by arrows. (*b*) Restriction map of the *ADH4*::Ty fragment isolated from a mutant with a Ty element at *ADH4*. (*c*) Restriction map of the *ADH4* fragment isolated from a strain which has no Ty insertion at *ADH4*. Restriction enzyme sites are abbreviated as follows: B, *Bam*HI; Bg, *Bgl*II; E, *Eco*RI; X, *Xho*I. The (S) indicates the *Sal*I site, which is the end of the Ty probe, and (H) indicates a *Hin*dIII site in the *ADH4* clone (other *Hin*dIII and *Sal*I sites are not indicated on this restriction map). (Reprinted, with permission, from Paquin and Williamson 1986.)

cerevisiae, have been identified, cloned, and sequenced. *ADH1* is the structural gene coding for ADHI, the classic fermentative isozyme of ADH (Ciriacy 1975a; Williamson et al. 1980). *ADH2*, the structural gene for ADHII, the glucose-repressed isozyme of ADH, is thought to be involved in the use of ethanol as a carbon and energy source (Ciriacy 1975a,b; Russell et al. 1983). *ADH3* is the gene encoding ADHIII, an ADH isozyme found only in the mitochondria (Ciriacy 1975a; Young and Pilgrim 1985). *ADH4* encodes ADHIV, a novel ADH identified and characterized during the course of this study (Paquin and Williamson 1986; Williamson and Paquin 1987). ADH activity is essential for fermentation. Growth in the presence of the respiratory inhibitor, antimycin A, requires fermentation and, thus, ADH activity. All the strains used in these experiments carry a deletion of the 5' region of *ADH1* (Williamson et al. 1983) and are, therefore, antimycin-A sensitive during growth on glucose. Several antimycin-A-resistant mutations derived from strains lacking ADHI activity had been shown to be due to Ty insertions adjacent to *ADH2*, which resulted in ADHII expression during growth on glucose (Williamson et al. 1981).

RESULTS

Ty transposition rates were estimated by first determining the mutation rate to antimycin-A resistance and then testing the antimycin-A-resistant mutants for Ty insertions by Southern blot analysis. Mutation rates were estimated using the P_o method (Lea and Coulson 1949; $P_o = e^{-m}$, where P_o equals the proportion of cultures without mutants and m is the mean number of mutations per culture; m is then divided by the mean number of cells per culture to give the mutation rate), a method that estimates mutation rates based on the number of cultures without mutants. This method was chosen because antimycin-A-resistant mutants grow faster than the parent antimycin-A-sensitive strains. Thus, any method that estimates mutation rates to antimycin-A resistance, based on the number of mutants per culture, will overestimate the mutation rate. For each mutation rate estimate, 16 to 18 cultures were grown at the experimental temperature for six to eight generations, plated on media containing antimycin A and glucose, and incubated for five days at 30°C. Since all antimyin-A-resistant mutants were selected under the same conditions, any differences in the numbers or types of mutations are due to the experimental treatments. The number of cells per culture was adjusted so that the mean number of mutations per culture was approximately one. Thus, an estimate of the number of cultures in which Ty transpositions to *ADH2* had occurred was obtained by testing only one antimycin-A-resistant mutant from each culture by Southern blot analysis for a Ty insertion at *ADH2*. Transposition rates were then calculated using the number of cultures without a mutant due to Ty insertion at *ADH2* as P_o.

At 30°C most of the mutations are not due to Ty insertions resulting in a mutation rate of approximately 5×10^{-9} and a transposition rate of approximately 10^{-10} (Table 1; Paquin and Williamson 1984). At low temperatures (15–20°C) the mutation rate jumped 50-fold to about 1×10^{-7} and the transposition rate jumped 100-fold to about 10^{-8}, indicating that Ty transposition is temperature sensitive.

However, even at 15°C only a third of the antimycin-A-resistant mutations were due to Ty insertions at *ADH2* (Table 1; Paquin and Williamson 1984), suggesting that other types of mutations were also increasing at low temperatures. We reasoned that Ty transposition to another site or sites in the genome could also result in antimycin A resistance and that the other mutants selected after growth at 15°C and 20°C might also be due to Ty transposition events. Therefore, we probed Southern blots of *Eco*RI-restricted DNA from antimycin-A-resistant mutants with an internal fragment of Ty (shown in Fig. 1; Paquin and Williamson 1986), in order to detect new Ty insertions. The Ty homologous bands on these Southern blots had one *Eco*RI site within the Ty element and one *Eco*RI site in adjacent, unique DNA and, therefore, vary in size depending on the location of *Eco*RI sites adjacent to the Ty elements. Twenty-nine bands homologous to Ty were seen in the antimycin-A-sensitive strains (Fig. 2). An additional band of the right size to be due to a Ty insertion at *ADH2* was observed in many of the antimycin-A-resistant mutants that were previously shown to have a Ty insertion at *ADH2* by Southern blot analysis using an *ADH2* probe. A new Ty band was not seen in DNA from all mutants carrying a Ty insertion at *ADH2* because some of the new bands comigrated with Ty bands present in the parental strains. In addition, about half the antimycin-A-resistant strains that did not carry a Ty insertion at *ADH2* had new bands homologous to Ty at other locations. The same three sizes of new bands (2.7, 2.9, and 3.1 kb) were seen repeatedly suggesting that either the antimycin-A-resistance was due to the new Ty insertions or that there were hot spots for Ty insertion. To determine whether these bands were linked to the antimycin-A-resistant phenotype, diploid mutants carrying new Ty bands were sporulated and the resultant tetrads dissected. The tetrads were tested for antimycin-A resistance and the presence of new Ty bands (Fig. 3; Paquin and Williamson 1986). In all four cases tested, the antimycin-A-resistant phenotype segregated with a new band homologous to Ty. Thus, the antimycin-A-resistant phenotype was linked to the new Ty insertions. When haploid strains carrying new Ty bands of different sizes were crossed to each other and the resultant diploids sporulated, all four spores were antimycin-A-resistant, indicating that the Ty insertions are closely linked.

Table 1
Antimycin-A-resistant Mutants from Strain 315-1D Grown on Glucose at 15°C, 20°C, or 30°C

Growth temp. (°C)	Mutation rate[a] to antimycin A resistance	Fraction of mutants with Ty insertions at ADH2	Transposition rate[b] to ADH2	Fraction of mutants with Ty insertions at ADH4	Transposition rate[b] to ADH4	Mutants with Ty insertions (%)
15	$1 \pm 1 \times 10^{-7}$	5/12	2×10^{-8}	7/12	5×10^{-8}	100%
20	$2 \pm 1 \times 10^{-7}$	5/18	1×10^{-8}	13/18	6×10^{-8}	100%
30	$7 \pm 7 \times 10^{-9}$	0/8	$<7 \times 10^{-10}$	0/8	$<7 \times 10^{-10}$	6%[c,d]
	$7 \pm 12 \times 10^{-9}$	0/5	$<9 \times 10^{-10}$	0/5	$<9 \times 10^{-10}$	
	$5 \pm 5 \times 10^{-9}$	0/11	$<3 \times 10^{-10}$	1/11	3×10^{-10}	
	$5 \pm 3 \times 10^{-9}$	1/12	2×10^{-10}	0/12	$<2 \times 10^{-10}$	

[a]Mutations per cell per generation ±95% confidence interval.
[b]Transpositions per cell per generation.
[c]One mutant that has a Ty insertion at a location other than ADH2 or ADH4 is not included in the percentage of mutants with Ty insertions.
[d]Each line represents an independent experiment.
(Reprinted, with permission, from Paquin and Williamson 1986.)

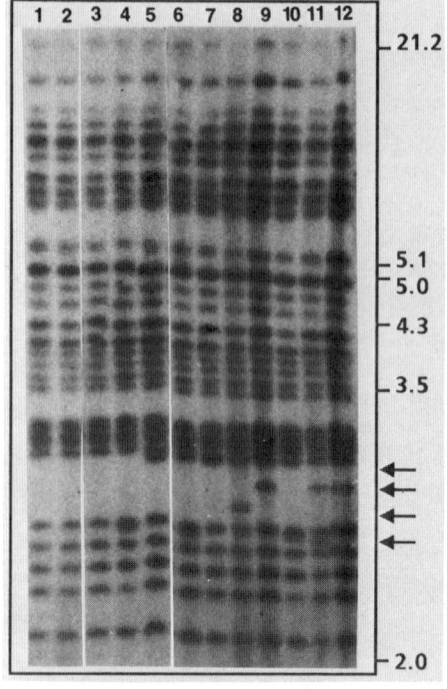

Figure 2
Yeast genomic DNA probed with a Ty fragment. DNA was isolated from various yeast strains, cut with EcoRI, electrophoresed, and transferred to nitrocellulose. The nick-translated fragment of Ty used as a probe is indicated in Fig. 1. Lanes 1, 2, and 3 contain DNA from the isogenic diploid CP2AB and two haploid progeny, CP2A and CP2B. Lanes 4–12 contain DNA from antimycin-A-resistant mutants derived from CP2AB. Lanes 5 and 7 contain DNA from mutants which have a Ty insertion adjacent to ADH2. A new 3.2-kb band can be seen in lane 5 but is obscured by another band in lane 7. Lanes 4, 6, and 8–12 contain DNA from antimycin-A-resistant mutants which do not have a Ty insertion adjacent to ADH2. Lanes 4, 6, and 11 have new bands of 2.7 kb, lane 8 has a new band of 2.9 kb, and lanes 9, 11, and 12 have new bands of 3.1 kb. Lane 11 has two new bands, and lane 10 has no new bands. Numbers on the right are in kilobases. (Reprinted, with permission, from Paquin and Williamson 1986.)

The next set of experiments were designed to determine which isozyme of ADH was being expressed in the antimycin-A-resistant mutants carrying the new Ty band. An antimycin-A-resistant mutant carrying a new Ty band was crossed to a strain carrying null alleles for *ADH1*, *ADH2*, and *ADH3*, and then the resultant diploid sporulated and dissected. Crude extracts of the tetrads and parent strains were run on cellulose acetate gels and stained for ADH activity. Neither ADHII nor ADHIII segregated with the antimycin-A-resistant phenotype. Instead a novel band, staining for ADH activity, appeared in the antimycin-A-resistant strains. Indeed, one segregant that had neither ADHII nor ADHIII activity was antimycin A resistant and had the new ADH activity band. Extracts from the strain carrying only the new ADH band had different substrate preferences than ADHI, ADHII, or ADHIII. This new ADH isozyme was named ADHIV (Paquin and Williamson 1986; Williamson and Paquin 1987).

The region adjacent to the new Ty band was then cloned and sequenced in order to further characterize ADHIV. DNA from a mutant carrying a new Ty band was restricted with *Eco*RI and run on a gel. An agarose slab containing DNA corresponding to the new Ty band was cut out of the gel, electroeluted,

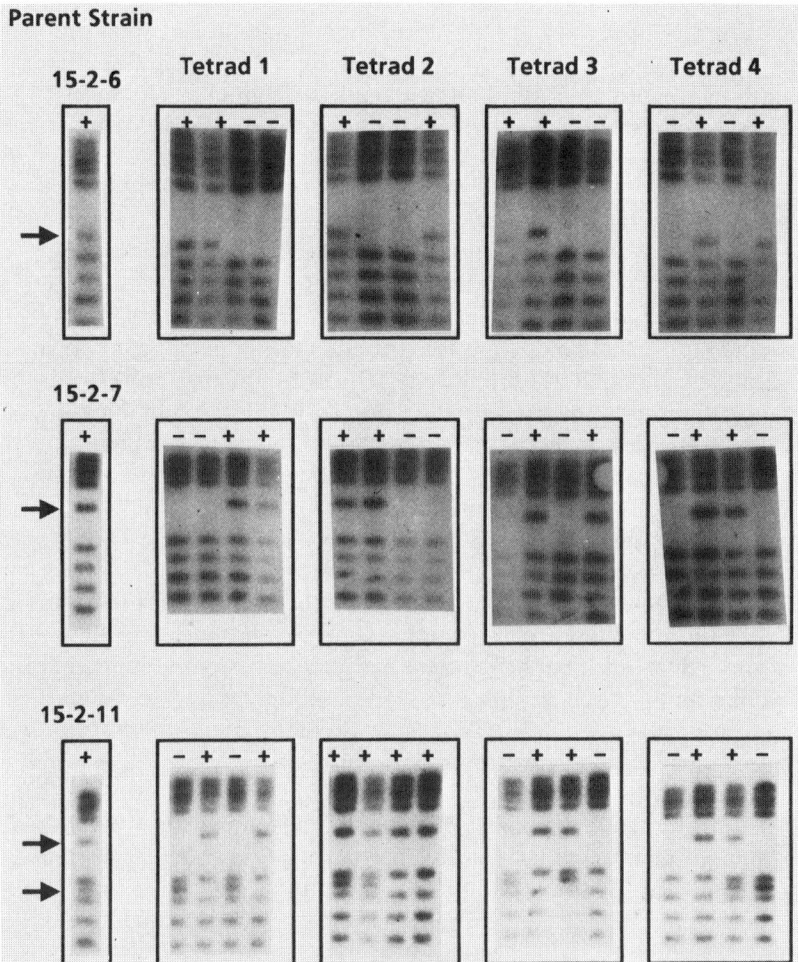

Figure 3

Segregation of new Ty bands with the antimycin-A-resistant phenotype. Southern blots of DNA cut with *Eco*RI from tetrads of antimycin-A-resistant mutants which do not have a Ty adjacent to *ADH2* but do have a new Ty band. The Ty hybridization probe is the same as in Fig. 2. Only the region of the Southern blots containing the new Ty bands (→) is shown. The ability to grow on plates containing antimycin A and glucose is indicated by a + sign and antimycin A sensitivity by a − sign. In every case the new Ty band segregates with the antimycin-A-resistant phenotype. In mutant 15-2-11 the antimycin A resistance segregates with the new 3.1-kb band but not with the new 2.7-kb band. In tetrad 2 of mutant 15-2-11, the new Ty band and the antimycin-A-resistant phenotype are seen in every spore. This could be due to either a mitotic crossover event before sporulation, a gene conversion event, or a false tetrad. (Reprinted, with permission, from Paquin and Williamson 1986.)

ligated into pUC8, and transformed into *Escherichia coli*. Colony hybridizations using the Ty fragment shown in Figure 1 identified a plasmid carrying homology to Ty. The restriction map of the insert suggests that it contains a region homologous to Ty and adjacent unique sequence DNA, as expected (Fig. 1). Southern blot analysis using the *Eco*RI/*Bam*HI fragment of this clone as a probe showed an 8-kb band in the parent antimycin-A-sensitive strains and the expected 2.7- , 2.9- , or 3.1-kb bands seen with the Ty probe in the antimycin-A-resistant strains. This confirmed that the clone was from the location where the new Ty bands were inserted. This unique sequence DNA adjacent to the Ty was then used as a probe to clone the wild-type sequence from a strain that does not carry a Ty in this location (Fig. 1). DNA sequencing identified an open reading frame (ORF), located in the appropriate location and orientation to be turned on by the Ty insertion. Northern blot analysis demonstrated the presence of a mRNA homologous to this region in antimycin-A-resistant mutants carrying a new Ty band that was not present in the parent strains. The ORF codes for a protein which is not homologous to any of the other three ADH isozymes of *S. cerevisae*. It is, however, homologous to an iron requiring ADH isozyme of *Zymomonas mobilis* (Williamson and Paquin 1987). Thus, this DNA contains the structural gene for ADHIV, *ADH4*. Sequencing of the location of the Ty insertions identified no homology to Ty sequences in the wild-type copy of *ADH4*, so the Ty insertions must be due to transposition events rather than homologous recombination.

Transposition rates to *ADH4* were estimated, using the method used to estimate transposition rates to *ADH2*. All of the antimycin-A-resistant mutants which were not due to Ty insertions at *ADH2* were tested by Southern blot analysis for Ty insertions at *ADH4*. Ty transposition to *ADH4* shows the same temperature sensitive change in transposition rate as transposition to *ADH2* (Table 1). Ty transpositions to *ADH2* or *ADH4* account for almost all of the antimycin-A-resistant mutants observed at low temperatures.

DISCUSSION

Ty transposition to two different loci both show a 100-fold increase in transposition rates at 15°C and 20°C. At these low temperatures, almost all the spontaneous mutations are due to Ty transposition events, whereas at 30°C only 2 out of 35 antimycin-A-resistant mutations are due to Ty transposition events. The increase in Ty transposition at low temperatures is not accompanied by an increase in Ty RNA as would be expected if RNA was limiting transposition (Paquin and Williamson 1984). Investigation of the effects of carbon source and mating type competence (two factors that effect Ty RNA levels) on Ty transposition rates to *ADH2* and *ADH4* showed only minor (approximately fivefold) changes in transposition rates. Crude extracts of Ty reverse transcriptase show a temperature sensitive profile (Garfinkel et al. 1985), which suggests that the increase in transposition rates was at least partially due to the temperature-sensitive reverse transcriptase.

Two other types of DNA rearrangements that result in antimycin-A resistance were isolated from the 208 antimycin-A-resistant mutants analyzed. One type of rearrangement was a duplication of *ADH2* (C.E. Paquin and V.M. Williamson, unpubl.). Another rearrangement was an amplification of *ADH4*. The amplified copies of *ADH4* are carried on a 42-kb linear, extrachromosomal palindrome with telomeric ends (Walton et al. 1986). Both of these DNA rearrangements occur at approximately the same frequency as Ty insertions at 30°C but do not increase in frequency at low temperatures. Thus, the only type of DNA rearrangement detected that increases at low temperatures is Ty transposition.

ACKNOWLEDGMENT

The work reported here was done at Atlantic Richfield Company Plant Cell Research Institute (ARCO PCRI, now The PCRI Inc.), 6560 Trinity Court, Dublin, California.

REFERENCES

Boeke, J.D., D.J. Garfinkel, C.A. Styles, and G.R. Fink. 1985. Ty elements transpose through an RNA intermediate. *Cell* **40:** 491.

Cameron, J.R., E.Y. Loh, and R.W. Davis. 1979. Evidence for transposition of dispersed repetitive families in yeast. *Cell* **16:** 739.

Ciriacy, M. 1975a. Genetics of alcohol dehydrogenase in *Saccharomyces cerevisiae*. I. Isolation and genetic analysis of *ADH* mutants. *Mutat. Res.* **29:** 315.

———. 1975b. Genetics of alcohol dehydrogenase in *Saccharomyces cerevisiae*. II. Two loci controlling synthesis of the glucose repressible ADHII. *Mol. Gen. Genet.* **138:** 157.

Clare, J. and P. Farabaugh. 1985. Nucleotide sequence of a yeast Ty element: Evidence for a novel mechanism of gene expression. *Proc. Natl. Acad. Sci.* **82:** 2829.

Errede, B., M. Company, J.D. Ferchak, C.A. Hutchison III, and W.S. Yarnell. 1985. Activation regions in a yeast transposon have homology to mating type control sequences and to mammalian enhancers. *Proc. Natl. Acad. Sci.* **82:** 5423.

Garfinkel, D.J., J. Boeke, and G.R. Fink. 1985. Ty element transposition: Reverse transcriptase and virus-like particles. *Cell* **42:** 507.

Lea, D.E. and C.A. Coulson. 1949. The distribution of the numbers of mutants in bacterial populations. *J. Genet.* **49:** 264.

Paquin, C.E. and V.M. Williamson. 1984. Temperature effects on the rate of Ty transposition. *Science* **226:** 53.

———. 1986. Ty insertions account for most of the spontaneous antimycin A resistance mutations during growth at 15°C of *Saccharomyces cerevisiae* strains lacking ADHI. *Mol. Cell. Biol.* **6:** 70.

Roeder, G.S. and G.R. Fink. 1980. DNA rearrangements associated with a transposable element in yeast. *Cell* **21:** 239.

Russell, D.W., M. Smith, V. Williamson, and E.T. Young. 1983. Nucleotide sequence of the yeast alcohol dehydrogenase II gene. *J. Biol. Chem.* **258:** 2674.

Walton, J., C. Paquin, K. Kaneko, and V. Williamson. 1986. Resistance to antimycin A in yeast by amplification of *ADH4* on a linear, 42 kb palindromic plasmid. *Cell* **46:** 857.

Williamson, V.M. 1983. Transposable elements in yeast. *Int. Rev. Cytol.* **83:** 1.

Williamson, V.M. and C. Paquin. 1987. Homology of *Saccharomyces cerevisiae ADH4* to an iron-activated alcohol dehydrogenase from *Zymomonas mobilis*. *Mol. Gen. Genet.* **209:** 374.

Williamson, V., D. Beier, and E.T. Young. 1983. Use of transformation and meiotic gene conversion to construct a yeast strain containing a deletion in the alcohol dehydrogenase I gene. In *Genetic engineering in eukaryotes* (ed. P.F. Lurkin and A. Kleinhofs), p. 21. Plenum Publishing, New York.

Williamson, V., E.T. Young, and M. Ciriacy. 1981. Transposable elements associated with constitutive expression of yeast alcohol dehydrogenase II. *Cell* **23:** 605.

Williamson, V.M., J. Bennetzen, E.T. Young, K. Nasmyth, and B. Hall. 1980. Isolation of the structural gene for alcohol dehydrogenase by genetic complementation in yeast. *Nature* **283:** 214.

Young, E.T. and D. Pilgrim. 1985. Isolation and DNA sequence of *ADH3*, a nuclear gene encoding the mitochondrial isozyme of alcohol dehydrogenase of *Saccharomyces cerevisiae*. *Mol. Cell. Biol.* **5:** 3024.

Effects of DNA Damage on Transcription and Transposition of Ty Retrotransposons of Yeast

KEVIN McENTEE AND VICTORIA A. BRADSHAW
Department of Biological Chemistry
School of Medicine
University of California-Los Angeles
Los Angeles, California 90024

OVERVIEW

Treatment of *Saccharomyces cerevisiae* with the mutagen/carcinogen 4-nitroquinoline-1-oxide (NQO) results in increased levels of transcripts homologous to a set of DNA damage responsive (Ddr) genes. The structural characterization of one such Ddr sequence (Ddr78A) demonstrates that it is related to the retrotransposon Ty. After NQO exposure for 4–6 hours, transcripts homologous to Ddr78A, as well as to a Ty*1* probe, accumulate in cells.

We have asked whether the increased transcript levels of these transposable elements is accompanied by increased transposition. We have isolated a series of mutants that constitutively express normally glucose-repressible alcohol dehydrogenase activities, *ADH2* or *ADH4*. Previous studies have demonstrated that most of the spontaneous constitutive mutations in these genes are due to Ty transposition. By Southern hybridization analysis of the *ADH2* and *ADH4* regions from more than 100 NQO-induced mutations, we present evidence that DNA damage stimulates Ty transposition to these loci by 2–13-fold.

INTRODUCTION

Ty elements are a family of dispersed repetitive sequences in *S. cerevisiae* that show considerable structural similarities to proviral forms of mammalian retroviruses. Typically, Ty elements contain highly conserved direct repeats of approximately 330–340 bp (δ-elements) which flank a central region of approximately 5.3 kb (ε-region). The major Ty transcript is 5.6 kb and initiates and terminates within δ sequences (Elder et al. 1983). DNA sequence analysis of two Ty elements indicate that there are two long, overlapping reading frames within the coding region of Ty and that the more distal sequences encode a reverse transcriptase (Clare and Farabaugh 1985; Hauber et al. 1985).

Ty elements transpose in yeast and transposition proceeds through an RNA intermediate. Boeke et al. (1985) have demonstrated that increased transcription of a single Ty element can increase transposition of this and other Ty elements. Moreover, in cells showing increased transposition, virus-like particles accumulate which contain both Ty RNA and reverse-transcriptase activity (Garfinkle et al. 1985).

Ty transposition can alter the regulation of adjacent genes by inserting in either orientation upstream from the start site of transcription (Williamson 1983). Ty transposition can lead to constitutive expression of the *ADH2* and *ADH4* genes which encode or control expression of normally glucose-repressible alcohol dehydrogenases (Williamson et al. 1981, 1983). Based on its ability to increase or decrease expression of genes in regions where it transposes, Ty represents a model system for investigating mutagenic effects of retrotransposons in eukaryotic cells.

During the isolation of Ddr genes in yeast, we obtained a set of DNA sequences that hybridized to a 5.6-kb transcript whose levels increased three- to tenfold following exposure to DNA damaging agents such as UV-irradiation or NQO. By Southern hybridization analysis we demonstrated that these elements were repetitive and that some of these Ddr sequences hybridized with a well-characterized Ty element (McClanahan and McEntee 1984). Recently, Rolfe et al. (1986) have shown that the levels of Ty homologous transcripts increase after UV-irradiation of yeast cells.

In this paper we present a characterization of one such Ddr sequence, Ddr78A. Moreover, we present evidence that the levels of Ty transposition to two genetic loci can be stimulated by exposure of yeast to low concentrations of the mutagen/carcinogen, NQO.

RESULTS

Structure of Ddr78A

The Ddr78A clone was isolated from a λ library of yeast genomic DNA. Interestingly, this clone hybridized to a 5.6-kb transcript that accumulated following DNA-damaging treatments of yeast, but the clone did not hybridize or hybridized weakly to the S13-Ty*1* clone (Cameron et al. 1979). DNA sequence analysis of the Ddr78A clone showed that it was 75% homologous to the sequence of the δ region of the Ty*912* element (Roeder and Fink 1980). Within the Ddr78A sequence we find several of the highly conserved regions that are likely to be responsible for expression and transposition of this element (Table 1). Thus, Ddr78A is likely to be expressed in yeast although we do not know whether the adjacent ϵ-region encodes functional Ty products.

Table 1
Conserved Regions of Ddr78A

Sequence	Proposed function
TGTTGGAATA	Unknown
TATATTATCATATACGGTGTT	Unknown
GGAAGCTGAA	Unknown
AACATATAAAA	'TATA' box
CCTCGAGGAGAACTTTT	Transcriptional start site
AACAATG	Transcription termination site (downstream δ)

See Williamson (1983).

Elevated Levels of Ddr78A Homologous Transcripts in NQO-Treated Yeast

We used Northern hybridization analysis to measure the levels of transcripts homologous to Ddr78A. Total RNA as well as oligo(dT) selected mRNA was isolated from yeast cells following exposure to NQO for 60 minutes or for 6 hours. The RNA was separated by electrophoresis in formaldehyde-agarose gels, transferred to nitrocellulose and hybridized as previously described (McClanahan and McEntee 1984). As shown in Figure 1, there is a substantial increase in the level of RNA homologous to Ddr78A after 6 hours of NQO treatment in both total and poly(A)-selected fractions. Although other Ddr genes show increased accumulation after 30–60 minutes, Ddr78A homologous transcripts do not accumulate after 60 minutes of exposure (Fig. 1). In other experiments, we have seen an increase in Ddr78A transcript levels after 4 hours of NQO exposure.

We also hybridized these RNAs with three other probes, a *HIS4* specific probe, pFW45, a ubiquitin (*UBI4*) probe, pUB200, and a probe homologous to histone H2A and H2B transcripts, *TRT1*. As seen in Figure 1 (A and B), there is little if any difference in the levels of the *HIS4* and *UBI4* specific RNAs in the control and drug-treated samples. Note that the level of histone H2A and H2B transcripts are less in the NQO-treated cell RNA samples at both 60 minutes (D) and 6 hours (C). Histone mRNA synthesis is closely coupled to DNA synthesis. Inhibition of DNA synthesis by UV or NQO as well as by hydroxyurea results in a dramatic immediate decrease in the level of H2B transcription (Lycan et al. 1987).

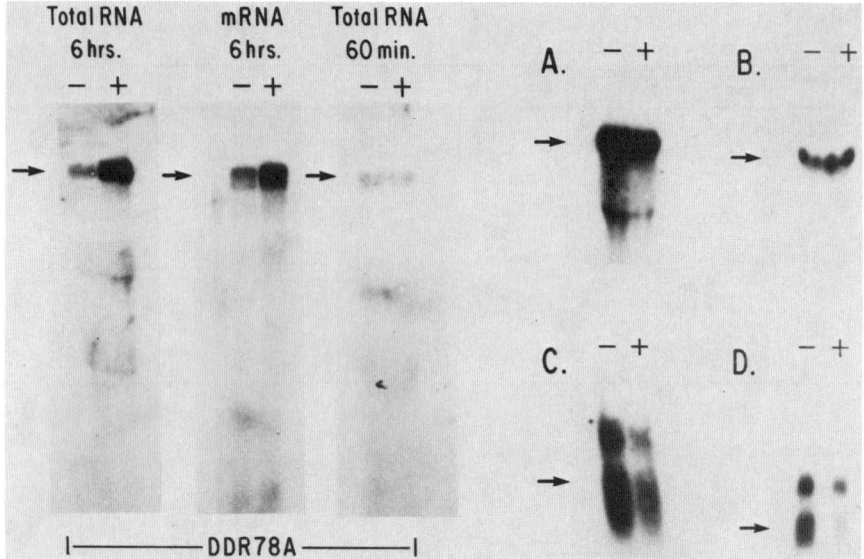

Figure 1
Increased levels of Ddr78A transcript in NQO-treated yeast. Total RNA and mRNA were prepared from yeast strain M12B after 6 hr of growth in the absence (−) or presence (+) of NQO (1.5 µg/ml). Total RNA was also prepared from cells after 60 min. Northern blots were hybridized with labeled Ddr78A probe. Total RNA Northern blots from M12B cells grown in the absence (−) or presence (+) of NQO (1.5 µg/ml) probed with (A) *HIS4* probe, pFW 45; (B) *UBI4* probe, pUB200; (C) and (D) histone H2A and H2B probe *TRT1*. In A through C, RNA was prepared after 6 hr of exposure, whereas in D the RNA was isolated after 60 min of treatment.

Elevated Levels of Ty Homologous Transcripts in NQO-treated Yeast

We examined the levels of Ty homologous RNAs in untreated and NQO-treated cells using the S13-Ty*1* clone to probe Northern filters of total and oligo(dT)-selected RNAs. The results are shown in Figure 2. An increase in the level of Ty homologous transcripts can be detected in mRNA prepared from DNA-damaged cells by oligo(dT) selection. The effect of NQO treatment is less evident when total RNA is examined (Fig. 2). Based on the relative hybridization intensities, we presume that the Ddr78A probe is hybridizing to a small subset of transcripts that are recognized by the Ty*1* probe or alternatively, that the Ddr78A probe is hybridizing to a set of RNA transcripts that are not recognized by the S13-Ty probe. As is the case with the Ddr78A homologous transcripts, the S13-Ty homologous transcripts accumulate after 4–6 hours of NQO treatment.

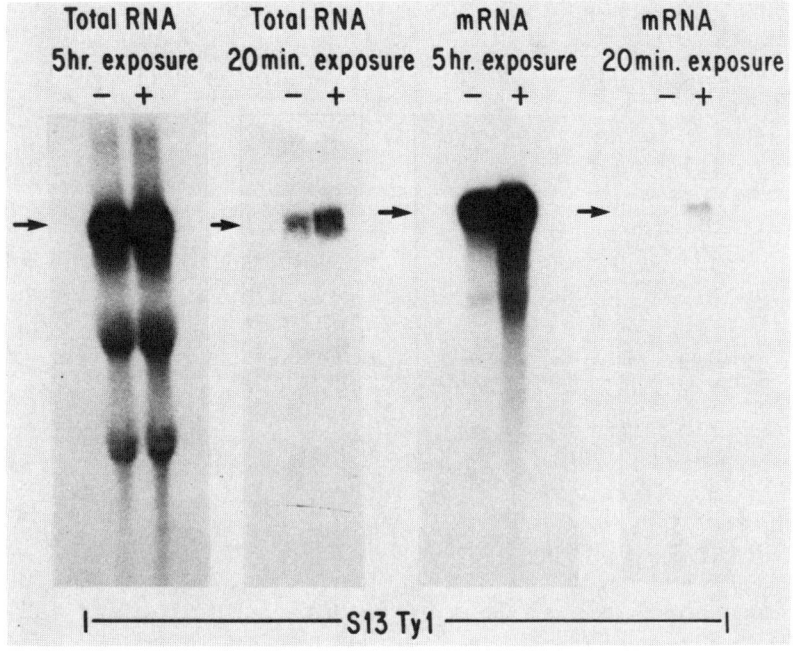

Figure 2
Elevated Ty transcripts in NQO-treated transcripts. Northern blots were prepared using total RNA and oligo-dT-selected RNA from strain M12B following growth in the absence (−) or presence (+) of NQO (1.5 μg/ml, 6 hr). Filters were hybridized with labeled S13-Ty*1* probe and the autoradiograms were exposed for the indicated times.

NQO Treatment Stimulates Ty Transposition to the ADH2 and ADH4 Loci

We have examined the transposition of Ty to two loci which control expression of two distinct glucose repressible alcohol dehydrogenases, *ADH2* and *ADH4*. Neither of these genes is expressed in cells growing on glucose, but mutants can be readily selected that produce increased levels of either *ADH2* or *ADH4*. Based on hybridization analysis and cloning of individual constitutive mutants, Williamson et al. (1981) have demonstrated that most if not all spontaneous mutations result from Ty transposition to the *ADH2* region or into upstream regulatory regions of the *ADH4* gene (Paquin and Williamson 1986).

We chose to examine the effects of NQO treatment on production of *ADH* constitutive mutants. For this analysis we employed strain 315-1D, which lacks the normally constitutive *ADH1* gene, and selected mutants which grew on glucose medium containing antimycin A, a respiration inhibitor. We examined the *ADH2* and *ADH4* chromosomal regions of several antimycin resistant clones by isolating DNA from small cultures, digesting it with *Hin*dIII and subjecting the DNA to Southern hybridization analysis using cloned *ADH2* and *ADH4* gene probes. We have used three criteria to determine whether a chromosomal rearrangement at either *ADH2* or *ADH4* is due to a Ty insertion: (1) the rearrangements alter the mobilities of characteristic upstream restriction fragments (in the case of *ADH2* it is a 3.4-kb *Hin*dIII fragment and in the case of *ADH4* it is a 3.0-kb *Hin*dIII fragment); (2) the size of the rearrangements are consistent with transposition of a 5.9-kb Ty element; (3) the rearranged DNA contains at least one *Xho*I restriction site which is absent from either of the wild-type *ADH2* and *ADH4* upstream regions but which is present and highly conserved in the δ-sequences of Ty elements (Williamson 1983).

As shown in Table 2, NQO exposure results in a dose-dependent increase in the frequency of antimycin-resistant mutants. At low concentrations of drug (0.1 and 0.2 μg/ml) there is a significant increase in the number of resistant cells with little cell killing. At high doses (0.5 μg/ml) the increase in the proportion of antimycin-resistant mutants is accompanied by substantial cell killing. The frequency of antimycin-resistant cells increases 17–120-fold over the spontaneous levels under the conditions of these experiments.

We have determined the fraction of antimycin-resistant mutants that result from Ty transposition to either *ADH2* or *ADH4* for both spontaneous and NQO-induced mutants. The results of these measurements are summarized in Table 2. These data confirm the results of Paquin and Williamson (1986) who demonstrated that most spontaneous, antimycin-resistant mutations result from Ty transposition to either *ADH2* or *ADH4*. Our values were obtained at 30°C, a temperature at which spontaneous transposition is reduced (Paquin and Williamson 1984).

Table 2
Effects of NQO on Ty Transposition to *ADH2* and *ADH4*

NQO (μg/ml)	Survival[a] (%)	Frequency of antimycin A-resistant mutants	Frequency of transposition[b]	
			ADH2	ADH4
0	100	1.0×10^{-5}	2.6×10^{-6}	4.1×10^{-6}
0.1	100	1.74×10^{-4}	1.0×10^{-5} (3.8×)	$<1 \times 10^{-5}$ (<2.4×)
0.2	65	4.1×10^{-4}	1.3×10^{-5} (5×)	1.3×10^{-5} (3×)
0.5	1.8	9.8×10^{-4}	3.1×10^{-5} (12×)	3.1×10^{-5} (7.6×)
0.5	3.5	1.2×10^{-3}	3.3×10^{-5} (12.6×)	$<3.3 \times 10^{-5}$ (<8×)

[a]Cells were exposed to the indicated concentration of NQO for 2 hr at 30°C in YPG medium. Survival was measured on YPD plates and values are expressed relative to the untreated (no NQO) control.
[b]Values are expressed as Ty transposition events per viable cell. The numbers in parenthesis are the ratios of the frequency of transposition after NQO exposure to the spontaneous frequency of transposition.

The data of Table 2 demonstrate that the frequency of Adh-constitutive mutations due to Ty transposition to either the *ADH2* or *ADH4* region increases following NQO exposure. For example, the spontaneous frequency of *ADH2*-constitutive mutations, resulting from Ty transposition was calculated to be 2.6×10^{-6} in untreated cells. After exposure to 0.1 µg/ml NQO, the frequency of Ty transposition to *ADH2* increased to 1.0×10^{-5}, or approximately a 3.8-fold higher level. Following treatment with 0.5 µg/ml NQO, the observed frequency of Ty transposition events to the *ADH2* locus was calculated to be 3.3×10^{-5} or an increase of 12-fold over spontaneous levels. These results indicate that Ty transposes to the *ADH2* and *ADH4* loci at increased frequencies when cells are treated with the mutagen/carcinogen, NQO. Similar effects on Ty transposition have been observed following UV-irradiation (V. Bradshaw, unpubl.).

DISCUSSION

Our results demonstrate that DNA damage produced by the chemical mutagen/carcinogen, NQO, can result in increased Ty transcript levels in yeast and can stimulate transposition of Ty elements to the *ADH2* and *ADH4* chromosomal regions. Ty transcript accumulation was measured after long exposure (6 hours) to high concentrations of NQO (1.5 µg/ml), whereas the transposition results were obtained with lower doses of NQO (0.1–0.5 µg/ml) following 2 hours of exposure. The differences in conditions were necessitated by the different sensitivities toward killing shown by the two strains: strain M12B is 3–5 times more resistant to killing than is strain 315-1D. Furthermore, we chose to examine chemical mutagenesis under conditions where cell survival was high.

Our data suggest that only a subset of Ty elements respond to DNA damage. We cloned a Ty-like sequence, Ddr78A, which is approximately 75% homologous to the δ region of Ty*912*. Northern data demonstrate significant accumulation of transcripts homologous to this DNA. Furthermore, we have detected increased levels of transcripts homologous to a well-characterized Ty*1* element, S13, although the relative increase is less. It is likely that some Ty elements are not responsive to DNA damage and are expressed at high levels.

The kinetics of Ddr78A transcript accumulation are considerably slower than those of other Ddr genes. After a 60 minute exposure to NQO, we observe maximal induction of other Ddr transcripts, whereas we observe an increase in the level of Ddr78A homologous transcripts (or Ty transcripts) after 4–6 hours of NQO exposure. We do not understand these kinetic differences, but we suggest there is a requirement for induction of other genes before Ty transcripts can accumulate.

Transposition of Ty to the *ADH2* and *ADH4* chromosomal regions is stimulated by exposure of cells to DNA-damaging agents. We have detected a 2–12-fold increase in the frequency of Ty transposition to these loci after exposure to either NQO or UV-irradiation (V. Bradshaw, unpubl.). In contrast to spontaneous antimycin-resistant mutants, the vast majority of NQO- or UV-induced antimycin-resistant mutants show no significant sequence alterations in the *ADH2* and *ADH4* upstream regions and are probably due to localized (i.e., base substitution) changes in *cis*-acting sequences or *trans*-acting genes. Further characterization of these mutants is in progress.

Although Ty transpositions represent a small fraction of DNA damage-induced mutations at these loci, nevertheless, these results have several important implications. First, they demonstrate that mutagens, both chemical and physical, may act by inducing either localized mutations (base substitutions, frameshifts) or larger genomic rearrangements through transposition. Second, these results are consistent with the concept of "genomic challenge" or "inducible evolution" in which environmental stress (in this case DNA damage) stimulates genomic rearrangements (Echols 1981; McClintock 1984). Finally, these results may provide insight into the mechanisms of chemical carcinogenesis, where it has been argued that chromosomal translocations and transpositions are important causal factors of malignancy (Cairns 1981; Klein 1981; Bishop 1987).

ACKNOWLEDGMENT

This work was supported by a research grant from the American Cancer Society to K.M.

REFERENCES

Bishop, J.M. 1987. The molecular genetics of cancer. *Science* **235**: 305.
Boeke, J.D., D.J. Garfinkle, C.A. Styles, and G.R. Fink. 1985. Ty elements transpose through an RNA intermediate. *Cell* **40**: 491.
Cairns, J. 1981. The origin of human cancers. *Nature* **289**: 353.
Cameron, J.R., E.Y. Loh, and R.W. Davis. 1979. Evidence for transposition of dispersed repetitive DNA families in yeast. *Cell* **16**: 739.
Clare, J. and P. Farabaugh. 1985. Nucleotide sequence of a yeast Ty element: Evidence for a novel mechanism of gene expression. *Proc. Natl. Acad. Sci.* **82**: 2829.
Echols, H. 1981. SOS functions, cancer and inducible evolution. *Cell* **25**: 1.
Elder, R.T., E.Y. Loh, and R.W. Davis. 1983. RNA from the yeast transposable element Ty1 has both ends in the direct repeats, a structure similar to retrovirus RNA. *Proc. Natl. Acad. Sci.* **80**: 2432.
Garfinkle, D.J., J.D. Boeke, and G.R. Fink. 1985. Ty element transposition: Reverse transcriptase and virus-like particles. *Cell* **42**: 507.

Hauber, J., P. Nelböck-Hochstetter, and H. Feldmann. 1985. Nucleotide sequence and characteristics of a Ty element from yeast. *Nucleic Acids Res.* **13:** 2745.

Klein, G. 1981. The role of gene dosage and genetic transpositions in carcinogenesis. *Nature* **294:** 313.

Lycan, D.E., M.A. Osley, and L.M. Hereford. 1987. Role of transcriptional and posttranscriptional regulation in expression of histone genes in *Saccharomyces cerevisiae*. *Mol. Cell. Biol.* **7:** 614.

McClanahan, T.A. and K. McEntee. 1984. Specific transcripts are elevated in *Saccharomyces cerevisiae* in response to DNA damage. *Mol. Cell. Biol.* **4:** 2356.

McClintock, B. 1984. The significance of responses of the genome to challenge. *Science* **226:** 792.

Paquin, C.E. and V.M. Williamson. 1984. Temperature effects on the rate of Ty transposition. *Science* **226:** 53.

―――. 1986. Ty insertions at two loci account for most of the spontaneous antimycin A resistant mutations during growth at 15°C of *Saccharomyces cerevisiae* strains lacking ADH1. *Mol. Cell. Biol.* **6:** 70.

Roeder, G.S. and G.R. Fink. 1980. DNA rearrangements associated with a transposable element in yeast. *Cell* **21:** 239.

Rolfe, M., A. Spanos, and G. Banks. 1986. Induction of yeast Ty element transcription by ultraviolet light. *Nature* **319:** 339.

Williamson, V.M. 1983. Transposable elements in yeast. *Int. Rev. Cytol.* **83:** 1.

Williamson, V.M., E.T. Young, and M. Ciriacy. 1981. Transposable elements associated with constitutive expression of yeast alcohol dehydrogenase II. *Cell* **23:** 605.

Williamson, V.M., D. Cox, E.T. Young, D.W. Russell, and M. Smith. 1983. Characterization of transposable element-associated mutations that alter yeast alcohol dehydrogenase II expression. *Mol. Cell. Biol.* **3:** 20.

Inducible Cellular Responses to DNA Damage

MICHAEL E. LAMBERT* and JAMES I. GARRELS
Cold Spring Harbor Laboratory
Cold Spring Harbor, New York 11724

OVERVIEW

Despite considerable efforts over the years to study specific cellular responses to DNA damage in mammalian cells, including increased expression of endogenous retroviruses, induction of DNA repair, growth arrest, and cell killing and mutagenesis, there are little systematic data on the overall patterns of change in cellular gene expression induced after exposure to DNA-damaging agents. In addition, very little is known about the functional significance of induced changes in gene expression or of the regulation of these inducible phenomena. To identify specific protein markers for inducible responses to DNA damage in mammalian cells, we have, therefore, utilized the QUEST system of high-resolution two-dimensional protein gel electrophoresis (HR2D) and quantitative analysis to study patterns of change in cellular protein synthesis induced in both rat and human cells in response to several classes of DNA damage. Our findings indicate that several proteins of known cellular function are rapidly induced during the acute phase of cellular response to DNA damage, including proliferating cell nuclear antigen (PCNA) or cyclin and human major histocompatibility class I (MHC-I) proteins.

INTRODUCTION

Although damage to the cellular DNA of certain prokaryotes transiently induces a highly regulated pattern of change in gene expression, often termed the "SOS" response (Witkin 1976), there is little direct evidence for the existence of analogous responses in eukaryotic cells. Based on the *recA*-dependent regulation of *lexA* repressible, DNA-damage-inducible genes (Walker 1984), several a priori criteria can be established, however, in defining such an analogous response in mammalian cells. First, this system would be dependent on the extent and persistence of damage in the genome. Second, it would display essential features of negative regulation, and third, it would be induced specifically by DNA damage and not merely by inhibition of growth.

*Present address: Department of Immunology, Research Institute of Scripps Clinic, La Jolla, California 92037

One inducible cellular response to DNA damage in mammalian cells that fits these criteria is the *trans*-activated replication of certain latent papovaviruses observed in semipermissive systems (Lambert et al. 1983, 1986a,b; Nomura and Oishi 1984; van der Lubbe et al. 1986; Ronai et al. 1987). Diverse classes of DNA-damaging agents have been found to enhance a large T antigen process of asynchronous viral DNA replication for both polyoma (Lambert and Weinstein 1987) and simian virus 40 (SV40) (Lavi 1981). This effect (1) is directly dependent on dose (Lambert and Weinstein 1987), (2) is not induced by growth arrest per se (Lambert et al. 1986b), and (3) is also inducible by treatment with low doses of either cycloheximide or actinomycin D (Ronai et al. 1987).

Other changes associated with DNA damage in eukaryotes include the increased transcription of endogenous retroviral elements in yeast (Rolfe et al. 1986; McEntee and Bradshaw, this volume), *Drosophila* (McDonald et al. 1987), and mammalian cells (Tennant and Rascati 1980), as well as the increased expression of several proteins that are themselves also induced by metabolic inhibition of growth or in nondividing cells (Rahmsdorf et al. 1982; Herrlich et al. 1986). DNA damage and heat shock can also induce common transcripts in both *Drosophila* (Vivino et al. 1986) and yeast (McClanahan and McEntee 1986), and specific types of DNA damage, such as ultraviolet light (UV), transiently induce the increased expression of several genes, including tissue plasminogen activator (Miskin and Reich 1980), as well as p53 (Maltzman and Czyzyk 1984), c-*fos*, and c-*myc* (Ronai et al. 1988). In addition to these specific effects, as well as those associated with adaptive responses to alkylation damage (Samson and Schwartz 1980), both chemical carcinogens and UV light induce increased expression at several cellular loci by altering methylation of cellular DNA (Lieberman et al. 1983; Wilson and Jones 1983).

The functional significance of most inducible cellular responses to DNA in mammalian cells is not known (Lambert et al. 1986a). There is evidence for an SOS-like processing system of targeted mutagenesis (Cornelis et al. 1982; Sarasin et al. 1982), although the regulation of its induction is only poorly understood. Studies of an UV extracellular factor secreted by damaged human fibroblasts (Schorpp et al. 1984) indicate that this factor can "communicate" the UV stress response to untreated cells. Overall, however, there is no set of proteins of known cellular function that serves as unique markers for a common inducible response to diverse types of DNA damage in mammalian cells.

As an approach to the systematic study of inducible cellular responses to DNA damage in mammalian cells, we have, therefore, used the QUEST system of HR2D, quantitative analysis, and data base development (Garrels 1983; Garrels et al. 1984) to compare the responses elicited by several different classes of DNA-damaging agents, in both human fibroblasts and epithelial cells, as well as in a well-characterized rodent cell line designated REF52 (McClure et al. 1982; Franza and Garrels 1984).

Induction of PCNA Expression in REF52 Cells following DNA Damage

We have used HR2D analysis to identify sets of cellular proteins induced individually and in common to several different DNA-damaging regimens in the rodent cell line REF52. Exponentially growing cultures of REF52 cells were treated with single low doses of either the metabolically activated form of benzo[a]pyrene (BP), benzo[a]pyrene-*trans*-7,8-dihydrodiol-9,10 epoxide (BPDE), or N-2 acetylaminofluorene (2-AAF), N-acetoxy-2 acetylaminofluorene (NAAF), or N-methyl-N''-nitro-N-nitrosoguanidine (MNNG). The principal DNA lesions induced by each agent are distinct, and repair of these lesions is mediated by largely separate DNA repair pathways. The chemical carcinogens BPDE and NAAF, for example, both form bulky DNA adduct lesions (at the N-2 and C-8 positions of guanine, respectively), whereas MNNG induces primary O^6-methyl alkylation damage. In this experimental series, treated cells were pulse-labeled with high-specific activity [^{35}S]methionine at intervals up to 24 hours after exposure, and the radiolabeled proteins were resolved by HR2D. Sets of autoradiographic images from individual experiments were then quantitated and the spot patterns were matched. The entire series was linked to form a data base in which the behavior of any single or set of polypeptides could be studied.

In one exercise, we used the QUEST system to select a set of protein spots that are induced early in common by MNNG, BPDE, or NAAF. The set was derived by establishing the set of proteins induced by twofold at 4 hours after MNNG treatment and then restricting this set to those that were also induced within the same time frame, to an equal degree, by BPDE and NAAF. The set was found to contain several previously identified cellular proteins. Included is PCNA, or cyclin, a nuclear, cell-cycle-specific phosphoprotein (36 kD, I.E.F. pH ~5.0) (Mathews et al. 1984) which has been shown to be serum inducible (MacDonald-Bravo and Bravo 1985) and Carcinogen-Inducible Protein 1 (CIN 1) (30 kD, I.E.F. pH ~5.1), which is the major BPDE-inducible spot seen in REF52 cells (Lambert et al. 1986a).

The inducibility of PCNA and a highly restricted number of coregulated polypeptides in response to specific classes of DNA-damaging agents suggests that these proteins may serve as a specific marker for a certain class of DNA damage. In this regard, MNNG has previously been shown to induce aphidicolin-sensitive unscheduled DNA repair synthesis (UDS) (Miller and Chinault 1982), suggesting at least partial dependence on either DNA polymerase-α or -δ. By contrast, bleomycin-induced UDS is highly aphidicolin-insensitive (Miller and Chinault 1982) and may be mediated instead by DNA polymerase-β. In the current study, we have in fact only observed PCNA induction by agents that induce aphidicolin-sensitive UDS, including MNNG, BPDE, and NAAF. By contrast, bleomycin and X ray, which induce aphidicolin-insensitive UDS, failed to induce PCNA. This correlation suggests the possibility that the induction of PCNA reflects involvement in long patch repair synthesis, a function that would be consistent with the postulated role in enhancing the processivity of DNA polymerase-δ during replicative elongation (Prelich et al. 1987). In order to test this correlation in the REF52 system, we have, therefore, also measured both PCNA synthesis and UDS immediately after MNNG-induced damage of quiescent REF52 cells. The results of such experiments have demonstrated a temporal correlation between [^{3}H]thymidine uptake in the presence of hydroxyurea (repair synthesis) and the rate of induced PCNA synthesis. Induction of PCNA synthesis by agents that induce aphidicolin-sensitive repair, such as MNNG, BPDE, and NAAF, is consistent with its involvement in long patch repair and with its postulated role as an auxiliary protein to DNA polymerase-δ. Recently, Linn et al. (S. Linn, pers. comm.) have also demonstrated that DNA polymerase-δ is integrally involved in UV-induced UDS in HeLa cells, and UV damage of human amnion cells has also been shown to rapidly induce a pattern of nuclear fluorescent PCNA staining similar to that observed at the onset of serum-induced DNA synthesis (Celis and Madsen 1986).

Induction of Human MHC-I Expression in Human Cells following DNA Damage

We have also used the QUEST system to study quantitative patterns of change in protein synthesis induced after exposure of human fibroblasts and keratinocytes to several classes of DNA damage. We compared the effects of ultraviolet light C (UVC) and BPDE, in the normal fetal lung fibroblast strain LI58, as well as three DNA-repair deficient strains, GM5509 (xeroderma pigmentosum group A) (Kraemer et al. 1975), GM1389 (xeroderma pigmentosum variant) (Cleaver et al. 1984), and GM1915 (a familial hypercholesterolemia hyperlipoproteinemia strain deficient for initiation of UVC- and BPDE-induced UDS (Joe et al. 1985). In addition, we studied the effects elicited in the SV40 DNA-transformed human keratinocyte line SVK14 (Taylor-Papadimitriou et al. 1982) by UVC, BPDE (MNNG), and 7,12 dimethyl-benz[a]anthracene. To test whether DNA-damage-inducible proteins are under negative regulation, we also compared the inducible patterns of change elicited by DNA damage with the changes induced by single low doses of the protein synthesis inhibitor, cycloheximide. Cell cultures were treated with single doses of each drug or modality and pulse-labeled with [^{35}S]methionine at intervals up to 24 hours after exposure, and the radiolabeled proteins were resolved by HR2D. Sets of autoradiographic images from individual experiments were then quantitated and the protein patterns were matched using the QUEST system.

Initially, we examined the effects of UVC and cycloheximide on MHC-I expression after exposure of quiescent, confluent cultures of LI58 cells. Shown in Figure 1 are the regions containing MHC-I from the HR2D patterns

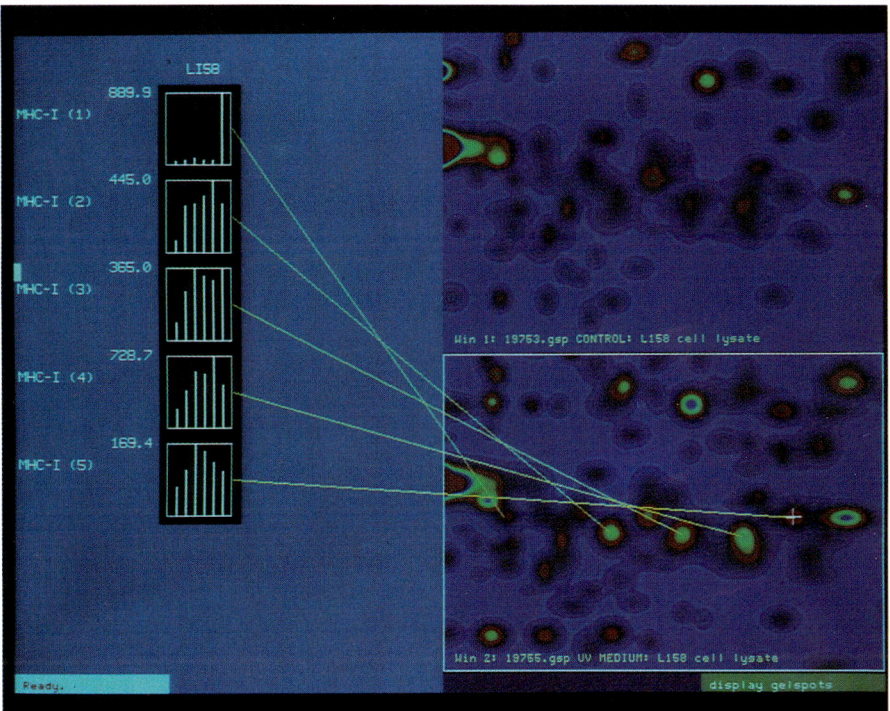

Figure 1
Induction of human MHC-I expression by DNA damage and cycloheximide in quiescent cultures of LI58 cells. For each MHC-I protein listed at the left, quantitative data are plotted on the right. Each bar represents a gel and the gels compared are listed below. The maximum ppm (counts loaded per gel) detected for a polypeptide for the entire experiment is displayed on the left. The height of each bar is scaled within each individual graph to the maximum detected value (shown as full scale). The synthetic image shown on the right is of a broad range ampholyte (pH 3–10), first dimension, 10% acrylamide slab gel, second-dimension gel. Basic proteins are to the right, acidic to the left. Order of samples (left to right): (bar *1*) Control cells; (bar *2*) UVC-treated cells (1 J/m^2); (bar *3*) UVC-treated cells (2 J/m^2); (bar *4*) UVC-treated cells (5 J/m^2); (bar *5*) UVC-treated cells (10 J/m^2); (bar *6*) cycloheximide-treated cells (1 µg/ml). Confluent cultures of L158 cells were treated as indicated and labeled at 4–8 hours after treatment with [^{35}S]methionine.

of total cell lysates prepared from either control or UVC-treated cells. As seen, UVC and cycloheximide both induced the synthesis of MHC-I within 4 hours. Induction of MHC-I in resting cultures of nondividing LI58 cells suggests that the increase in synthetic rate is not a secondary consequence of DNA damage or cycloheximide-induced arrest of DNA synthesis but is, instead, a direct consequence of treatment with either regimen. The identity of the MHC-I products was determined by direct immunoprecipitation of MHC-I products from cycloheximide-induced LI58 cells, using the anti-framework MHC-I rabbit sera K-455 (M.E. Lambert et al., in prep.).

To test further whether MHC-I induction in human fibroblasts by DNA damage is dependent on specific DNA lesions or on DNA repair, we examined the effects of both UVC and BPDE in exponentially growing cultures of four human fibroblast strains, including LI58 and the repair defective strains, GM5509, GM1389, and GM1915 (M.E. Lambert et al., in prep.). Both UVC and BPDE induced MHC-I expression in LI58 cells within 4 hours, indicating that the triggering signal for induction is not restricted to DNA lesions induced by UVC damage. UVC treatment of exponentially growing LI58 cells induced higher levels of MHC-I induction as compared to UVC treatment of quiescent LI58 cells. The UDS-defective lines, GM5509 and GM1915, and the UDS-competent, postreplication repair-defective line, GM1389, exhibited highly altered patterns of MHC-I induction. In particular, higher average ratios of MHC-I induction in the repair-defective lines suggest a positive correlation between the persistence of unrepaired DNA damage in the genome and the extent of MHC-I induction.

Finally, we examined patterns of MHC-I expression induced in the SV40 DNA-transformed human keratinocyte line, SVK14, after exposure to several different DNA-damaging regimens (M.E. Lambert et al., in prep.). Although the overall number of K-455 reactive MHC-I products detected in total cell lysate patterns of this transformed cell type was lower than for any of the primary fibroblast strains tested, all DNA-damage regimens tested, as well as cycloheximide, induced MHC-I. Induction of MHC-I was also observed after treatment with ultraviolet light A (UVA, 345–440 nm), although the dose used was greater than 1000 times that of UVC. The metabolic inhibitor hydroxyurea (5 mM) alone did not induce MHC-I overall, again indicating that arrest of DNA synthesis is insufficient for MHC-I induction. Time course studies demonstrated a persistence of these effects to 24 hours after exposure.

These data demonstrate that MHC-I induction is part of the cellular stress response elicited in human fibroblasts and keratinocytes by several classes of DNA-damaging agents. The ability of cycloheximide to induce MHC-I suggests that this induction may serve as a probe for the action of a stress response pathway in mammalian cells under negative regulatory control.

Recent studies indicate that the mouse major histocompatibility genes, *H-2K* and *D* mRNA, are transcriptionally induced by protein synthesis inhibitors, including cycloheximide (R. Zeff and S. Nathenson, pers. comm.), and that RNA synthesis inhibitors can synergize with γ-interferon in the transcriptional induction of MHC-I in human K562 leukemia cells (Chen et al. 1987). These findings are consistent with the hypothesis that transcription of MHC-I genes may be regulated, in part, by labile transcriptional repressors. The precise molecular mechanisms, however, underlying DNA damage and cycloheximide induction of MHC-I, the specific MHC-I alleles induced, and the possible effects on posttranscriptional modifications have not been determined. Cycloheximide could act through differential inhibition of the synthesis of specific repressors or through increased stabilization of specific transcripts, although this may not be equivalent in mode of action to that of DNA damage.

The MHC-I complex itself represents a set of genes that serves an important function in cellular immunity, and evidence for DNA-damage inducibility of these proteins raises several new possibilities. Two sets of extensively polymorphic genes encoding cell surface molecules are harbored in the MHC (for review, see Steinmetz and Hood 1983). The class I genes encode polypeptide chains that are glycosylated, membrane-integrated, and have apparent molecular weights of approximately 45,000. Three loci, called A, B, and C, encode the polymorphic molecules that are ubiquitously expressed and which present foreign molecules to cytotoxic T lymphocytes reviewed in Möller 1985a,b). The murine MHC, which has been examined in greater detail than its human counterpart, contains, in addition to the polymorphic class I genes, a set of some 30 nonpolymorphic class I genes (see Nathenson et al. 1986). The latter do not seem to be engaged in presenting foreign molecules to the T lymphocytes, their constitutive expression seems to be restricted to the thymus, and their function is unknown. As mentioned above, the majority of the MHC-I antigens are nonpolymorphic, display a very restricted, constitutive tissue expression, and are not normally recognized by the α/β-T-cell receptor-bearing cytotoxic lymphocytes (see Nathenson et al. 1986). In view of their great structural similarity with the polymorphic class I MHC antigens, however, it seems reasonable to suggest that the molecular mode of operation of nonpolymorphic class I antigens will be very similar to that of the polymorphic ones. The fact that the expression of nonpolymorphic class I antigens can be induced by a variety of DNA-damaging agents (M.E. Lambert et al., in prep.) raises the possibility that these molecules may serve as markers for damaged cells such that the immune system may intervene. It is conceivable that double-negative, γ/δ-T-cell receptor-bearing cytotoxic lymphocytes may be effector cells in such a process (Brenner et al. 1987).

Overall, therefore, the expression of human MHC-I genes after exposure to cellular DNA damage and to protein synthesis inhibitors provides an excellent opportunity to study the molecular and cellular mechanisms underlying the inducibility of a putative SOS-like function in mammalian cells.

CONCLUSION

Overall, our studies in REF52 cells and in human fibroblasts and epithelial cells demonstrate that DNA-damage-induced stress responses in mammalian cells can be defined in a highly precise and quantitative manner in terms of inducible changes in patterns of protein synthesis. The overall complexity of the changes observed presumably reflects the pleiotropic effects of DNA damage on induction of endogenous retroviral expression, as well as of distinct DNA repair pathways, and of the highly regulated process of induced growth arrest.

REFERENCES

Brenner, M.B., J. McLean, H. Scheft, J. Riberdy, S.-L. Ang, J.G. Seidman, P. Devlin, and M.S. Krangel. 1987. Two forms of the T-cell receptor γ protein found on peripheral blood cytotoxic T lymphocytes. *Nature* **325:** 689.

Celis, J.E. and P. Madsen. 1986. Increased nuclear cyclin/PCNA antigen staining of non S-phase transformed human amnion cells engaged in nucleotide excision DNA repair. *FEBS Lett.* **209(2):** 277.

Chen, E., R.W. Carr, and G.D. Ginder. 1987. Negative and positive regulation of human-leukocyte antigen class I gene transcription in K562 leukemia cells. *Mol. Cell. Biol.* **7:** 4572.

Cleaver, J., W.C. Charles, and S.H. Hong. 1984. Efficiency of repair of pyrimidine dimers and psoralen monoadducts in normal and xeroderma pigmentosum human cells. *Photochem. Photobiol.* **40:** 621.

Cornelis, J.J., Z.Z. Su, and J. Rommelaire. 1982. Direct and indirect effects of ultraviolet light on the mutagenesis of parvovirus H-1 in human cells. *EMBO J.* **1:** 693.

Franza, B.R. and J.I. Garrels. 1984. Transformation-sensitive proteins of REF-52 cells detected by computer-analyzed two-dimensional gel electrophoresis. *Cancer Cells* **1:** 137.

Garrels, J.I. 1983. Quantitative two-dimensional gel electrophoresis of proteins. *Methods Enzymol.* **100:** 411.

Garrels, J.I., J.T. Ferrar, and C.B. Burwell IV. 1984. The QUEST system for computer-analyzed two-dimensional gel electrophoresis of proteins. In *Two-dimensional gel electrophoresis of proteins: Methods and applications* (ed. J.E. Celis and R. Bravo), p. 37. Academic Press, New York.

Herrlich, P., P. Angel, H.J. Rahmsdorf, U. Mallick, A. Poting, L. Hieber, C. Lucke-Huhle, and M. Schorpp. 1986. The mammalian genetic stress response. *Adv. Enzyme Regul.* **25:** 485.

Joe, C.O., J.O. Norman, T.R. Irvin, and D.L. Busbee. 1985. DNA polymerase activity in a repair-deficient human cell line. *Biochem. Biophys. Res. Commun.* **128:** 754.

Kraemer, K., H.G. Coon, R.A. Petinga, S.F. Barrett, A.E. Rahe, and J.H. Robbins. 1975. Genetic heterogeneity in xeroderma pigmentosum: Complementation groups and their relationship to DNA repair rates. *Proc. Natl. Acad. Sci.* **72:** 59.

Lambert, M.E. and I.B. Weinstein. 1987. Nitropyrenes are inducers of polyoma viral-DNA synthesis. *Mutat. Res.* **183:** 203.

Lambert, M.E., J.I. Garrels, J. McDonald, and I.B. Weinstein. 1986a. Inducible cellular responses to DNA damage in mammalian cells. In *Antimutagenesis and anticarcinogenesis mechanisms* (ed. D.M. Shankel et al.), p. 291. Plenum Press, New York.

Lambert, M.E., S. Gattoni-Celli, P. Kirschmeier, and I.B. Weinstein. 1983. Benzo[a]pyrene induction of extrachromosomal viral DNA synthesis in rat cells transformed by polyoma virus. *Carcinogenesis* **4:** 587.

Lambert, M.E., S. Pellegrini, S. Gattoni-Celli, and I.B. Weinstein. 1986b. Carcinogen induced asynchronous replication of polyoma DNA is mediated by a *trans*-acting factor. *Carcinogenesis* **7:** 1011.

Lavi, S. 1981. Carcinogen mediated amplification of viral DNA sequences in simian virus 40-transformed Chinese hamster ovary cells. *Proc. Natl. Acad. Sci.* **78:** 6144.

Lieberman, M.W., L.R. Beach, and R. D. Palmiter. 1983. Ultraviolet radiation-induced metallothionein-I gene activity is associated with extensive DNA methylation. *Cell* **35:** 207.

MacDonald-Bravo, H. and R. Bravo. 1985. Induction of the nuclear protein cyclin in serum-stimulated quiescent 3T3 cells is independent of DNA synthesis. *Exp. Cell Res.* **156:** 455.

Maltzman, W. and L. Czyzyk. 1984. UV irradiation stimulates levels of p53 cellular tumor antigen in nontransformed mouse cells. *Mol. Cell. Biol.* **4:** 1689.

Mathews, M.B., R.M. Bernstein, B.R. Franza, and J.I. Garrels. 1984. Identity of the proliferating cell nuclear antigen and cyclin. *Nature* **309:** 374.

McClanahan, T. and K. McEntee. 1986. DNA damage and heat shock dually regulate genes in *Saccharomyces cerevisiae*. *Mol. Cell. Biol.* **6:** 90.

McClure, D.B., M.J. Hightower, and W.C. Topp. 1982. Effect of SV40 transformation on the growth-factor requirements of the rat embryo cell line REF52 in serum-free medium. *Cold Spring Harbor Conf. Cell Proliferation* **9:** 345.

McDonald, J.F., D.J. Strand, M.E. Lambert, and I.B. Weinstein. 1987. The responsive genome: Evidence and evolutionary implications. In *Development as an evolutionary process* (ed. R. Rauff and E. Rauff), p. 239. A.R. Liss, New York.

Miller, M.R. and D.N. Chinault. 1982. Evidence that DNA polymerases α and β participate differentially in DNA repair synthesis induced by different agents. *J. Biol. Chem.* **257:** 46.

Miskin, R. and E. Reich. 1980. Plasminogen activator: Induction of synthesis by DNA damage. *Cell* **19:** 217.

Möller, G., ed. 1985a. *Molecular genetics of class I and II MHC antigens 1. Immunol. Rev.* **84**.

———. 1985b. *Molecular genetics of class I and II MHC antigens 2. Immunol. Rev.* **85**.

Nathenson, S.G., J. Geliebter, G.M. Pfaffenbach, and R.A. Zeff. 1986. Murine major histocompatibility complex class-I mutants: Molecular analysis and structure–function implications. *Annu. Rev. Immunol.* **4:** 471.

Nomura, S. and M. Oishi. 1984. UV irradiation induces an activity which stimulates simian virus 40 rescue upon cell fusion. *Mol. Cell. Biol.* **4:** 1159.

Prelich, G., C.K. Tan, M. Kostura, M.B. Mathews, A.G. So, K.M. Downey, and B. Stillman. 1987. Functional identity of proliferating cell nuclear antigen and a DNA polymerase-δ auxiliary protein. *Nature* **326:** 517.

Rahmsdorf, H.J., U. Mallick, H. Ponta, and P. Herrlich. 1982. A B-lymphocyte-specific high turnover protein: Constitutive expression in resting B cells and induction of synthesis in proliferating cells. *Cell* **29:** 459.

Rolfe, M.A., A. Spanos, and G. Banks. 1986. Induction of yeast Ty element transcription by ultraviolet light. *Nature* **319:** 339.

Ronai, Z.A., E. Okin, and I.B. Weinstein. 1988. Ultraviolet light induced expression of oncogenes in rat fibroblasts and human keratinocyte cells. *Oncogene* **2:** 201.

Ronai, Z.A., M.E. Lambert, M.D. Johnson, E. Okin, and I.B. Weinstein. 1987. Induction of asynchronous replication of polyoma DNA in rat cells by ultraviolet irradiation and the effects of various inhibitors. *Cancer Res.* **47:** 4565.

Samson, L. and J.L. Schwartz. 1980. Evidence for an adaptive DNA repair pathway in CHO and human skin fibroblast cell lines. *Nature* **287:** 861.

Sarasin, A., F. Bourre, and A. Benoit. 1982. Error prone replication of ultraviolet-irradiated simian virus 40 in carcinogen treated monkey kidney cells. *Biochimie* **64:** 815.

Schorpp, M., U. Mallick, H.J. Rahmsdorf, and P. Herrlich. 1984. UV-induced extracellular factor from human fibroblasts communicates the UV response to nonirradiated cells. *Cell* **37:** 861.

Steinmetz, M. and L. Hood. 1983. Genes of the major histocompatibility complex in mouse and man. *Science* **222:** 727.

Taylor-Papadimitriou, J., P. Purkis, E.B. Lane, I.A. McKay, and S.E. Chang. 1982. Effects of SV40 transformation on the cytoskeleton and behavioural properties of human keratinocytes. *Cell Differ.* **11:** 169.

Tennant, R.W. and R.J. Rascati. 1980. Mechanisms of cocarcinogenesis involving indogenous retroviruses. In *Modifiers of chemical carcinogenesis* (ed. T. Slaga), p. 185. Raven Press, New York.

van der Lubbe, J.L.M., P.J. Abrahams, C.M. van Drunen, and A.J. van der Eb. 1986. Enhanced induction of SV40 replication from transformed rat cells by fusion with UV-irradiated normal and repair-deficient human fibroblasts. *Mutat. Res.* **165:** 47.

Vivino, A.A., M.D. Smith, and K.W. Minton. 1986. A DNA damage-responsive *Drosophila melanogaster* gene is also induced by heat shock. *Mol. Cell. Biol.* **6:** 4767.

Walker, G.C. 1984. Mutagenesis and inducible responses to deoxyribonucleic acid damage in *Escherichia coli*. *Microbiol. Rev.* **48:** 60.

Wilson, V.L. and P.A. Jones. 1983. Inhibition of DNA methylation by chemical carcinogens in vitro. *Cell* **32:** 329.

Witkin, E.M. 1976. Ultraviolet mutagenesis and inducible DNA repair in *Escherichia coli*. *Bacteriol. Rev.* **40:** 869.

Induction of VL30 Element Expression as a Response to Anoxic Stress

GARTH R. ANDERSON,* DANIEL L. STOLER,*
JOSEPH P. SCOTT,[†] AND BECKY K. FARKAS*
*Molecular and Cellular Biology
Roswell Park Memorial Institute
Buffalo, New York 14263
[†]Oil City Hospital
Oil City, Pennsylvania 16301

OVERVIEW

VL30 elements have structural properties of transposons and, when pseudotyped by retroviruses, have been shown to be transposable. These elements are particularly interesting in that they represent one of two sets of rat cellular sequences transduced into the genomes of the Kirsten and Harvey murine sarcoma viruses. Anoxic stress induces normal rat fibroblasts to transcribe exceptionally high levels of functional mRNA, corresponding to the VL30 sequences. This stress is relatively mild, with no reduction in cell viability associated with the induction response. One anoxic stress response protein, of 34 kD and with lactate dehydrogenase (LDH) activity, appears likely to be encoded by VL30 sequences.

Induction of VL30 RNA is not seen with a variety of other respiratory poisons, with the exception of a low level response to cyanide. The induction by cyanide, however, appears different, being part of a general response to highly toxic conditions. The induction of VL30 seen with cyanide is substantially less than with anoxia and, unlike anoxia, also results in production of endogenous type C leukemia virus.

INTRODUCTION

The Kirsten and Harvey murine sarcoma viruses are acute transforming retroviruses that share the unusual property of containing transposable element sequences as one of two cell-derived inserts. Along with a well-characterized *ras* oncogene, these viruses each contain major fractions of a rat VL30 element (Anderson and Robbins 1976; Ellis et al. 1980). Although *ras* function plays a major role in the oncogenic activity of these viruses, the VL30 sequences also contribute to their transforming potential (Wei et al. 1980).

Banbury Report 30: Eukaryotic Transposable Elements as Mutagenic Agents
© Cold Spring Harbor Laboratory. 0-87969-230-8/88. $1.00 + .00

The VL30 element family remains poorly understood, although major salient features have been defined. In mouse and rat, there are 50 to 100 copies of VL30 per haploid genome. These represent related but nonidentical elements which by restriction mapping fall into at least four broad families (Keshet et al. 1980). The VL30 elements, or virus-like 30S RNA, are of approximately 5.4 kb in length and have long terminal repeat (LTR) sequences at each end (Keshet and Shaul 1981). Packaging signals and a tRNA primer binding site enable native VL30 RNA molecules to be packaged as pseudotypes by budding retroviruses and then be efficiently reverse-transcribed to DNA upon a new round of infection. This process results in transposition of the VL30 element into new sites (Carter et al. 1986).

The Kirsten and Harvey virus genomes each contain approximately 3 kb of a rat VL30 element (Young et al. 1980). The presence of VL30 sequences alongside *ras* oncogenes in these two independently isolated sarcoma viruses suggests that such association represents more than coincidence. Other acute transforming retroviruses with two separate cell-derived elements are known, including avian erythroblastosis virus with *erbA* and *erbB* genes (Vennstrom and Bishop 1982; Weinberger et al. 1986) and the MillHill no.2 virus with *mil* and *myc* genes (Jansen et al. 1984). In these cases, the two inserts represent two cooperating oncogenes, one from the growth factor/receptor/signal transducer family and one from the nuclear/nucleic acid binding family (Land et al. 1983).

Ten years ago, we first observed that anoxic stress induces normal rat fibroblasts to transcribe large amounts of an RNA homologous to sequences contained in the Kirsten and Harvey sarcoma virus genomes (Anderson and Matovcik 1977). These observations have led to our finding that the VL30 elements are a major anoxic response system, associated with a specific polypeptide that is likely to be involved in the oncogenic activity of the Kirsten and Harvey viruses. This polypeptide has been found in human cancer (Anderson et al. 1983), suggesting human VL30 elements may also be active in cancer.

RESULTS

Before it was known that the Kirsten and Harvey sarcoma viruses had two separate rat-derived inserts, we had observed that anoxic stress of subconfluent cultures causes a profound induction of RNA hybridizing to a Kirsten murine sarcoma virus cDNA probe (Anderson and Matovcik 1977). Hybridization to subclones specific to each of the two rat inserts, *ras* and VL30, has more recently revealed that it is VL30 sequences alone that are induced by anoxia (Table 1). This RNA is polyadenylated and polysome-

Table 1
Anoxic Stress Induces VL30 But Not *ras* Expression

Cells	Relative RNA	
	VL30	*ras*
FRE	1	1
FRE, anoxic	620	0.2

Anoxia was for 24 hr. *ras* concentrations were measured by hybridization to a HiHi3 probe, and VL30 by hybridization to the Young pVL30 rat probe. Crt analysis indicates that following anoxia approximately 10^4 VL30 RNA molecules are present.

associated (Anderson et al. 1979), and on Northern blots is of the same size (5.4 kb) as authentic VL30 RNA (G.R. Anderson et al., in prep.).

Transcription of the VL30 RNA begins within 3 hours after cultured cells are placed under anoxia; the true lag period is probably shorter, reflecting time required for oxygen to be removed from the culture media. Anoxia is a surprisingly mild stress for fibroblasts, with cells remaining viable for approximately 3 days. Cell division does not occur while under anoxia, although proliferation resumes upon return to aerobic culture conditions (Fig. 1). This cessation of proliferation is mirrored by a cessation of DNA synthesis, although RNA and protein synthesis continue at nearly normal rates for 16-24 hours.

The relative mildness of the anoxic stress induction of VL30 is also seen in the ability of vigorous (presumably anaerobic) exercise to induce VL30 transcription in vivo in muscle tissue (Anderson et al. 1983). Here, VL30 RNA is readily detectable within 6 hours following exercise, and expression remains high if animals are exercised vigorously once daily.

High-resolution one- and two-dimensional gel analyses have revealed that multiple proteins are inducible by anoxic stress, including two "glucose-regulated proteins" at 94 and 78 kD, induced preferentially in confluent cultures, and a major anoxia specific protein at 34 kD which induces preferentially when subconfluent cultures are so stressed (Anderson et al. 1979, 1988; Sciandra et al. 1984). Antisera to the rat 34-kD anoxic stress protein detects it as an antigen in a wide variety of nonrat cells expressing rat VL30 sequences as a result of infection with Kirsten sarcoma virus (Fig. 2). The 34-kD antigen immune coprecipitates with a 21-kD antigen, which at the peptide level appears related to but distinct from known *ras* p21 gene products.

The expression of the p34 antigen is not an automatic result of neoplastic transformation, since it is not seen in SV40 or polyoma transformed cells. However, this antigen has also been found in a methylcholanthrene-transformed mouse cell line and is induced by Rous sarcoma virus transformation. The possible role of VL30 expression in these additional cases has not been directly examined.

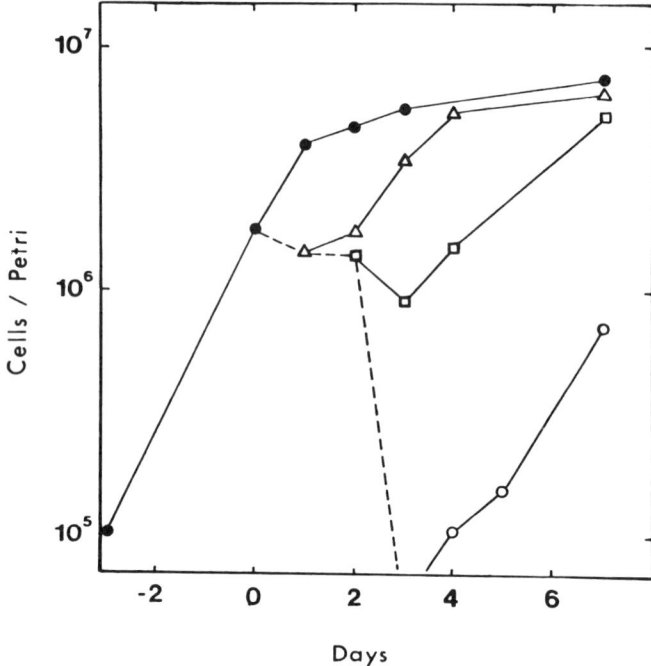

Figure 1
Viability of FRE rat cells subjected to various periods of anoxic stress. Fischer rat cells were grown in 65 mm Petri dishes. Beginning at time 0, individual dishes were subjected to 24 (△), 48 (□), or 72 (○) hr of continuous anoxic cultures. After return to an aerobic atmosphere, cells were media changed and cell counts taken as indicated. (●) Aerobically cultured controls. All cells subjected to 96 hr or more of anoxic stress were killed.

Extensive biochemical characterization of the 34-kD anoxic stress protein reveals it has LDH activity (LDH_k). LDH_k is distinguished from conventional LDH isozymes by binding of flavin mononucleotide with inhibition by flavin-adenine dinucleotide, inhibition by the cellular alarmone Gp4G, and a slightly smaller subunit molecular weight and higher isoelectric point. This protein binds to certain nucleic acids but not others, in no clear pattern as yet (Anderson et al. 1988). In human cancer, the 34-kD anoxic stress protein has been found at high levels in a wide variety of tumor types, ranging from 5- to 500-fold higher concentrations than in adjoining normal tissue (Anderson et al. 1983). The 34-kD anoxic stress protein has also been found frequently in the serum of patients, particularly those with metastatic disease, and appears of practical value as a cancer marker (Polonis et al. 1984; Manly et al. 1987).

The induction of VL30 RNA in response to anoxic stress is not paralleled by a response to a wide variety of respiratory poisons (Table 2). The only

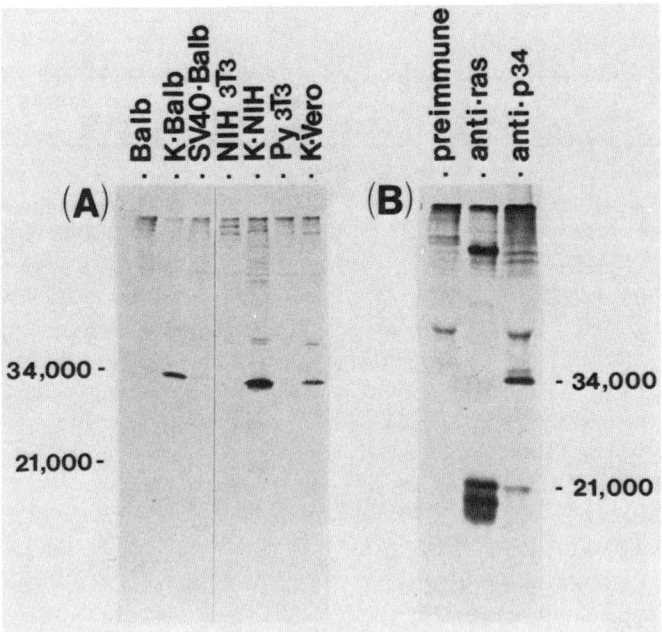

Figure 2
Anoxic stress protein $p34/LDH_k$ as an antigen in Kirsten sarcoma virus transformed cells. ^{35}S-labeled extracts were reacted with polyclonal rabbit antisera to $asp34/LDH_k$ and immune precipitates were analyzed on SDS-PAGE. (A) Analysis of extracts of uninfected BALB 3T3 and NIH 3T3 mouse cells, and Kirsten sarcoma cells (K), SV40, polyoma (PY) transformed cells. K·Vero: Kirsten sarcoma virus transformed Vero monkey cells. (B) Longer exposures of Kirsten transformed Vero extracts reacted with preimmune, anti-*ras*, and anti-*asp*34 sera, showing the 21-kD antigen associating with *asp*34.

respiratory poison which shows significant induction of VL30 is treatment with cyanide, but the response is far less than that seen with anoxia. The cyanide response further differs from anoxic stress in that total RNA synthesis nearly completely ceases, indicative of a more generally toxic condition (Fig. 3). Cyanide treatment also results in induction of endogenous type C leukemia virus RNA, resulting in a burst of type-C virus particle production (Table 3) not seen with anoxia (Whitaker-Dowling et al. 1979).

DISCUSSION

Induction of transposable elements in response to stress is a widespread phenomenon seen throughout the evolutionary tree. Induction of rat VL30 elements in response to anoxic stress of cultured fibroblasts is unusual, however, both in its magnitude and in its specificity for this one particular

Table 2
Induction of Rat VL30 by Respiratory Blocks and Other Agents

Agent[a]	Concentrations tested	Maximal induction VL30 RNA[b]
3-Acetyl pyridine	0.01–20 μg/ml	1.4
Amobarbital	0.1–3 × 10^{-4} M	1.6
Anoxia	0–20% O_2	980
Sodium arsenate	0.1–4 × 10^{-5} M	1.0
Sodium azide	0.01–8 × 10^{-2} M	4.0
Butylated hydroxytoluene	1 × 10^{-4} M	0.9
Canavanine	200 μg/ml	1.0
Chloramphenicol	50 μg/ml	1.0
Sodium cyanide	0.01–3 × 10^{-2} M	8.5
Diamide	0.4–4 × 10^{-5} M	1.1
Dinitrophenol	0.03–1 × 10^{-3} M	2.5
Dithiothreitol, oxidized	0.3–1 × 10^{-3} M	1.0
Glutathione, oxidized	0.3–3 × 10^{-3} M	1.0
Glutathione, reduced	1–9 × 10^{-3} M	0.8
Gramicidin	0.3–20 μg/ml	1.3
Oligomycin	0.4–10 μg/ml	1.6
n-Propylgallate	0.01–1 × 10^{-3} M	1.6
Retinoic acid	10^{-6} M	1.0
Rotenone	0.4–4 × 10^{-5} M	1.6
Thioglycollate	1–9 × 10^{-3} M	1.1
D-α-tocopherol	0.1–10 U/ml	0.8

[a]Fischer rat cells were treated as indicated for a 24 hr period.
[b]Induction is expressed relative to the concentration of VL30 RNA in untreated FRE cells. The value shown is the maximal induction seen following a 24 hr treatment.

relatively mild stress. Tissue anoxia is a normal physiological stress seen in muscle with vigorous exercise, and induction of VL30 RNA also occurs in vivo in exercised muscle. The observation that anoxically induced VL30 sequences are polyadenylated and polysome-associated indicates that transla-

Table 3
Induction of Endogenous Type-C Virus by Cyanide

Days[a]	Virus production[b]
0	2.0
1.5	8.7
3.0	7.0
4.5	3.6

Uninfected FRE rat cells were treated 6 hr with 9 mM NaCN, pH 7.5. Cultures were media changed and then assayed for virus particle associated reverse transcriptase as per Whitaker-Dowling et al. 1979.
[a]After cyanide treatment.
[b]Reverse-transcriptase, cpm × 10^{-3}.

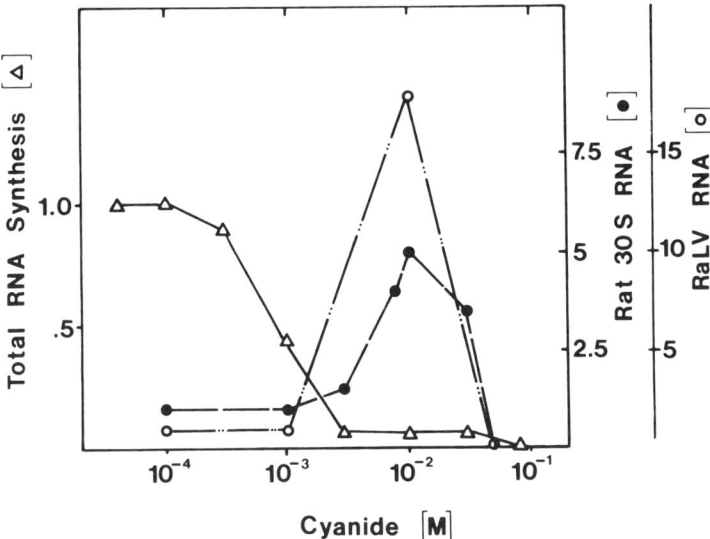

Figure 3
Dose dependency of the cyanide induction response. Uninfected FRE cells were treated for 24 hr with sodium cyanide, at the concentrations indicated. Total cell RNA was then isolated and analyzed by nucleic acid hybridization for (●) VL30 rat 30S RNA, (○) endogenous leukemia virus RaLV-RNA, and (△) total RNA synthesis monitored in parallel cultures labeled with ^{3}H uridine.

tion of VL30 RNA to anoxic stress gene products is evidently a part of the total response.

VL30 sequences have been found frequently transcribed at high levels in animal cancers (Anderson and Robbins 1976; Singh et al. 1985; Dragani et al. 1987). VL30 sequences are more directly associated with the primary events of neoplastic transformation by the fact that they constitute one of the two cell-derived inserts of the Kirsten and Harvey murine sarcoma viruses. Their role in these viruses has yet to be clearly defined, although two major possibilities exist. In the simplest case, VL30 sequences may encode important gene products directly. One such likely gene product is anoxic stress protein $p34/LDH_k$. A second possibility, not necessarily incompatible with the first, is that expression of virally incorporated VL30 sequences may inhibit expression of closely related host VL30 elements that would otherwise suppress the transforming activity of the viral *ras* oncogene. Consistent with this, we have observed that anoxic induction of host VL30 sequences does not occur in Kirsten-sarcoma-virus-transformed rat cells.

Although absolute proof that the anoxia-inducible VL30 encodes anoxic stress response protein $p34/LDH_k$ will require a match of protein and nucleic acid sequencing results, a provocative model can be constructed based on the known properties of this protein. Four potentially relevant features of this protein are its LDH activity, its nucleic acid binding activity, its coprecipitation with a p21, and its specific inhibition by guanine nucleotides (G), including Gp4G. The LDH activity of this protein has obvious utility as an anoxic stress response protein, providing an alternate means for the oxidation of nicotinamide-adenine dinucleotide (NADH) (Fig. 4). The lactate so produced as a byproduct may also be reflected in the high lactate production seen in most human cancers (Warburg et al. 1924). The nucleic acid binding activity of this protein is consistent with the potential for autoregulation of this (and possibly other) gene expression. Whereas $asp34/LDH_k$ binds nucleic acid, NADH effectively releases it. Under anoxic stress NADH levels can be expected to rise; this could dissociate it from binding transcriptional regulatory sites. Transcription and translation would then provide a means to oxidize the excess NADH. If a VL30 indeed encodes the $asp34$, and this protein autoregulates its own expression, a likely binding site would be in the VL30 LTR. Last and most speculatively, the coprecipitation of a p21 antigen with the $asp34/LDH_k$, and its regulation by guanine nucleotides, raise the possibility that this protein might function in a G-protein system. The other gene product of the Kirsten and Harvey sarcoma virus, ras p21, clearly is a G_α homolog (Gibbs et al. 1984; Hurley et al. 1984; Sweet et al. 1984). G_α proteins are known to be modulated by cellular G_β proteins of approximately

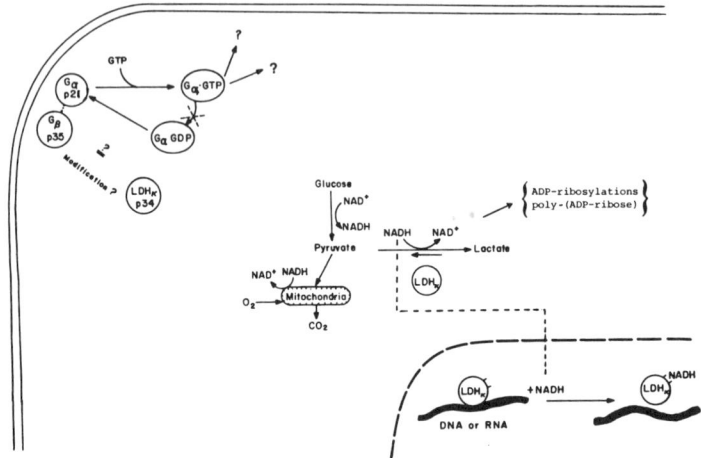

Figure 4

Anoxic stress protein $p34/LDH_k$ may function at multiple levels in the anoxic stress response: metabolically as an alternate oxidation route for NADH, by binding nucleic acid and perhaps autoregulating its own expression, and possibly by interacting in a G-protein regulatory role.

35 kD (Gilman 1984). If the VL30 sequences of these viruses encode a modified or defective G_β homolog, interference with suppression by authentic cellular G_β proteins would be expected, resulting in enhancement of the transforming potential of the viral *ras* gene product. Whether or not $asp34$/ LDH_k truly can function as a G_β, however, remains to be proven.

Two very real questions remain as to why the anoxic stress response system need be part of a transposable element. Indeed, does transposition itself play a role in the anoxic stress response, and is there a functional relevance in the ability of VL30 elements to be packaged into type-C retroviruses, with the potential to be dispersed extracellularly?

ACKNOWLEDGMENTS

We thank Roy Heinaman and Bill Kovacik for technical assistance. We thank Marcia Held for her expert typing of this manuscript.

REFERENCES

Anderson, G.R. and B.K. Farkas. 1988. The major anoxic stress response protein p34 is a distinct lactate dehydrogenase. *Biochemistry* (in press).

Anderson, G.R. and L.M. Matovcik. 1977. Expression of murine sarcoma virus genes in uninfected rat cells. *Science* **197**: 1371.

Anderson, G.R. and K.C. Robbins. 1976. The rat sequences of the Kirsten and Harvey murine sarcoma virus genomes: Nature, origin and expression in rat tumor RNA. *J. Virol.* **17**: 335.

Anderson, G.R., K.R. Marotti, and P.A. Whitaker-Dowling. 1979. A candidate rat-specific gene product of the Kirsten murine sarcoma virus. *Virology* **99**: 31.

Anderson, G.R., V.R. Polonis, J.K. Petell, R.A. Saavedra, K.F. Manly, and L.M. Matovcik. 1983. LDH_k in human cancer. In *Isozymes: Current topics in biological and medical research* (ed. M.C. Rattazzi et al.), vol. 11, p. 155. Alan Liss, New York.

Carter, A.T., J.D. Norton, Y. Gibson, and R.J. Avery. 1986. Expression and transmission of a rodent retrovirus-like VL30 gene family. *J. Mol. Biol.* **188**: 105.

Dragani, T.A., G. Manenti, G. Della Porta, and I.B. Weinstein. 1987. Factors influencing the expression of endogenous retrovirus-related sequences in the liver of B6C3 mice. *Cancer Res.* **47**: 795.

Ellis, R.W., D. DeFeo, J. Maryak, H. Young, T. Shih, E. Chang, D. Lowy, and E. Scolnick. 1980. Dual evolutionary origin for the rat genetic sequences of Harvey murine sarcoma virus. *J. Virol.* **36**: 408.

Gibbs, J.B., I.S. Sigal, M. Poe, and E. Scolnick. 1984. Intrinsic GTPase activity distinguishes normal and oncogenic *ras* p21 molecules. *Proc. Natl. Acad. Sci.* **81**: 5704.

Gilman, A.G. 1984. G proteins and dual control of adenylate cyclase. *Cell* **36**: 577.

Hurley, J.B., M.I. Simon, D.B. Teplow, J.D. Robinshaw, and A.G. Gilman. 1984.

Homologies between signal transducing G proteins and *ras* gene products. *Science* **226:** 860.

Jansen, H.W., R. Lurz, K. Bister, T. Bonner, G. Mark, and U. Rapp. 1984. Homologous cell-derived oncogenes in avian carcinoma virus MH2 and murine sarcoma virus 3611. *Nature* **307:** 281.

Keshet, E. and Y. Shaul. 1981. Terminal direct repeats in a retrovirus-like repeated mouse gene family. *Nature* **289:** 83.

Keshet, E., Y. Shaul, J. Kaminchik, and H. Aviv. 1980. Heterogeneity of virus-like genes encoding retrovirus-associated 30S RNA and their organization within the mouse genome. *Cell* **20:** 431.

Land, H., L.F. Parada, and R.A. Weinberg. 1983. Tumorigenic conversion of primary embryo fibroblasts requires at least two cooperating oncogenes. *Nature* **304:** 596.

Manly, K.F., N. Petrelli, G.R. Anderson, L.J. Emrich, L. Herrera, and A. Mittelman. 1987. Expression of an unusual isozyme of lactate dehydrogenase in the serum of cancer patients and comparison with carcinoembryonic antigen. *Cancer Res.* **47:** 6156.

Polonis, V.R., G.R. Anderson, J. Brzykcy, A.O. Vladutiu, and K.F. Manly. 1984. An unusual oxygen-sensitive lactate dehydrogenase isozyme associated with Kirsten murine sarcoma virus in human serum. *Cancer Res.* **44:** 2236.

Sciandra, J.J., J.R. Subjeck, and C.S. Hughes. 1984. Induction of glucose-regulated proteins during anaerobic exposure and of heat-shock proteins after reoxygenation. *Proc. Natl. Acad. Sci.* **81:** 4843.

Singh, K., S. Saragosti, and M. Botchan. 1985. Isolation of cellular genes differentially expressed in mouse NIH 3T3 cells and a SV40-transformed derivative: Growth specific expression of VL30 genes. *Mol. Cell. Biol.* **5:** 2590.

Sweet, R.W., S. Yokoyama, T. Kamata, J.R. Feramisco, M. Rosenberg, and M. Gross. 1984. The product of *ras* is a GTPase and the T24 oncogenic mutant is deficient in this activity. *Nature* **311:** 273.

Vennstrom, B. and J.M. Bishop. 1982. Isolation and characterization of chicken DNA homologous to the two putative oncogenes of avian erythroblastosis virus. *Cell* **28:** 135.

Warburg, O., K. Posener, and E. Negalein. 1924. Metabolism of carcinoma cells. *Biochem. Z.* **152:** 309.

Wei, C.-M., D.R. Lowy, and E.M. Scolnick. 1980. Mapping of transforming region of the Harvey murine sarcoma virus genome by using insertion-deletion mutants constructed *in vitro*. *Proc. Natl. Acad. Sci.* **77:** 4674.

Weinberger, C., C.C. Thompson, E.S. Ong, R. Lebo, D.J. Groul, and R.M. Evans. 1986. The c-erb-A gene encodes a thyroid hormone receptor. *Nature* **324:** 641.

Whitaker-Dowling, P.A., K.R. Marotti, and G.R. Anderson. 1979. Apparent post-transcriptional block to anaerobic induction of endogenous leukemia virus. *J. Virol.* **32:** 234.

Young, H.A., M.A. Gonda, D. DeFeo, R.W. Ellis, K. Nagashima, and E.M. Scolnick. 1980. Heteroduplex analysis of cloned rat endogenous replication-defective 30S retrovirus and Harvey murine sarcoma virus. *Virology* **107:** 89.

DNA Modification of *Mu* Elements in Maize

VICKI LYNN CHANDLER
Institute of Molecular Biology
University of Oregon
Eugene, Oregon 97403

OVERVIEW

The increased mutation rate in *Mutator* stocks of maize is associated with the transposition of *Mu* elements. The high mutation rate is heritable, although *Mutator* stocks occasionally lose their mutagenic activity. Southern blot analysis of *Mu* elements demonstrates a correlation between loss of activity and an inhibition of digestion of *Mu* elements by methylation-sensitive restriction enzymes. This modification is stable through several generations of outcrosses. Most stocks of maize are non-*Mutator*, yet they usually contain a few copies of intact *Mu* elements. We have characterized an intact *Mu1*-like element from the non-*Mutator* line B37 that is identical to *Mu1*, an element known to transpose in *Mutator* stocks. In B37 genomic DNA this element is modified such that certain methylation-sensitive restriction enzymes will not cut it. Modification of this element is lost when it is placed in an active *Mutator* stock by genetic crosses. We hypothesize that DNA modification of transposable elements may serve as a heritable, but potentially reversible, mechanism for regulating transposable elements. One source of genomic instability may be the liberation of transposable elements through the demethylation of cryptic elements.

INTRODUCTION

McClintock (1952) first described the presence of mobile genetic elements that interact with structural genes to cause genetic instabilities in maize. Subsequently, elements have been characterized in many prokaryotic and eukaryotic organisms. A strong body of evidence suggests that transposable elements cause genetic variation both within the lifetime of an organism and during evolution. They produce mutations by inserting within or near genes and by causing chromosomal rearrangements. Frequently, transposable element-induced mutations are unstable, giving rise to reversion events or additional, secondary mutations. The clonal nature of maize development, combined with visual markers such as the purple anthocyanin pigments, permits the analysis of changes in the timing, frequency, and pattern of gene expression at the mutated locus (for review, see Fedoroff 1983).

Banbury Report 30: Eukaryotic Transposable Elements as Mutagenic Agents
© Cold Spring Harbor Laboratory. 0-87969-230-8/88. $1.00 + .00

About ten years ago, Robertson (1978) described a *Mutator* stock of maize which had a high mutation rate, producing new mutant alleles at frequencies ranging from 10^{-3} to 10^{-5}. Mutator activity is labile and occasionally lost when *Mutator* plants are inbred or outcrossed (Robertson 1983). For example, when a *Mutator* stock is crossed to a standard stock of maize with a normal mutation rate, approximately 10% of the F_1 progeny lose the high mutation rate while 90% retain it. This apparent multifactor segregation, in combination with the observation that about one-third of the mutations isolated from *Mutator* stocks are unstable, suggests that *Mutator* activity is the result of an active transposable element family.

Consistent with this hypothesis, several related transposable elements have been isolated from *Mutator* plants by their insertion into well-characterized genes. A 1.4-kbp insertion, *Mu1*, was cloned from *Mutator*-induced mutations at *Adh1* (Bennetzen et al. 1984) and *A1* (O'Reilly et al. 1985). A second element, *Mu1.7*, which differs from *Mu1* primarily by an extra 380-bp segment (L. Taylor and V. Walbot, in prep.) transposed into the *bronze-1* (*bz1*) locus (Taylor et al. 1986). Approximately 10–60 copies of these elements are found in *Mutator* plants, and most mutant alleles that have been examined with molecular probes contain insertions of *Mu1*- or *Mu1.7*-like elements (Lillis and Freeling 1986). Additional *Mu* elements have been studied that share terminal inverted repeats with *Mu1* and *Mu1.7*, but have unrelated internal sequences (V. Chandler, unpubl.). The best characterized of these elements, termed *Mu3*, is 1.8-kbp long and was isolated as a result of its insertion into the *Adh1* gene (C. Chen et al., in prep.). It is not known which of these elements encodes factors required for their transposition or if other factors are necessary. However, *Mu* elements do not interact genetically or share sequence homology with other types of maize transposable elements such as *Ac/Ds* and *Spm* (Robertson and Mascia 1981).

Standard non-*Mutator* stocks of maize whose pedigrees show no *Mutator* ancestors contain sequences with homology to *Mu* elements (Chandler et al. 1986), although the number of *Mu1* and *Mu1.7* elements is lower than that found in *Mutator* stocks. There is no genetic evidence of a high mutation rate in these non-*Mutator* stocks, suggesting that the elements are not transposing at detectable frequencies. To address the regulation of *Mutator* transposition, we have isolated unstable mutations in several loci required for purple anthocyanin pigment production (Walbot et al. 1986). We have assayed *Mutator* activity in these stocks by following the segregation of the unstable allele. In this paper, we review a correlation between modification of certain methylation-sensitive restriction sites within *Mu* elements and decreased transmission of the unstable phenotype. We also describe the methylation state of *Mu* elements in non-*Mutator* stocks and discuss the implications of our results on the regulation of *Mu* transposable elements.

RESULTS

Unstable alleles that result from *Mu* element insertion into loci required for anthocyanin pigmentation are identified as exceptional variegated kernels by crossing a purple, homozygous dominant stock containing *Mutator* activity to a colorless, homozygous recessive tester. All the F_1 kernels should be purple except for rare mutant kernels that are colorless (stable) or variegated (unstable). The variegated phenotype is the result of the restoration of genic activity during somatic growth, producing colored revertant sectors on the colorless mutant background. Plants derived from colorless or variegated kernels are crossed to the recessive tester. For example, a plant heterozygous for a mutant allele at the *B* locus (*b-Peru-mu/b*) crossed by a homozygous recessive tester (*b/b*) shows the expected 1:1 segregation of variegated (*b-Peru-mu/b*) to colorless (*b/b*) kernels (Fig. 1A). The activity of the element at the locus is monitored by following the segregation of the variegated phenotype through subsequent generations. Plants segregating fewer variegated kernels than expected have been identified. An example of an ear with a sector that lacks variegated kernels is shown in Figure 1B. This allele's capacity for somatic reversion has been lost in an heritable manner, since the colorless kernels remain colorless in subsequent crosses to tester. Similar observations have been made with *Mutator*-induced mutations at other genes. The best characterized is the *bz2-mu1* mutation in which approximately 20–30% of the plants segregate fewer spotted kernels than expected when self-pollinated or outcrossed to testers (Chandler and Walbot 1986; Walbot 1986). Thus, the activity of the element at *bz2-mu1* is labile and is occasionally lost upon self-pollination and outcrossing.

To investigate the molecular basis for the inactivity of *bz2-mu1*, the restriction pattern of *Mu* elements were examined in plants showing the distorted transmission of the mutant phenotype and in plants segregating the variegated allele normally (Chandler and Walbot 1986). A total of thirteen enzymes have been tested: *Taq*I, *Bst*EII, *Bst*NI, *Eco*RI, *Msp*I, *Hpa*II, *Bgl*I, *Ava*II, *Hinf*I, *Ava*I, *Cla*II, *Mlu*I, and *Tth*111-I. All of the 52 progeny from ears with a decreased transmission of the variegated allele contained *Mu* elements, but some or all of the elements were no longer restrictable with enzymes sensitive to 5-methyl-deoxycytosine. In contrast, 132 progeny from ears showing normal segregation of the variegated phenotype contained *Mu* elements that were cleavable by all enzymes tested. An example of a Southern blot with DNA samples from both types of individuals is shown in Figure 2. The modification appears to be specific for the *Mu* elements and immediately adjacent DNA since Southern blot hybridizations with probes specific for *Adh1*, *bz1*, and a repetitive nuclear DNA clone detect no differences in DNA modification between active and inactive stocks (Chandler and Walbot 1986).

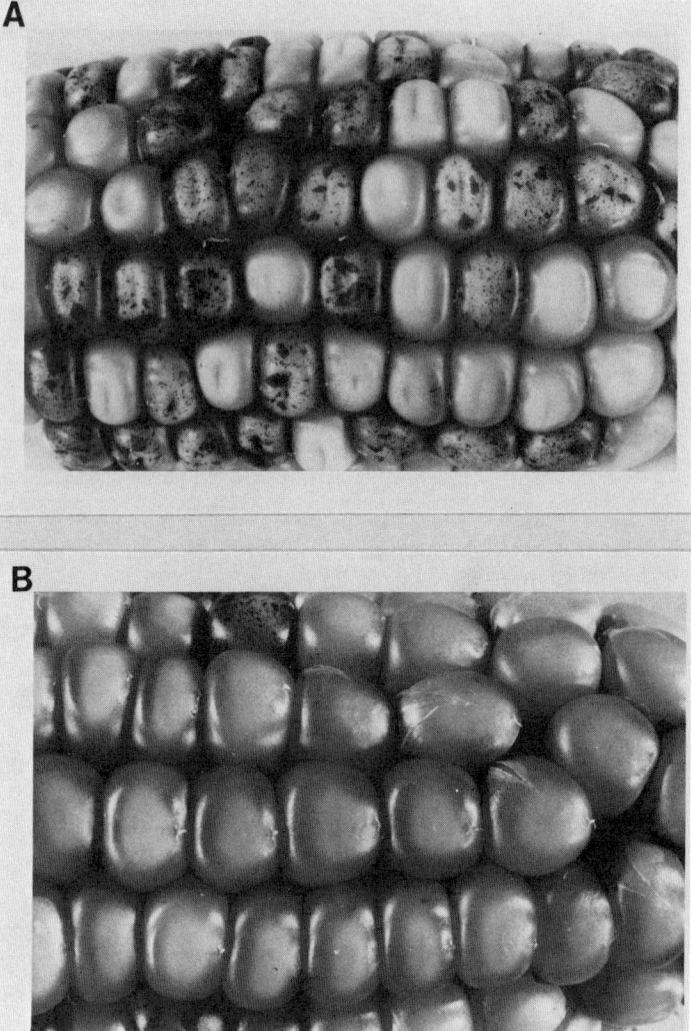

Figure 1
Photographs of a *Mutator*-induced unstable allele at the *B-Peru* locus. (*A*) Ear showing normal segregation of the *b-Peru-mu* allele. A heterozygous *b-Peru-mu/b* individual was crossed with a homozygous b tester *(b/b)* plant. The variegated kernels have the genotype *b-Peru-mu/b*, the colorless kernels are *b/b*. (*B*) Ear from a similar cross to that shown in panel *A*, except that the variegated *b-Peru-mu* phenotype did not segregate normally. A sector is observed on the ear that contains fewer variegated kernels than the 50% expected.

Kinetic studies suggest that DNA modification can occur in all the *Mu* elements within one generation; however, plants that contain both modified and unmodified elements have been identified, suggesting that the modifica-

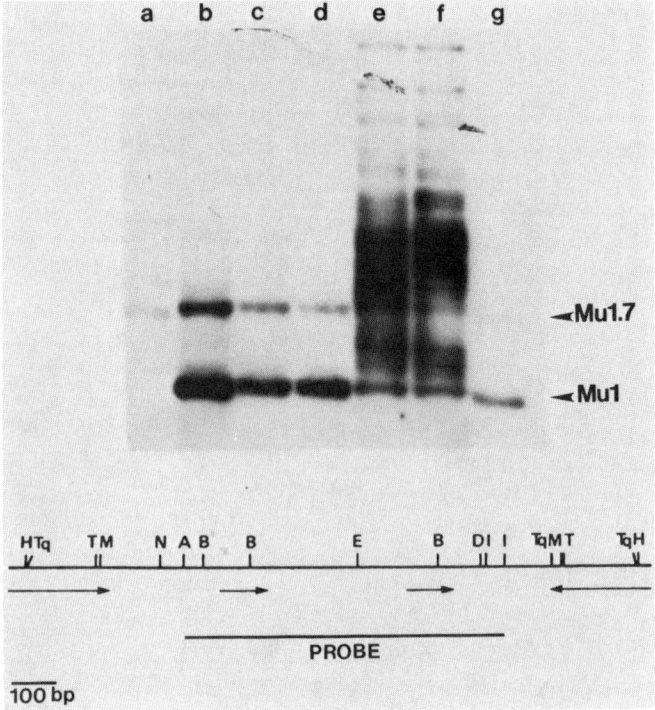

Figure 2
HinfI digests of DNA from active and inactive *Mutator* stocks. DNA was prepared from immature cobs of plants with normal segregation of the *bz2-mu1* allele (active) or decreased transmission of the *bz2-mu1* allele (inactive). Five µg of DNA from each individual was digested with HinfI, electrophoresed on 0.8% agarose, transferred to genetran, and hybridized with an internal *Mu1* probe as described previously (Chandler et al. 1986). A map of *Mu1* is drawn below the autoradiogram with the HinfI sites (H) and the probe indicated. Lane *g* contains samples of cloned *Mu1* and *Mu1.7* elements diluted to approximately three and one copies per haploid genome, respectively. Lane *a* is DNA from a non-*Mutator* stock with a few fragments homologous to *Mu* elements. The active *Mutator* stocks (*b–d*) contain the HinfI fragments expected of *Mu1* and *Mu1.7*, whereas the inactive *Mutator* stocks contain mostly larger restriction fragments (lanes *e,f*). Restriction enzyme cleavage sites are shown: H, HinfI; Tq, TaqI; T, Tth111-I; M, MluI; N, NotI; A, AvaI; B, BglI; E, BstEII; D, DdeI; I, BstNI.

tion can occur in a progressive manner (Chandler and Walbot 1986; V. Chandler, unpubl.).

Most maize stocks contain sequences very similar to *Mu* elements, and yet they show no genetic activity characteristic of *Mu* transposition. The lack of activity in non-*Mutator* stocks could be the result of nucleotide changes that have rendered the elements defective. Alternatively, the elements may be intact, but their low numbers, the absence of regulatory factors, or DNA modification may prevent their transposition at a detectable frequency. To investigate these possibilities an intact *Mu1*-like element, termed *Mu1.4-B37* was cloned from the non-*Mutator* inbred line B37 (Chandler et al. 1986). The sequence of *Mu1.4-B37* is virtually identical to that of *Mu1* with only a single nucleotide difference (V. Chandler et al., in prep.). This nucleotide difference is silent in regard to altering any protein that may be encoded by the major *Mu1* open reading frames (Barker et al. 1984). *Mu1.4-B37* is not found in the same genomic location in most other maize lines and is flanked by 9-bp direct repeats, suggesting that it transposed into its current genomic location in B37.

The comparison of *Mu1.4-B37* and *Mu1* suggests that they should be equally capable of transposition since they are identical. However, B37 has a low number of *Mu1*-like elements and a normal mutation rate, suggesting that *Mu1.4-B37* is inactive. Southern blot analysis was done to test if one reason for its inactivity might be DNA modification (Chandler et al. 1986). As shown in Figure 3 (lanes a and b), the *Hin*fI sites in the *Mu1.4-B37* termini are modified in B37 DNA such that digestion with *Hin*fI yields fragments larger than those predicted from the cloned sequences (see map in Fig. 3). Complete digestion is expected to yield 1.3-kbp and 0.95-kbp fragments homologous to the *Mu1* and flanking probes, respectively. Instead, both the *Mu* and flanking probes hybridize strongly to the same 2.55-kbp fragment and, to a lesser extent, to a 2.25-kbp fragment. This demonstrates that the *Hin*fI sites in the *Mu* element are not cleaved and the *Mu* and flanking sequences are now fused. The 2.55-kbp fragment corresponds to inhibition at both of the *Hin*fI sites in the *Mu1.4-B37* element, and the 2.25-kbp fragment represents cleavage of the left site and inhibition of the right site (see map in Fig. 3). The 2.25-kbp fragment is much less than one copy per genome, suggesting that in most cells neither site within the *Mu* element is cleaved. Both the *Mu* and right flanking probes hybridize to a few additional fragments. The 1.65-kbp and 1.85-kbp *Hin*fI fragments homologous to the *Mu1* probe (Fig. 3, lane a) result from other *Mu*-homologous sequences in the genome (L. Talbert and V. Chandler, in prep.). The right flanking DNA probe also hybridizes to one additional fragment unlinked to *Mu1.4-B37* (lanes b, e, and f).

The correlation between DNA modification and lack of activity in *Mutator* stocks suggests that a prerequisite for activation of *Mu1.4-B37* may be the loss of its DNA modification. To test whether modification of the *Hin*fI sites in *Mu1.4-B37* is stable in the presence of *Mutator* activity, the *Mu1.4-B37* element was introduced into an active *Mutator* background by genetic crosses. Individual B37 plants containing a single homozygous *Mu1.4-B37* element were crossed to several active *Mutator* plants homozygous for the *bz2-mu1* allele. F_1 progeny were self-pollinated to generate F_2 individuals, and *Mutator* activity assessed by following segregation of the variegated *bz2-mu1* allele. DNA was isolated from F_1 and F_2 individuals, and accessibility of the *Mu1.4-B37 Hin*fI sites was examined using Southern blot analysis.

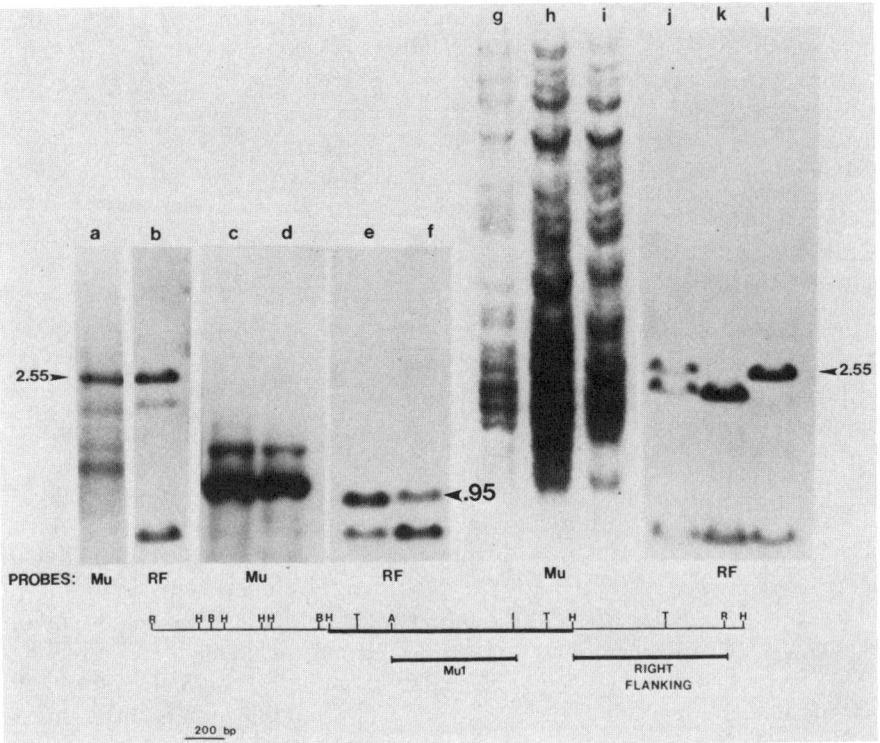

Figure 3
Modification of *Mu1.4-B37*. DNA samples were prepared from leaves or immature, unfertilized ears, digested with *Hin*fI and analyzed on Southern blots. The *Mu1* and *Mu1.7* elements in each plant were monitored using the internal *Mu1* probe (*Mu*), while the individiual *Mu1.4-B37* element was monitored using the right flanking probe (RF). Both probes are indicated in the restriction map of the cloned *Mu1.4-B37* element and its flanking DNA. Lanes *a* and *b* are DNA samples from the non-*Mutator* B37 parent hybridized with the *Mu* probe (*a*) or the right flanking probe (*b*). Lanes *c–f* are DNA samples from two progeny of a cross between the non-*Mutator* B37 parent and an active *Mutator* plant hybridized with the *Mu* probe (*c, d*) or the right flanking probe (*e, f*). These individuals segregated the *bz2-mu1* allele normally. Lanes *g–l* are DNA samples from three F$_2$ progeny that failed to segregate the *bz2-mu1* allele normally. Lanes *g–i* are hybridized with the *Mu* probe and lanes *j–l* with the right flanking probe. Restriction enzyme sites are shown on the map: R, *Eco*RI; H, *Hin*fI; B, *Bam*HI; T, *Tth*111-I; A, *Ava*I; and I, *Bst*NI.

As expected, all of the *Mu1* and *Mu1.7* elements are digestable with *Hin*fI in active *Mutator* plants (Fig. 3, lanes c,d). This is also true for the *Mu1.4-B37* element as evidenced by the loss of the 2.55-kbp and 2.25-kbp fusion fragments and the appearance of the 0.95-kbp right flanking fragment (Fig. 3, lanes e,f). This demonstrates that the modification state of the *Mu1.4-B37* element can change depending on the genetic background of the stock.

Furthermore, in F_2 *Mutator* stocks that lost activity, *Mu1.4-B37* again becomes modified (Fig. 3, lanes j–l) as do all the *Mu* elements in the plant (Fig. 3, lanes g–i). Only the 2.25-kbp or the 2.55-kbp fragments are observed, showing that *Hin*fI sites relatively close to the element are cleavable. The amount of the 2.55-kbp and 2.25-kbp fusion fragments varies among different F_2 individuals, demonstrating that the modification of the left *Mu1.4-B37* *Hin*fI site does not always occur.

DISCUSSION

Several lines of evidence suggest that modified *Mu* elements are nonfunctional. We have demonstrated a correlation between an increased level of DNA modification of *Mu* elements and loss of *Mutator* activity as measured by somatic reversion (Chandler and Walbot 1986). Bennetzen (1987) showed that deoxycytosine modification of the internal *Eco*RII site in *Mu* elements correlates with loss of mutagenic activity in highly inbred *Mutator* stocks (Robertson 1986). In addition, when inactive *Mutator* stocks containing modified elements are outcrossed to non-*Mutator* stocks the number of *Mu* elements decreases dramatically (V. Chandler, unpubl.). This is in sharp contrast to active lines where the number of elements tends to remain high, presumably because of new transposition events (Alleman and Freeling 1986). Since the presence of modified *Mu* elements is only correlated with lack of activity, we do not know whether DNA modification is the primary mechanism of inactivation, is directly involved in the maintenance of the inactive state, or if inactive elements become more easily modified. Also, the loss of *Mutator* activity is not always associated with DNA modification, suggesting that other mechanisms of regulation must also exist (Bennetzen 1987; V. Chandler, unpubl.).

All *Mutator* maize stocks descend from a single mutant stock described by Robertson (1978). However, most non-*Mutator* stocks of maize contain a few elements closely related to *Mu1* and *Mu1.7* elements. We have shown that an element from a non-*Mutator* line, *Mu1.4-B37*, is identical to *Mu1*. With other transposable element families, such as the *Ac/Ds* elements, it appears that sequences surrounded by the *Ac/Ds* terminal repeats can be induced to transpose if *Ac* is present to provide the necessary enzymes for excision and transposition (Döring and Starlinger 1984). If this is the case for *Mu* elements, *Mu1.4-B37* should be capable of transposition if supplied with the necessary transacting factors. Our sequencing data suggest that *Mu1.4-B37* should be as capable as *Mu1* of coding transacting factors.

Mu1.4-B37 normally exists in a modified state in the inbred line B37, and we hypothesize that this may contribute to its apparent inactivity. The strong correlation between DNA modification of *Mu* elements and the loss of activity in *Mutator* stocks suggests that the activation of *Mu1.4-B37* may require the loss of its DNA modification. We show here that this occurs when *Mu1.4-B37* is crossed into an active *Mutator* stock. Additionally, in *Mutator* stocks that subsequently lose activity, *Mu1.4-B37* again becomes modified as do all the *Mu* elements in the stock. Thus, the modification state of *Mu1.4-B37* reflects the state of all the *Mu* elements in a *Mutator* line, demonstrating that *Mu1.4-B37* is capable of responding to factors in a manner similar to other *Mu* elements. Factor(s) in active *Mutator* plants may inhibit the modification of *Mu* elements. This could result from the presence of a demethylation activity in active *Mutator* stocks or by inhibition of the methylase.

Modification of *Mu* elements, whether carried out by enzymes encoded by the elements themselves or by other genes in the genome, provides a heritable but potentially reversible mechanism for regulating transposable element activity. *Mu* elements are not the only transposable elements whose activities may be regulated by DNA modification. The activity of the prokaryotic transposable element IS*10* is regulated by adenosine methylation (Roberts et al. 1985), and transposition of *Ac* and *Spm* elements in maize is correlated with hypomethylation at certain restriction sites (Schwartz and Dennis 1986; Chomet et al. 1987). DNA modification of transposable elements may serve as a general cellular mechanism for regulating a potentially deleterious activity, and elements that move at high frequency may have evolved strategies for escaping the modification systems.

A common source of genomic instability may be the liberation of previously silent transposable element systems. Environmental and genetic stresses have been correlated with increased mutability in plants and have been hypothesized to activate transposable elements (McClintock 1978; Burr and Burr 1981). Treatments that evoke DNA repair pathways may create hemimodified elements, which in the next round of replication give rise to hypomodified elements. Thus, cryptic elements may become transiently demethylated, which might lead to their activation. It is interesting to note that the *Ac* transposable element was first detected after chromosome breakage, an event that might be expected to stimulate repair synthesis (McClintock 1984).

ACKNOWLEDGMENTS

I thank Devon Turks for her excellent technical assistance, Virginia Walbot for helpful discussions during the course of this work, and Luther Talbert for comments on the manuscript. This research was supported by a National Science Foundation grant (DCE-8451656) and matching funds from Pioneer Hi-Bred International, CIBA-GEIGY, and Northwest Area Foundation to V.L.C.

REFERENCES

Alleman, M. and M. Freeling. 1986. The *Mu* transposable element of maize: Evidence for transposition and copy number regulation during development. *Genetics* **112:** 107.

Barker, R.F., D.V. Thompson, D.R. Talbot, J. Swanson, and J.L. Bennetzen. 1984. Nucleotide sequence of the maize transposable element *Mu1*. *Nucleic Acids Res.* **12:** 5955.

Bennetzen, J. 1987. Covalent DNA modification and the regulation of Mutator element transposition in maize. *Mol. Gen. Genet.* **208:** 45.

Bennetzen, J.L., J. Swanson, W.C. Taylor, and M. Freeling. 1984. DNA insertion in the first intron of maize *Adh1* affects message levels: Cloning of progenitor and mutant *Adh1* alleles. *Proc. Natl. Acad. Sci.* **81:** 4125.

Burr, B. and F. Burr. 1981. Transposable elements and genetic instabilities in crop plants. *Stadler Genet. Symp.* **13:** 115.

Chandler, V. and V. Walbot. 1986. DNA modification of a maize transposable element correlates with loss of activity. *Proc. Natl. Acad. Sci.* **83:** 1767.

Chandler, V., C. Rivin, and V. Walbot. 1986. Stable non-Mutator stocks of maize have sequences homologous to the *Mu1* transposable element. *Genetics* **114:** 1007.

Chomet, P.S., S. Wessler, and S. Dellaporta. 1987. Inactivation of the maize transposable element *Activator* (*Ac*) is associated with its DNA modification. *EMBO J.* **6:** 295.

Döring, H.-P. and P. Starlinger. 1984. Barbara McClintock's controlling elements: Now at the DNA level. *Cell* **39:** 253.

Fedoroff, N.V. 1983. Controlling elements in maize. In *Mobile genetic elements* (ed. J. Shapiro), p. 1. Academic Press, New York.

Lillis, M. and M. Freeling. 1986. *Mu* transposons in maize. *Trends Genet.* **2:** 183.

McClintock, B. 1952. Chromosome organization and genic expression. *Cold Spring Harbor Symp. Quant. Biol.* **16:** 13.

———. 1978. Mechanisms that rapidly reorganize the genome. *Stadler Genet. Symp.* **10:** 25.

———. 1984. The significance of responses of the genome to challenge. *Science* **226:** 792.

O'Reilly, C., N.S. Shepherd, A. Pereira, Z. Schwarz-Sommer, I. Bertram, D.S. Robertson, P.A. Peterson, and H. Saedler. 1985. Molecular cloning of the *al* locus of *Zea mays* using the transposable elements *En* and *Mu1*. *EMBO J.* **4:** 877.

Roberts, D., B.C. Hoopes, W.R. McClure, and N. Kleckner. 1985. IS10 transposition is regulated by DNA adenosine methylation. *Cell* **43:** 117.

Robertson, D.S. 1978. Characterization of a Mutator system in maize. *Mutat. Res.* **51:** 21.

———. 1983. A possible dose dependent inactivation of Mutator (*Mu*) in maize. *Mol. Gen. Genet.* **191:** 86.

———. 1986. Genetic studies on the loss of *Mu* mutator activity in maize. *Genetics* **113:** 765.

Robertson, D.S. and P.N. Mascia. 1981. Tests of 4 controlling-element systems for mutator activity and their interaction with *Mu* mutator. *Mutat. Res.* **84:** 283.

Schwartz, D. and E. Dennis. 1986. Transposase activity of an Ac controlling element in maize is regulated by its degree of methylation. *Mol. Gen. Genet.* **205:** 476.

Taylor, L., V. Chandler, and V. Walbot. 1986. Insertion of 1.4 kb and 1.7 kb *Mu* elements into the *bronze*-1 gene of *Zea mays* L. *Maydica* **31:** 31.

Walbot, V. 1986. Inheritance of mutator activity in *Zea mays* as assayed by somatic instability of the *bz2-mu1* allele. *Genetics* **114:** 1293.

Walbot, V., C. Briggs, and V. Chandler. 1986. Properties of mutable alleles recovered from Mutator stocks of *Zea mays* L. In *Genetics, development, and evolution* (ed. J.P. Gustafson et al.), p. 115. Plenum Press, New York.

Inducers/Regulators of Retroviral Element Expression

A Possible Role of Retrotransposons in Carcinogenesis

I. BERNARD WEINSTEIN,* WENDY L. HSIAO,* LING-LING HSIEH,*
TOMMASO A. DRAGANI,[†] CARL PERAINO,[‡] AND MARTIN BEGEMANN*
*Columbia University Comprehensive Cancer Center and
Institute of Cancer Research
New York, New York 10032
[†]Instituto Nazionale per lo Studio e la Cura dei Tumori
20133 Milan, Italy
[‡]Division of Biological and Medical Research
Argonne National Laboratory
Argonne, Illinois 60439

OVERVIEW

Carcinogenesis involves complex changes in gene expression and the structure of cellular DNA. The DNA of mammalian cells, including humans, contains thousands of copies of endogenous retrovirus-like elements whose structures closely resemble transposable elements found in lower organisms. Because these elements might play a role in multistage carcinogenesis, we have carried out detailed studies on their expression during the course of development of carcinogen-induced and spontaneously arising rodent tumors. These studies indicate that carcinogen-induced liver tumor formation in both mice and rats is associated with a striking increase in the expression of endogenous retrovirus-like elements in carcinogen-induced liver tumors. In related studies we found that agents that induce liver cell proliferation cause the transient expression of these sequences. Studies employing cycloheximide and transient expression assays suggest that *trans*-acting repressor and activator proteins normally regulate the transcription of these sequences. Further studies of this phenomenon may lead to insights into the more widespread aberrations in gene expression often seen in tumors. Studies are also in progress to determine whether the expression of retrovirus-like sequences in tumors also leads, via reverse transcription, to gene transposition and insertion mutations and thus contributes to the process of tumor progression.

INTRODUCTION

We have previously postulated that multistage carcinogenesis involves, in addition to point mutations in cellular oncogenes, other complex changes in the genome, including gene transposition (Weinstein et al. 1984). Indeed, there is evidence that mammalian retroviruses resemble the transposable element copia in *Drosophila* and the transposable element Ty in yeast *Saccharomyces cerevisae*. The similarities include the following: (1) all three types of elements are flanked on their 5' and 3' sides by LTR-like sequences (termed δ in Ty) that promote transcription of adjacent genes; (2) they can all produce mutations by transpositions and insertion into nonhomologous DNA sequences; (3) the latter transpositions occur via an RNA transcript and reverse transcription; (4) the reverse transcriptase activity is encoded within the retrovirus-like sequence (*pol* gene) of all three types of elements; and (5) they generate five base pair duplications at the 5' and 3' sides of their sites of insertion into cellular DNA (for review, see Weinstein et al. 1984; Baltimore 1985). In view of the frequent chromosome rearrangements seen in fully malignant tumors, it is also of interest that transposition of the Ty element of yeast is frequently associated with chromosomal rearrangements.

A few examples now exist in which exogeneous retroviruses produce insertional mutagenesis in murine cells (Breindl et al. 1984; King et al. 1985). In addition, mutations in the coat color in mice can be due to insertion of a retrovirus genome into a locus determining coat color (Jenkins et al. 1981). The most direct evidence that endogenous retroviruses can undergo transposition and cause insertion mutations in mammalian cell are the examples in which the intracisternal A particle (IAP) sequence has been found inserted either into an immunoglobulin gene (Kuff et al. 1984) or the c-*mos* gene (Rechavi et al. 1982; Gattoni-Celli et al. 1983) in murine myelomas. It is of interest that these myelomas produce an abundance of IAP particles, which would be expected to increase the frequency of IAP transposition if, as seems to be the case, transposition occurs via reverse transcription. Also of interest is the finding that the constitutive production of interleukin 3 by the WEHI-38 murine leukemia cell line is due to the transposition of an endogenous IAP-long terminal repeat (LTR) sequence upstream of the interleukin 3 gene (Ymer et al. 1985). Because of their similarities, it has recently been proposed that copia, Ty, and endogenous retroviruses in mammalian cells be classified "retrotransposons" (Baltimore 1985; Boeke et al. 1985).

The LTR sequences are of particular interest because they contain transcription control signals including a TATAAA box, the CCAATC sequence, and enhancer sequences. They can potentiate the activity of other promoters even when the LTR itself is located 3' to the adjacent gene. Furthermore, the LTRs have structural similarities to similar elements of yeast and *Drosophila* (for review, see Weinstein et al. 1984; Baltimore 1985). The normal mouse genome contains at least five families of endogenous LTR-like sequences. Taken together, this means that a normal mouse cell contains over 2000 LTR-like sequences. Although it was often assumed that this was peculiar to rodents, there is increasing evidence that the normal human genome also contains several families and a very large number of copies of retrovirus-like elements. Thus, these elements represent a major repertoire for potential aberrations in the control of gene expression during carcinogenesis in rodents and in humans.

Carcinogen-transformed cells often display a switch-on in the transcription of endogenous retrovirus-like elements (Weinstein et al. 1984). Since these sequences also code for reverse transcriptase, once their expression is switched on, the RNA transcripts could undergo reverse transcription, and the resulting DNA copies might insert to new sites in the genome, thus producing insertion mutations and abnormalities in gene expression (Fig. 1). An epigenetic event (i.e., activation of transcription of endogenous retrovirus-like elements) could, therefore, lead to a stable genetic event and, thus, play a

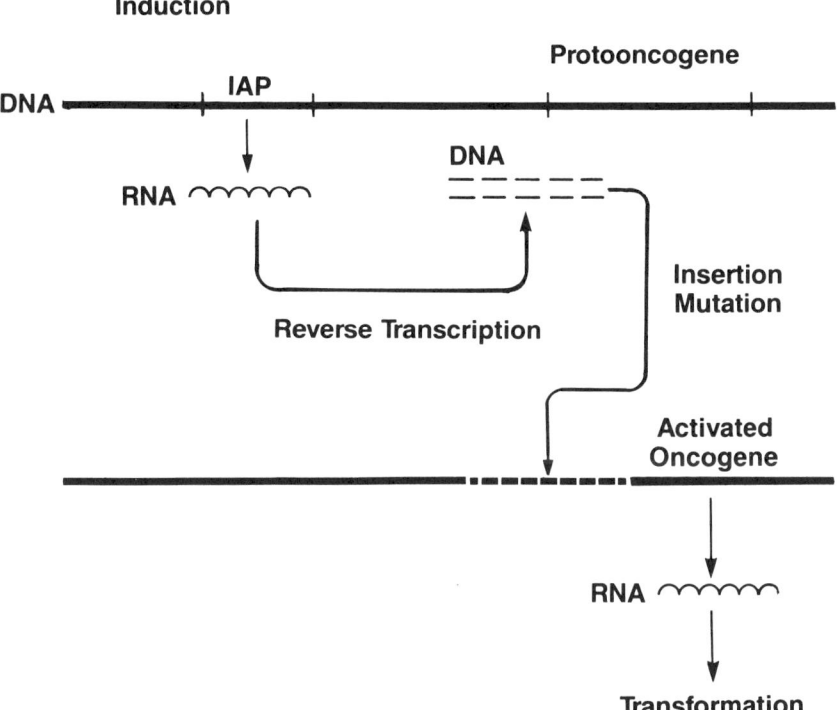

Figure 1
Hypothetical mechanism by which transient activation of transcription of an endogenous IAP, perhaps in response to exposure to environmental chemicals, could lead to transposition of this sequence, and, thus, a stable genetic change. This type of mechanism could explain activation of the oncogene c-*mos* in chemically induced murine plasmacytomas. (For additional details, see Weinstein et al. 1984.)

role during multistage carcinogenesis. DNA-damaging agents might also activate the transcription and transposition of retrotransposons. Thus, recent studies in yeast indicate that transcription of the Ty element is influenced by a number of factors including mating type, carbon source, and DNA-damaging agents (UV, etc.), and the rate of Ty transposition is increased 100-fold by exposure to elevated temperature (Paquin and Williamson 1984; Boeke et al. 1985). Transcription of the retrovirus-like transposable element copia is also enhanced by certain environmental stresses in *Drosophila* (Strand and McDonald 1985) and also when this element is integrated into the genome of rodent cells (Lambert et al. 1986; McDonald et al. 1987). For a review of evidence that various forms of environmental stress can induce the expression of transposition of mobile elements in a variety of both prokaryotic and eukaryotic organisms, see McDonald et al. (1987). We have, therefore, been carrying out a series of studies on the expression of endogenous retrovirus-like sequences in normal and carcinogen-transformed rodent cells.

RESULTS

Activation of Transcription of Endogenous Retrovirus DNA Sequences in Carcinogen-transformed Cells and Tumors

In our early studies we found that C3H 10T1/2 mouse fibroblasts, transformed by either chemical carcinogens or radiation, express a set of poly(A)$^+$ RNAs that contain sequences homologous to the LTR sequences of Moloney murine leukemia virus (Mo-MLV) (Kirschmeier et al. 1982). These RNAs were not detected or were present at very low levels in the corresponding untransformed cells. Since our initial findings were obtained in a cell culture system, it was necessary to demonstrate that they were relevant to carcinogenesis in the intact animal. We found that RNA sequences homologous to a Mo-MLV-LTR probe are also expressed in primary squamous cell skin carcinomas induced in mice by treatment with dimethylbenz(a)anthracene (DMBA) plus 12-*O*-tetradecanoyl-phorbol-13-acetate or repeated applications of DMBA alone, but these sequences were not expressed in the epidermis of normal mice (Housey et al. 1985). We have also found increased expression of endogenous retrovirus-related sequences in murine hepatocellular adenomas and carcinomas, which were induced by nitrosodiethylamine (NDEA) in male B6C3F mice or that occurred spontaneously in male C3HF mice (Dragani et al. 1986). In all of these tumors there were increased levels of RNAs homologous to Mo-MLV and IAP probes when compared to normal mouse liver. Furthermore, the retrovirus-like VL30 sequence was also expressed at increased levels in several but not all of the tumors. Some of these findings are illustrated in Figure 2. More recently, we found increased

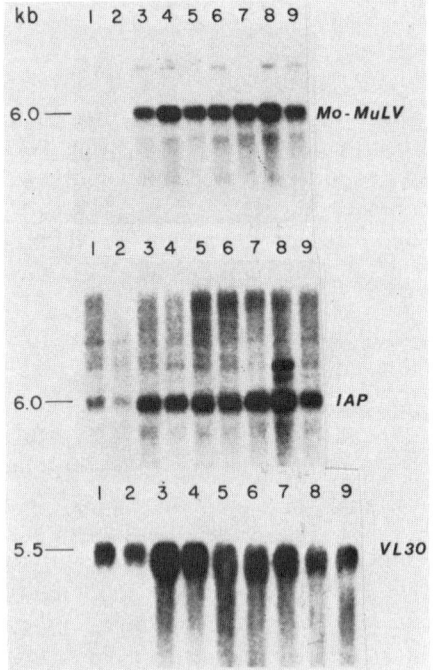

Figure 2
Analysis of Mo-MLV, IAP, and LV30 transcripts in murine B6C3F$_1$ hepatocellular tumor and normal livers. (Lanes *1*–*2*) Controls (normal liver); (lanes *3*–*6*) NDEA-induced adenomas; (lanes *7*–*9*) NDEA-induced carcinomas. (For additional details, see Dragani et al. 1986.)

expression of endogenous rat leukemia virus- and rat 30S-related sequences in rat hepatocellular adenomas and carcinomas induced by diethylnitrosamine and phenobarbital (Hsieh et al. 1987). Parallel assays indicated that these liver tumors often expressed increased levels of c-*myc* transcripts. Activation of c-*myc* does not, however, appear to be the cause for the increased expression of the retrovirus-related sequences, because the level of expression of this oncogene did not parallel the levels of expression of the retrovirus-related sequences, nor did the level of expression of the latter sequences correlate with the expression of the c-Ha-*ras* or c-*fos* proto-oncogene (Hsieh et al. 1987).

In view of the above results, we have also examined the effects of various factors on the expression of endogenous retrovirus-like sequences in non-neoplastic rodent liver (Dragani et al. 1987). The expression of RNA transcripts from three families of endogenous retrovirus-related sequences was investigated during liver cell proliferation in B6C3 mice. Treatment with a single dose of the liver mitogen and promoter of mouse hepatocarcinogenesis 1,4-bis[2-(3,5-dichloropyridyloxy)]benzene (TCPOBOP), or with carbon tetrachloride (CCl$_4$) induced liver cell proliferation at days 2 and 3 after treatment. Both of these treatments led to a marked increase in Mo-MLV-related 6 kb RNAs, which were most abundant at day 1 after TCPOBOP treatment and at day 2 after CCl_4. At days 1 and 2 after TCPOBOP and at days 1, 2, and 3 after CCl_4, IAP-related 6-kb RNAs were markedly increased. VL30-related transcripts were slightly decreased after TCPOBOP, but they were markedly increased after days 1 and 2 following CCl_4. The livers of 15-day-old untreated mice contained about a threefold higher level of Mo-MLV-related RNAs than adult liver. IAP-related 6-kb transcripts were present at threefold greater abundance in 7-day-old than in 15-day-old or adult liver. RNAs homologous to VL30 were detected at about the same levels in infant as well as adult livers.

Inhibition of protein synthesis by the administration of cycloheximide to adult mice caused a marked increase in the amount of Mo-MLV-, IAP-, and VL30-related RNAs in the livers of treated mice, suggesting the existence of labile proteins that normally regulate the abundance of these transcripts (Dragani et al. 1987). We postulate that the amounts of these putative regulatory proteins vary during both normal development and carcinogenesis and also in response to specific agents that induce liver cell proliferation. Consistent with this hypothesis are recent studies on rat liver regeneration, in which we have found that at 24 hours following partial hepatectomy, coinciding with the peak in DNA synthesis, there was a sixfold increase in the abundance of transcripts homologous to the endogenous rat-leukemia virus-like sequences, and by 72 hours the level of these transcripts returned to normal (L.L. Hsieh et al., unpubl.).

Mechanisms Underlying the Constitutive Expression of Endogenous Retrovirus Sequences in Tumor Cells

The above studies provide evidence that increased expression of endogenous-retrovirus-related sequences appears to be a hallmark of the carcinogenic process, and that it can occur at a relatively early stage since it was seen in mouse and rat liver adenomas. There are several possible mechanisms that might explain this phenomenon. These include the following: (1) synthesis by the tumor cells of a *trans*-acting transcription factor that recognizes and binds to specific enhancer/promoter sequences, present in the LTR regions of endogenous retrovirus-like sequences, and enhances their transcription; (2) hypomethylation of the endogenous retrovirus sequences; (3) alterations in the chromatin structure in the region of these sequences; and (4) point mutations, sequence amplifications, or gross rearrangements of these sequences.

We have found that the transient exposure of C3H 10T1/2 cells to 5-azacytidine, which leads to extensive hypomethylation of cellular DNA, causes a marked increase in the expression of both IAP and MLV-related

DNA sequences (Hsiao et al. 1986). Thus, DNA methylation appears to play a role in controlling the expression of these sequences. This may be relevant since disturbances in DNA methylation can occur during chemical carcinogenesis (Hsiao et al. 1986).

At the same time, we think that mechanism (1) (see above) also plays a role, in view of the increasing evidence that eukaryotic cells contain specific *trans*-acting proteins that act on specific enhancer and promoter sequences, to either enhance or inhibit the transcription of flanking structural genes during differentiation and virus infection. We have, therefore, recently carried out a series of transient expression experiments, using the construct prcm chloroamphenicol acetyltransferase (CAT)-1, obtained from M. Horowitz (Horowitz et al. 1984), in which an LTR sequence derived from an endogenous murine IAP sequence is linked to the bacterial gene CAT. We found that when this construct was transfected into the carcinogen-transformed cell line, MCA 10T1/2, there was about a tenfold greater expression of CAT activity than obtained in parallel experiments with normal C3H 10T1/2 cells (M. Begemann et al., unpubl.). Studies with ^{32}P-labeled DNA indicate that this difference is not due to greater uptake of the transfected DNAs by the transformed cells. Competition studies, in which an excess of plasmid DNA containing only the IAP LTR sequence was cotransfected with the IAP LTR-CAT construct, also provide evidence that factors required for the expression of IAP-LTR-linked sequences are present in limited amounts in normal 10T1/2 cells but are present in excess amounts in carcinogen-transformed derivatives. Thus, it would appear that the increased expression of endogenous retrovirus-related sequences in carcinogen-transformed rodent cells is due, at least in part, to altered levels of proteins that control the transcription of specific DNA sequences. This phenomenon may provide insights into the basis of the widespread alterations in gene expression that are often found in tumor cells (Weinstein et al. 1984). We are currently employing the DNA band shift gel electrophoresis method to determine whether transformed cells differ from normal cells in terms of the presence of proteins that bind to specific LTR sequences.

SUMMARY

As discussed above, and as is seen in Figure 1, it is possible that the persistent expression of high levels of endogenous retrovirus-like sequences in tumor cells could lead, via reverse transcription, to gene transposition, thus producing insertion mutations or LTR-enhanced gene transcription. These complex genomic events could play an important role in the process of multistage carcinogenesis.

ACKNOWLEDGMENTS

The authors are grateful to Mrs. Nancy Mojica for valuable assistance in the preparation of this manuscript. This research was supported by National Institutes of Health grant CA-02111 and funding from the National Foundation for Cancer Research (to I.B.W.); the U.S. Department of Energy, Office of Health and Environmental Research, under contract no. W-31-109-End-38; the National Toxicology Program, National Institute of Environmental Health Sciences under Interagency Agreement no. YO1-ES-29901 (to C.P.); and an Associaziole Italiana per le Finalized Project "Oncology" Consiglio Nacionale della Ricerche (C.N.R.), Rome Italy (to T.D.).

REFERENCES

Baltimore, D. 1985. Retroviruses and retrotransposons: The role of reverse transcription in shaping the eukaryotic genome. *Cell* **40**: 481.

Boeke, J.D., D.J. Garfinkel, C.A. Styles, and G.R. Fink. 1985. Ty elements transpose through an RNA intermediate. *Cell* **40**: 491.

Breindl, M., K. Harbersand, and R. Jaenisch. 1984. Retrovirus-induced lethal mutation in collagen I gene of mice is associated with an altered chromatin structure. *Cell* **38**: 9.

Dragani, T.A., G. Manenti, G. Della Porta, and I.B. Weinstein. 1987. Factors influencing the expression of endogenous retrovirus-related sequences in the liver of B6C3 mice. *Cancer Res.* **47**: 795.

Dragani, T.A., G. Manenti, G. Della Porta, S. Gattoni-Celli, and I.B. Weinstein. 1986. Expression of retroviral sequences and oncogenes in mouse hepatocellular tumors. *Cancer Res.* **46**: 1915.

Gattoni-Celli, S., W.L. Hsiao, and I.B. Weinstein. 1983. Re-arranged c-mos locus in a MOPC-21 murine myeloma cell line and its persistance in hybridomas. *Nature* **306**: 795.

Horowitz, M., S. Luria, G. Rechavi, and D. Givol. 1984. Mechanism of activation of the mouse c-mos oncogene by the LTR of an intracisternal A-particle gene. *EMBO J.* **3**: 2937.

Housey, G.H., P. Kirschmeier, S.J. Garte, F. Burns, W. Troll, and I.B. Weinstein. 1985. Expression of long terminal repeat (LTR) sequences in carcinogen-induced murine skin carcinomas. *Biochem. Biophys. Res. Commun.* **127**: 392.

Hsiao, W.L., S. Gattoni-Celli, and I.B. Weinstein. 1986. Effects of 5-azacytidine on the expression of endogenous retrovirus-related sequences in C3H 10T1/2 cells. *J. Virol.* **57**: 1119.

Hsieh, L.L., W.-L. Hsiao, C. Peraino, R.R. Maronpot, and I.B. Weinstein. 1987. Expression of retroviral sequences and oncogenes in rat liver tumors induced by diethylnitrosamine. *Cancer Res.* **47**: 3421.

Jenkins, N., N.G. Copeland, B.A. Taylor, and B.K. Lee. 1981. Dilute(d) coat colour mutation of DBA12J mice is associated with the site of integration of an ecotropic MuLV genome. *Nature* **293**: 370.

King, W., M.D. Patel, L.I. Lobel, S.P. Goff, and M.C. Nguyen-Huu. 1985. Insertion mutagenesis of embryonal carcinoma cells by retroviruses. *Science* **228:** 554.

Kirschmeier, P., S. Gattoni-Celli, D. Dina, and I.B. Weinstein. 1982. Carcinogen and radiation transformed C3H 10T1/2 cells contain RNA's homologous to the long terminal repeat sequences of a murine leukemia virus. *Proc. Natl. Acad. Sci.* **79:** 2773.

Kuff, E.L., A. Feenstra, K. Lueders, L. Smith, R. Hawley, N. Hozumi, and M. Shulman. 1984. Intracisternal A-particle genes as movable elements in the mouse genome. *Proc. Natl. Acad. Sci.* **80:** 1992.

Lambert, M.E., J.I. Garrels, J. McDonald, and I.B. Weinstein. 1986. Inducible cellular responses to DNA damage in mammalian cells. In *Antimutagenesis and anticarcinogenesis* (ed. Shankel et al.), p. 291. Plenum Press, New York.

McDonald, J.F., D.J. Strand, M.E. Lambert, and I.B. Weinstein. 1987. The responsive genome: Evidence and evolutionary implications. *MBL Lect. Biol.* **8:** 239.

Paquin, C.E. and V.M. Williamson. 1984. Temperature effects on the rate of Ty transposition. *Science* **226:** 53.

Rechavi, G., A. Givol, and L. Canaani. 1982. Activation of a cellular oncogene by DNA rearrangement: Possible involvement of an IS-like element. *Nature* **300:** 607.

Strand, D. and J. McDonald. 1985. Copia is transcriptionally responsive to environmental stress. *Nucleic Acids Res.* **13:** 4401.

Weinstein, I.B., S. Gattoni-Celli, P. Kirschmeier, M. Lambert, W.-L. Hsiao, J. Backer, and A. Jeffrey. 1984. Multistage carcinogenesis involves multiple genes and multiple mechanisms. In *Cancer cells. The transformed phenotype* (ed. A. Levine et al.), vol. 1, p. 229. Cold Spring Harbor Laboratory Press, Cold Spring Harbor, New York.

Ymer, S., W.O.J. Tucker, C.J. Sanderson, A.J. Hapel, and I.G. Young Campbell. 1985. Constitutive synthesis of interleukin-3 by the leukemia cell line WEHI-3B is due to retroviral insertions near the gene. *Nature* **317:** 255.

Retrovirus-like Particles, Reoviruses, and c-*src* Expression in Malignant Tumors of *Drosophila melanogaster*

ELISABETH GATEFF
Institute of Genetics
Johannes Gutenberg University
6500 Mainz, Federal Republic of Germany

OVERVIEW

Mutated, recessive differentiation genes of *Drosophila melanogaster* cause malignant transformation of specific tissues and cell types in homozygous mutant animals. Retrovirus-like particles (RLPs; previously virus-like particles, VLPs) were regularly observed in mutant tumor cells. The presence of RLPs and copia transcripts was studied in the tumorous lethal(2)giant larvae (*l(2)gl*) neuroblasts in situ and after transplantation in adult abdomens (Ryo et al. 1984). Neither RLPs nor copia transcripts were detected in the mutant neuroblasts in situ. However, large amounts of copia transcripts were found in RLPs from *l(2)gl* neuroblasts subcultured in the abdomens of wild-type flies. Karyotypic abnormalities increased dramatically with each new transfer generation in subculture in vivo.

Moderate amounts of RLPs were also found in the blood cells in situ in two blood tumor mutants: lethal(1)malignant blood cell neoplasm, *l(1)mbn*, and lethal(3)malignant blood cell neoplasm, *l(3)mbn*. The number of RLPs increased drastically after in vitro subculture of *l(3)mbn* blood cells and the blood cells of the mutant lethal(2)malignant blood neoplasm, *l(2)mbn*. In addition to large amounts of RLPs, a new respiratory enteric orphan virus (reovirus) designated as *Drosophila* F virus (DFV) and a well-known picorna virus, the *Drosophila* C virus (DCV), were found in in-vitro-subcultured *l(2)mbn* and *l(3)mbn* blood cells. Compared to the tumorous blood cells in situ, the in-vitro-subcultured blood cells showed slightly elevated c-*src* expression.

INTRODUCTION

Mutations in specific, recessive-lethal genes of *Drosophila* cause the development of malignant tumors in cells and tissues such as neuroblasts of the adult optic centers in the larval brain, the imaginal disks, the blood cells in the hematopoietic organs, and the gonial cells (Gateff 1978a,b,c,1982). These genes are expressed autonomously in the respective cells and most probably initiate and/or direct the realization of specific differentiation programs. In the homozygous mutant state, their function is abolished, resulting in malignant transformation.

The fine structure of the tumor cells of four mutants was studied and in all cases VLPs were found within their nuclei (Akai et al. 1967; Shrestha and Gateff 1982a,b,1986). An attempt to isolate and characterize these particles from tumorous blood cell lines in vitro failed. Instead, the DFV and DCV were found in large amounts in the cytoplasm of the cells (Gateff et al. 1980).

First indications that the VLPs may be retrovirus-like came from the studies of Flavell and Ish Horowicz (1981), who found in in-vitro-cultured *Drosophila* cells, containing large amounts of VLPs, circular copies of the transposable element copia. Subsequent studies by Shiba and Saigo (1983) demonstrated copia transcripts in the VLPs isolated from in-vitro-subcultured cell. Since copia structurally resembles retroviruses, the VLPs were renamed RLPs. Proof that the RLPs in the neuroblasts of the tumor mutant *l(2)gl* also contain copia transcripts was shown by Ryo et al. (1984). These results and related findings will be reviewed and discussed.

RESULTS

RLPs were described repeatedly in the nuclei of various *Drosophila* cells (for review, see Gateff 1978c). Studying the fine structure of the wild-type eye-antennal imaginal disk and sublines derived from it, in culture in adult abdomens, Akai et al. (1967) and Gateff et al. (1974) found numerous RLPs in the cells of in-vivo-subcultured imaginal disk sublines. With each new subculture generation their numbers increased. Wild-type imaginal disk cells, in contrast, were practically devoid of RLPs. RLPs were also found in the tumorous cells of the following four recessive-lethal tumor mutants: *l(2)gl* (Akai et al. 1967), *l(1)mbn*, (Shrestha and Gateff 1982a,b), *l(2)mbn*, and *l(3)mbn* (Shrestha and Gateff 1986).

The *l(2)gl* mutation is lethal at the end of larval life. The lethality is caused by a malignant neuroblastoma developing in the presumptive optic centers of the larval brain and tumors of the imaginal disk (Gateff and Schneiderman 1969, 1974). Comparative fine-structure studies of wild-type and *l(2)gl* neuroblasts in situ revealed very small amounts of RLPs in the *l(2)gl* brain cells, whereas wild-type brain cells were completely devoid of RLPs (Akai et al. 1967). However, after transplantation of mutant neuroblasts into the abdomens of wild-type flies and their serial subculture for many transfer generations, the amount of RLPs increased dramatically. Furthermore, Gateff and Schneiderman (1969, 1974) and Ryo et al. (1984) reported various karyotypic abnormalities in the *l(2)gl* neuroblasts after prolonged in vivo subculture.

The finding of copia transcripts in RLPs from the permanent *Drosophila* cell line GM_2 (Shiba and Saigo 1983) stimulated the investigation by Ryo et al. (1984), designed to analyze the nature of the RLPs found in the *l(2)gl* neuroblasts. In a first approach, the titer of RLPs in *l(2)gl* neuroblasts from the 27th transfer-generation in vivo was compared with that from the embryo-derived cell line, GM_2. This comparison showed an eightfold increase of RLPs in the *l(2)gl* neuroblasts over that in the GM_2 cells. Northern blots with RLP-RNA from *l(2)gl* neuroblasts, subcultured in vivo for 27 generations, were hybridized with a copia probe. Whereas the *l(2)gl* neuroblasts in situ were negative, the subcultured *l(2)gl* neuroblasts showed extremely high levels of copia RNA.

The above studies demonstrated that the particles, found in large amounts in the in-vivo-subcultured *l(2)gl* neuroblasts, contain copia transcripts. Furthermore, under the stress of in-vivo-subculture, copia transcripts increased dramatically in the mutant neuroblasts where they became packed into protein coats. T. Shiba (unpubl. data in Ryo et al. 1984) indicates that the increased amount of copia RNA was not due to a higher expression rate of copia sequences but, instead, related to an increased number of copia elements per genome.

The stress of subculture in vivo also influences the stability of the neuroblast karyotype. In an exhaustive analysis, Ryo et al. (1984) showed increasing amounts of karyotypic abnormalities in the *l(2)gl* neuroblasts during four consecutive subculture generation in vivo in wild-type adults. In situ, 2% of the *l(2)gl* neuroblasts showed tetraploid karyotypes. Already, during the first in vivo transfer generation, 22% of the investigated cells were tetraploid or aneuploid. In the fourth transfer generation, 49% of the cells exhibited karyotypic abnormalities. Among these, 5% were tetraploid, and the remaining 44% showed aneuploidy. Whether the instability of the karyotype in in-vivo-subcultured neuroblasts correlates to the increase of RLPs remains to be shown. These studies indicate that the stress conditions of in vivo subculture, together with the *l(2)gl* mutation, are in some way involved in the increase of copia element numbers, RLPs, and karyotypic instability.

Electron micrographic studies of the blood cells in the above-mentioned three blood tumor mutants, *l(1)mbn, l(2)mbn,* and *l(3)mbn*, revealed numerous RLPs in situ as well as after culture in vitro. In these mutants the phagocytic blood cell precursors, the so-called proplasmatocytes, in the hematopoietic organs are transformed (Gateff 1978a; Shrestha and Gateff 1982b, 1986). The tumorous cells grow excessively and kill the host at the end of larval life. In the wild type, in contrast, hematopoiesis is strictly regulated, and blood cells are produced as part of the developmental program and as guards against infection or parasitism (Rizki 1978; Shrestha and Gateff 1982a). The two blood cell types found in the hemolymph, the phagocytic plasmatocytes, and the crystal cells, originate in the hematopoietic organs located along the dorsal heart vessel. In these organs, immature blood cells, designated as prohemocytes, give rise to proplasmatocytes and procrystal cells that differentiate into plasmatocytes. Plasmatocytes are round cells that possess the capacity for phagocytosis. They can change their morphology by becoming elongated, exhibiting numerous cytoplasmic processes. This morphological variant is called podocyte. Podocytes in turn transform into extremely flat cells that are referred to as lamellocytes. Their function is to encapsulate foreign matter. In all three mutants, the precursor cells of the plasmatocytes, e.g., the proplasmatocytes, are malignantly transformed, whereas crystal cells are not affected.

Fine-structure studies of wild-type blood cells revealed the complete absence of RLPs in any of the differentiation stages of the plasmatocyte and crystal cell lines (Table 1; Shrestha and Gateff 1982a). In the blood tumor mutants, $l(1)mbn$ and $l(3)mbn$ in situ, on the other hand, RLPs were found in the nuclei as well as in the cytoplasm of the different blood cells (Table 1; Shrestha and Gateff 1982b, 1986). The two mutants differed considerably, however, in the amount of RLPs. The various differentiation stages of the $l(1)mbn$ blood cells showed up to 20 times more RLPs than their counterparts in the $l(3)mbn$ mutant. Table 1 further compares the content of RLPs in the $l(2)mbn$ and $l(3)mbn$ blood cells subcultured for more than 20 generations in vitro. Blood cells in vitro are more uniform and show the appearance of plasmatocytes. All cells contain RLPs. The $l(2)mbn$ blood cells, however, exhibit eight times more RLPs than the $l(3)mbn$ blood cells. Comparison of the amount of RLPs in the blood cells in situ with that after culture in vitro shows a drastic increase of RLPs after subculture in vitro.

The above observations permit the general conclusion that the stress created by the mutations and/or by the subculture of the mutant cells, either in vivo or in vitro, causes the production of RLPs. Furthermore, with progress of time in culture, in vivo or in vitro, RLPs increase often dramatically.

Our attempt to isolate and characterize the RLPs from in-vitro-subcultured $l(2)mbn$ and $l(3)mbn$ blood cells was unsuccessful. Nevertheless, our efforts resulted in the discovery of DFV and of the already well-known DCV (Gateff et al. 1980).

Both viruses contain RNA genomes and replicate exclusively in the cytoplasm. Since the two viruses were found neither in mutant blood cells in situ nor in the wild-type stock from which the mutants originated, the question concerning the origin of the viral genomes and their activation after subculture in vitro, resulting in virus-particle production, remains to be solved. Apparently, in the mutant cells, after subculture in vitro, the control mechanisms existing in situ are abolished and, thus, both the replication of the viral genome and viral particle production ensue. Nothing is known concerning the

Table 1
RLPs in the Nuclei and the Cytoplasm of Wild-type, l(1)mbn, l(2)mbn, and l(3)mbn Blood Cells In Situ and after Culture In Vitro

	RLPs in situ						RLPs in vitro			
	wild-type[a]		l(1)mbn[b]		l(3)mbn		l(2)mbn[c]		l(3)mbn[c]	
Cell type	nucleus	cytoplasm	nucleus	cytoplasm	nucleus	cytoplasm	nucleus	cytoplasm	nucleus	cytoplasm
Prohemocyte	–	–	38	4	2.2	–				
Proplasmatocyte	–	–	104	7	3.2	–				
Plasmatocyte	–	–	90	14	30.4	0.8	851	84	105	36
Podocyte	–	–	43	7	2.9	1.1				
Lamellocyte	–	–	134	5	7.1	1.6				
Procrystal cell	–	–	46	10	0.33	–				
Crystal cell	–	–	+	+	1.1	–				

The numbers are mean values from counts of RLPs in ten ultrathin sections from ten different cells.
[a] – indicates RLPs not present.
[b] + indicates crystal cells were not found.
[c] In culture in vitro no distinction can be made between different blood cell stages. All blood cells have generally the appearance of plasmatocytes.

interactions between cellular and viral factors, leading either to viral suppression or replication.

A final question addressed by A. Barnekow and E. Gateff (unpubl.) was whether c-*src* is overexpressed in the tumorous blood cells in situ and/or in vitro when compared to the wild-type. The results are presented in Table 2. The profile of pp60$^{c\text{-}src}$ kinase activity during wild-type *Drosophila* development is similar to that described by Schartl and Barnekow (1984) for normal vertebrate development. The highest kinase activity was detected 2–3 hours before the hatching of the larva from the egg. In the *l(3)mbn* mutant larvae in situ, no increased pp60$^{c\text{-}src}$ kinase activity over that of the third instar wild-type larvae could be measured. This implies that the mutation is not affecting pp60$^{c\text{-}src}$ expression. However, after subculture in vitro, pp60$^{c\text{-}src}$ kinase activity increased significantly. In both mutant cell lines the exponentially growing cells exhibited higher kinase activity than the cells in the stationary phase. Obviously, the in vitro conditions seem to be the primary cause for the increased kinase activities in the *l(2)mbn* and *l(3)mbn* blood cells.

Table 2

pp60$^{c\text{-}src}$ Kinase Activity during Wild-type Development and in *l(2)mbn* and *l(3)mbn* Mutant Blood Cells In Situ and In Vitro

	Cpm/mg protein (S.D. 5%)
Developmental stages of wild-type	
Embryo	
3 hr	250
5 hr	250
18 hr	800
21 hr	600
Third instar larva	250
Pupa	300
Adult	400
Blood cells from third instar mutant larvae in situ and in vitro	
l(2)mbn in vitro[a] (exponential growth)	1000
l(2)mbn in vitro (stationary phase)	650
l(3)mbn in situ	200
l(3)mbn in vitro (exponential growth)	700
l(3)mbn in vitro (stationary phase)	370

[a]The *l(2)mbn* mutant animals lost their tumor phenotype. For this reason measurements in situ could not be made. For antisera, preparations of cell extracts, immunoprecipitation, and protein determination see Barnekow and Bauer (1984).

DISCUSSION

The results presented here certainly pose more questions than they answer. The most pertinent question concerns the correlation between transposons and the induction of recessive-lethal mutations in genes responsible for tumor development. The structural analysis of mutated *Drosophila* genes revealed that, in general, approximately 70% carried insertions of one of the four groups of transposable elements. Data are available for only one of the *Drosophila* tumor genes, namely the brain and imaginal disk tumor mutant *l(2)gl*. The molecular analysis of more than 50 *l(2)gl* alleles disclosed that only 2 of them had a foreign DNA insert (Mechler et al. 1985). One of the inserts was identified as the copia-like element B104 (Lützelschwab et al. 1986). The insert in the second allele is not yet characterized. The remaining *l(2)gl* alleles represent smaller or larger deletions of part of the gene, the entire gene, or additional sequences in centromeric direction (Mechler et al. 1985; U. Schacker and B.M. Mechler, pers. comm.). For the remaining recessive tumor genes, molecular data are not yet available. The *l(2)gl* case demonstrates, however, the insertions of a transposable element are in most cases not the cause for the mutation. Nevertheless, the possibility cannot be excluded that some of the *l(2)gl* deletions may have originated from the excision of a P element. Excision of copia has not yet been demonstrated (Finnegan and Fawcett 1986).

P-element mutagenesis performed on five independent, recessive-lethal tumor genes revealed a high proportion of recessive-lethal mutations. From 15,278 P-element-mutagenized third chromosomes, 1317 had a recessive-lethal mutation. Among these, one P element was found in the temperature-sensitive brain tumor gene lethal(3) brain tumor[ts] *(l[3]bt*[ts]*)*. A second P element was associated with a visible deletion removing the gene (T. Löffler and E. Gateff, unpubl.). Twelve further P-element insertions were found in the vicinity of the blood tumor mutant *l(3)mbn* located in the region 64F–65D (G. Becker and E. Gateff, unpubl.). No P element insertions were found for the tumor genes on the first and second chromosomes (P. Gräf et al., unpubl.).

The second question addresses the possibility of a causal relationship between a tumor gene mutation and the increase of RLPs. In the case of *l(2)gl* neuroblasts in situ, small amounts of RLPs were found in the nuclei (Akai et al. 1967), indicating that the mutation may relate to copia amplification and RLP production. Similar observations were made in the tumorous blood cells in the *l(1)mbn* and *l(3)mbn* mutants. However, tissue culture alone can also cause RLP overproduction. This was the case with serially subcultured wild-type imaginal disks in vivo (Gateff et al. 1974). Table 1 shows that RLPs increased on the average 8–20 times in in vitro subcultured *l(2)mbn* and *l(3)mbn* blood cells, compared with the cells in situ. Furthermore, the considerable numerical difference of RLPs in the *l(1)mbn* and *l(3)mbn* mutants in situ indicates that additional factors play a role in suppressing or promoting RLPs. In the light of the above data, it appears that both a mutation and the physiological stress of tissue culture are separately and/or mutually involved in copia expression and transposition.

A further question concerns the karyotypic abnormalities observed in the *l(2)gl* neuroblasts after in vivo subculture (Ryo et al. 1984). Are they the result of copia transposition or is the stress environment of in vivo culture causing the observed phenomenon? In situ, the karyotype of *l(2)gl* neuroblasts is normal and copia is not expressed. In culture in vivo, on the other hand, 50% of the cells showed aberrant karyotypes after four transfer generations, it is safe to conclude that the mutation is not associated with the karyotypic abberrations. Copia is expressed during normal development. Nevertheless, RLPs are extremely rare. The Ty transposable element in yeast, which is stucturally similar to retroviruses and copia, is associated with VLPs that appear in the cytoplasm after induction of transposition (Garfinkel et al. 1985). In *Drosophila* a similar mechanism may underlay RLP production in the mutant tumor cells.

The presence of a reovirus and a picorna virus, both RNA viruses, in addition to the RLPs in the *l(2)mbn* and *l(3)mbn* tumorous blood cells cultured in vitro, indicates that the in vitro culture conditions promote the growth of these viruses. In mutant blood cells in situ the two viruses could not be detected (Gateff et al. 1980). It is possible, however, that at the time of the induction of the mutations the wild-type flies were virus-infected but subsequently lost the viruses, whereas in the mutants they persisted and became extremely prolific only after subculture in vitro.

The overexpression of $pp60^{c\text{-}src}$ in the exponentially growing mutant blood cells is a further phenomenon, observed only in the in-vitro-cultured blood cells. Mutant blood cells in situ show normal levels of $pp60^{c\text{-}src}$. This indicates that the mutations are not causing $pp60^{c\text{-}src}$ overexpression, rather the adverse in vitro conditions are the cause (Table 2). The presence of RLPs, on the other hand, seem to correlate with the mutations (Table 1).

In conclusion, copia overexpression and RLP production in *l(2)gl* neuroblasts is not causally related to the gene mutation. On the other hand, RLP production seems to relate to the *l(1)mbn* and *l(3)mbn* tumor mutations. In all instances, subculture in vivo or in vitro causes the increase of RLPs over that in situ. Subculture in vitro is also responsible for $pp60^{c\text{-}src}$ overexpression in exponentially growing tumorous blood cells. Finally, the *l(2)gl* mutation is not responsible for the karyotypic abnormalities in the neuroblasts observed after subculture in vivo.

ACKNOWLEDGMENTS

This research was supported by a Deutsche Forschungsgemeinschaft grant SFB 46. The author thanks Dr. B.M. Mechler and A. Barnekow for critical discussions.

REFERENCES

Akai, H., E. Gateff, L.E. Davis, and H.A. Schneiderman. 1967. Virus-like particles in cultured and mutant tissues of *Drosophila*. *Science* **157**: 810.

Barnekow, A. and H. Bauer. 1984. The differential expression of the cellular *src*-gene product pp60$^{c\text{-}src}$ and its phosphokinase activity in normal chicken cells and tissues. *Biochim. Biophys. Acta* **782**: 94.

Finnegan, D.J. and D.H. Fawcett. 1986. Transposable elements in *Drosophila melanogaster*. *Oxford Surv. Euk. Genes* **3**: 1.

Flavell, A.J. and D. Ish-Horowicz. 1981. Extrachromosomal circular copies of the eukaryotic transposable element copia in cultured *Drosophila* cells. *Nature* **292**: 591.

Garfinkel, D.J., J.D. Bocke, and G.R. Fink. 1985. Ty element transposition: Reverse transcriptase and virus-like particles. *Cell* **42**: 507.

Gateff, E. 1978a. Malignant neoplasms of genetic origin in *Drosophila melanogaster*. *Science* **200**: 1448.

———. 1978b. The genetics and epigenetics of neoplasms in *Drosophila*. *Biol. Rev.* **53**: 123.

———. 1978c. Malignant and benign neoplasms of *Drosophla melanogaster*. In *The genetics and biology of* Drosophila (ed. M. Ashburner and T.R.F. Wright), vol. 2B, p. 182. Academic Press, London.

———. 1982. Cancer genes and development—The *Drosophila* case? *Adv. Cancer Res.* **37**: 33.

Gateff, E. and H.A. Schneiderman. 1969. Neoplasms in mutant and cultured wild-type tissues of Drosophila. *Natl. Cancer Inst. Monogr.* **31**: 365.

———. 1974. Developmental capacities of benign and malignant neoplasms of *Drosophila*. *Wilhelm Roux's Archiv. Entwicklungsmech. Org.* **176**: 23.

Gateff, E., H. Akai, and H.A. Schneiderman. 1974. Correlations between developmental capacity and structure of tissue sublines derived from eye-antennal imaginal disks of *Drosophila melanogaster*. *Wilhelm Roux's Archiv. Entwicklungsmech. Org.* **173**: 89.

Gateff, E., L. Gissmann, N. Plus, H. Pfister, I. Schroeder, R. Shrestha, and H. zur Hausen. 1980. Characterization of two tumorous blood cell lines of *Drosophila melanogaster* and the viruses they contain. In *Invertebrate systems in vitro* (ed. E. Kurstak et al.), p. 517. Elsevier/North-Holland, Amsterdam.

Lützelschwab, R., G. Müller, B. Wälder, O. Schmidt, R. Fürbass, and B. Mechler. 1986. Insertion mutation inactivates the expression of the recessive oncogene *lethal(2)giant larvae* of *Drosophila melanogaster*. *Mol. Gen. Genet.* **204**: 58.

Mechler, B.M., W. McGinnis, and W.J. Gehring. 1985. Molecular cloning of *lethal(2)giant larvae*, a recessive oncogene of *Drosophila melanogaster*. *EMBO J.* **4:** 1551.

Rizki, M. 1978. The circulatory system and associated cells and tissues. In *The genetics and biology of* Drosophila (ed. M. Ashburner and T.R.F. Wright), vol. 2b, p. 397. Academic Press, London.

Ryo, H., T. Shiba, S. Kondo, and E. Gateff. 1984. Chromosomal aberrations and retrovirus-like particles by *in vitro* transplantation in neoplastic brain cells of a *Drosophila* mutant strain. *Gann* **75:** 22.

Schartl, M. and A. Barnekow. 1984. Differential expression of the cellular *src* gene during vertebrate development. *Dev. Biol.* **105:** 415.

Shiba, T. and K. Saigo. 1983. Retrovirus-like particles containing RNA homologous to the transposable element copia in *Drosophila melanogaster*. *Nature* **302:** 119.

Shrestha, R. and E. Gateff. 1982a. Ultrastructure and cytochemistry of the cell types in the larval hematopoietic organs and hemolymph of *Drosophila melanogaster*. *Dev. Growth Differ.* **24:** 65.

———. 1982b. Ultrastructure and cytochemistry in the tumorous hematopoietic organs and the hemolymph of the mutant *lethal(1)malignant blood neoplasm (l(1)mbn)* of *Drosophila melanogaster*. *Dev. Growth Differ.* **24:** 83.

———. 1986. Ultrastructure and cytochemisy of the tumorous blood cells in the mutant *lethal(3)malignant blood neoplasm* of *Drosophila melanogaster*. *Invertebr. Pathol.* **48:** 1.

HTLV-I, HIV, and Human T Cell Growth

WARNER C. GREENE, MITCHELL DUKOVICH, YUJI WANO,
AND MIRIAM SIEKEVITZ
Howard Hughes Medical Institute
Department of Medicine
Duke University School of Medicine
Durham, North Carolina 27710

OVERVIEW

The biological effects of human T-lymphotropic virus type I (HTLV-I) and human immunodeficiency virus (HIV) retroviral infection on interleukin 2 (IL-2) and IL-2 receptor gene expression in human $CD4^+$ T cells were studied. The HTLV-I-derived *trans*-activator gene product (*tat* I), was shown to stimulate the cellular promoter regulating IL-2 receptor (Tac) gene expression and to provide one of the two mitogenic signals required for activation of the IL-2 promoter. These findings raise the possibility that HTLV-I-induced T-cell transformation may involve an early period of uncontrolled autocrine T-cell proliferation mediated by the inappropriate expression of this T-cell growth factor and receptor gene. In contrast to HTLV-I, HIV infection of $CD4^+$ lymphocytes is frequently associated with cytopathic effects and clinically with the development of the acquired immune deficiency syndrome (AIDS). In viable $CD4^+$ T cells infected with HIV, this virus does not alter mitogen-induced IL-2 receptor gene expression but does markedly inhibit IL-2 gene activation. The lack of production of this growth factor, which is required for the growth of $CD4^+$ T cells, may contribute to the cytopathic effects of this retrovirus and play a role in the progressive depletion of the $CD4^+$ subset of T cells observed in AIDS. Additional studies of $CD4^+$ T cells latently infected with HIV have revealed that mitogenic activation of these cells triggers HIV gene replication. This response appears to reflect properties encoded by the viral long terminal repeat (LTR), as four different types of T-cell mitogens were shown to be capable of activating this viral element. Deletion analysis of the HIV LTR has permitted identification of important positive and negative regulatory sequences that modulate the overall level of viral gene expression following mitogen stimulation.

INTRODUCTION

HTLV-I retroviral infection may lead to the neoplastic transformation of $CD4^+$ T cells and to the development of the adult T-cell leukemia (ATL)

(Poiesz et al. 1980, 1981; Yoshida et al. 1982). The mechanism by which this oncogenic human retrovirus induces T-cell transformation remains undefined at present. However, unlike many other acutely transforming animal retroviruses, HTLV-I does not contain a recognized oncogene (Seiki et al. 1983; Wong-Staal and Gallo 1985), nor does this virus appear to produce cis-activation of select cellular proto-oncogenes by insertional mutagenesis, as the sites of proviral integration vary markedly from tumor to tumor (Seiki et al. 1984). ATL cell lines, however, are uniformly characterized by the deregulated display of large numbers of IL-2 receptors (Tac antigen) (Depper et al. 1984; Waldmann et al. 1984; Krönke et al. 1985) and occasionally by the production of IL-2 (Arya et al. 1984). These findings have raised the theoretical possibility that the inappropriate expression of these cellular proteins that regulate T-lymphocyte growth may contribute to the neoplastic transformation induced by HTLV-I. In terms of the mechanism of IL-2 receptor activation, HTLV-I encodes a 40-kD *tat* I protein that markedly enhances the expression of other viral genes in a *trans* manner and is obligately required for viral replication (Haseltine et al. 1984; Kiyokawa et al. 1984; Sodroski et al. 1984; Seiki et al. 1985; Shimotohno et al. 1985). These transcriptional properties of this protein have suggested the possibility that the *tat* I gene product might also be capable of activating the expression of certain cellular genes including IL-2 and the IL-2 receptor. In the following sections, we discuss our data demonstrating that the *tat* I protein activates the cellular promoters regulating IL-2 and IL-2 receptor gene expression.

HIV also displays selective tropism for $CD4^+$ T cells (Dalgleish et al. 1984; Klatzmann et al. 1984). Recent studies have implicated the CD4 receptor as the cellular binding site for this retrovirus (Dalgleish et al. 1984; Klatzmann et al. 1984; McDougal et al. 1986). Unlike HTLV-I, HIV infection often leads to marked cytopathic effects on this subset of T cells and clinically to a severe and uniformly fatal immunodeficiency termed AIDS (Broder and Gallo 1984). Recent investigations suggest that, under some circumstances, the virus may latently infect $CD4^+$ T cells (Hoxie et al. 1985; Folks et al. 1986; Zagury et al. 1986). In the following sections we discuss our studies of the role of mitogenic stimulation in the control of latent versus productive HIV infection and the marked inhibitory effects that this virus exerts on mitogen-induced IL-2 gene expression.

RESULTS AND DISCUSSION

The *tat* I Protein of HTLV-I Activates IL-2 Receptor and IL-2 Promoter Expression

To investigate functional effects of the *tat* I gene product on cellular gene expression, a full-length cDNA encoding this protein was isolated and inserted into an expression vector containing the HTLV-I-LTR. A second pX-derived cDNA encoding a 21-kD non-*trans*-activating protein derived from a different reading frame termed Δ *tat* I was similarly isolated, inserted, and used as a control in these studies. The *tat* I cDNA expression plasmid was shown to produce large quantities of the *tat* I protein following transfection in a variety of cells as measured by *trans*-activation of a cotransfected plasmid containing the HTLV-I-LTR linked to the chloramphenicol acetyl transferase (CAT) gene. In contrast, the Δ *tat* I control plasmid produced no *trans*-activation of the HTLV-I LTR. To analyze potential effects of the *tat* I protein on IL-2 receptor promoter expression, the *tat* I plasmid was cotransfected with a plasmid containing a 1.4-kb segment of DNA corresponding to the 5' flanking region of the Tac gene linked to the CAT receptor gene. The *tat* I, but not Δ *tat* I plasmid, consistently induced a threefold to sixfold increase in IL-2 receptor promoter activity in Jurkat T cells (Siekevitz et al. 1987). These changes were slightly greater than that observed following activation of the IL-2 receptor promoter with phytohemagglutinin (PHA), phorbol-12-myristate-13-acetate (PMA), or combinations of these mitogens. Inoue et al. (1985), Cross et al. (1987), Maruyama et al. (1987) and have similarly found evidence for *tat* activation of IL-2 receptor gene expression.

Transfection of other T cell lines revealed that mitogen and *tat* I inducibility of the IL-2 receptor promoter were not always coordinately linked. For example, in MT-1 cells, the *tat* I gene product activated the receptor promoter, whereas mitogenic stimulation was without effect. Interestingly, in YT cells (a natural killer-like cell line), PHA and PMA activation produced a marked increase in IL-2 receptor promoter activity, whereas the *tat* I protein was ineffective. These findings suggest that an additional cellular factor may be required for the action of the *tat* protein. It is possible that this putative cofactor is uniquely distributed within the $CD4^+$ subset of T cells, thus perhaps explaining the restricted range of cells transformed by HTLV-I.

The effects of the *tat* I protein on IL-2 promoter activation were also examined (Siekevitz et al. 1987). In contrast to the IL-2 receptor, two mitogenic signals (e.g., PHA + PMA) are required for endogenous IL-2 gene and transfected IL-2 promoter activation in Jurkat T cells. The *tat* I protein alone produced little or no stimulation of the IL-2 promoter in Jurkat T cells but was able to markedly synergize with the single ineffective mitogenic signals (PHA, PMA, or OKT3) leading to marked stimulation (>60-fold) of the IL-2 promoter. In contrast, the Δ *tat* I control plasmid was without effect in these studies. The stimulatory effects of the *tat* I plasmid on IL-2 receptor and IL-2 promoter activation appeared specific as this viral protein did not activate the promoters controlling β-actin or c-*myc* cellular gene expression nor enhanced the activity of the LTRs of the Rous sarcoma virus, visna virus, or gibbon ape leukemia virus in the presence or absence of mitogenic stimuli.

The finding that the *tat* I protein activates the IL-2 receptor promoter provides a plausible explanation for the constitutive high-level expression of IL-2 receptors observed in HTLV-I transformed T cell lines. Furthermore, the ability of the *tat* I protein to provide one of the two cellular signals required for IL-2 gene expression may also predispose these virally infected T cells to produce IL-2 following antigenic stimulation. Ordinarily, this single signal might be an ineffective stimulus for IL-2 production. The coproduction of IL-2 and IL-2 receptors could in turn mediate an early period of uncontrolled polyclonal T-cell growth. In vitro analysis of T-cell transformation induced by HTLV-I has in fact revealed an early stage of vigorous polyclonal T-cell proliferation. The autocrine proliferation of these T cells induced by *tat* I may also facilitate the occurrence of second stage events required for the completion of viral transformation and lead to the outgrowth of a clonal population of leukemic T cells which proliferate independently of IL-2. In support of this autocrine model of leukemogenesis, Arima et al. (1986) have recently identified primary ATL tumor cells whose proliferation, at least in part, involves the simultaneous display of IL-2 receptors and production of IL-2. In summary, it seems likely that the *tat* I protein plays an important early role in the induction of HTLV-I associated leukemia by amplifying viral replication and by deregulating the expression of the cellular genes that normally regulate T-cell growth, including IL-2 and the IL-2 receptor.

HIV Inhibits IL-2 Gene Activation

Mitogen-induced activation of IL-2 and IL-2 receptor gene expression was studied in purified $CD3^+/CD4^+$ normal human T lymphocytes and clonal populations of $CD3^+/CD4^+$ Jurkat leukemic T cells highly infected with HIV (M. Dukovich et al., in prep.). For analysis, $CD3^+/CD4^+$ human T cells were purified by affinity rosetting techniques. At varying times after HIV infection, potential changes in PHA-induced and PMA-induced IL-2 and IL-2 receptor gene expression were studied by a cytoplasmic dot-blotting technique. HIV gene expression was first detected six days following exposure of the cells to virus. Eight days following infection, a marked inhibition of IL-2 gene expression was observed that progressively increased throughout the remaining 14 days of analysis. In contrast, no changes in mitogen-induced IL-2 receptor mRNA levels were detected. The progressive fall in IL-2 mRNA levels was paralleled by the disappearance of CD4 receptor from the surface of these cells, indicative of HIV infection. Northern blotting studies of HIV-infected T cells isolated 14 days after viral infection revealed greater than 90% inhibition of PHA-induced and PMA-induced IL-2 gene expression. In contrast, mitogen-induced Tac gene expression was unaltered. Flow microfluorometric analysis of the infected-cell versus control-cell populations revealed that greater than 95% of the cells in both populations were $CD3^+$, indicating their T cell lineage; however, $CD8^+$ cytotoxic lymphocytes were not detectable in either the infected or uninfected cultures.

These findings did not distinguish between the possibilities that HIV infection resulted in the selective elimination of the IL-2 producing subpopulation of the $CD4^+$ lymphocytes, produced an intrinsic defect in IL-2 gene expression unrelated to cell death or resulted in the outgrowth of a different population of cells. To address the issue of cell heterogeneity, clonal populations of Jurkat T cells were homogeneously infected with HIV and activated with varying mitogens. As with normal T cells, a profound defect in IL-2 gene activation was observed in the surviving cloned population of these leukemia cells following HIV infection. In summary, these varying lines of data suggest that HIV infection of $CD4^+$ T lymphocytes may lead to a marked impairment of mitogen-activated IL-2 gene expression but fails to alter IL-2 receptor gene expression. It is possible that the lack of IL-2 production may play a role in the cytopathic effects induced by HIV as well as contribute to the progressive decline in the number of $CD4^+$ lymphocytes and to the emergence of T-cell immunodeficiency characteristic of AIDS.

Latent versus Productive HIV Infection: The Role of Immune Stimulation

Recent studies indicate that the interaction of HIV with $CD4^+$ T cells may lead to a period of latent viral infection (Hoxie et al. 1985; Folks et al. 1986; Zagury et al. 1986). For example, HIV-infected $CD4^+$ cells or cell lines can be maintained under select conditions of culture for variable periods of time in the absence of detectable virus production. The intracellular events that trigger the transition from latent to productive viral infection remain poorly defined at present; however, the state of T-cell activation may be important. The greater understanding of this process might facilitate the development of specific agents capable of blocking HIV activation and the cytopathic effects associated with the replication of this virus. We have recently demonstrated that mitogen activation of Jurkat T cells latently infected with HIV leads to a marked increase in HIV gene transcription, as measured in nuclear runoff assays (Siekevitz et al. 1987). Similarly, using purified $CD4^+$ populations of normal lymphocytes, the addition of mitogenic agents was found to augment the levels of HIV-mRNA. To investigate whether T-cell activation stimulated HIV gene expression by altering the activity of the HIV-LTR, plasmids containing this viral element linked to the CAT gene were prepared and used for transient transfection assays in Jurkat T cells. Activation of these transfected cells with PHA or PMA produced a marked increase in HIV-LTR-CAT activity. In addition, increases in HIV-LTR-CAT activity were also obtained following activation with the calcium ionophore, ionomycin, and the *tat* I gene product of HTLV-I. These different mitogenic agents were each shown to produce synergistic activation of the HIV-LTR in the presence of the HIV-derived *tat* III protein. For example, combinations of PMA and *tat* III led to a greater than 9000-fold increase in HIV-LTR-CAT activity. RNase protection studies confirmed an absolute increase in HIV-LTR-CAT mRNA and demonstrated that transcription was appropriately initiated at the expected capping site within the HIV-LTR.

To further evaluate the *cis*-acting sequences that mediate mitogen-induced changes in HIV-LTR activity, a series of Bal 31 deletion mutants prepared by Dr. Steve Josephs (National Institutes of Health) were analyzed. Transfection of these various deletion constructs revealed the presence of multiple positive and negative regulatory elements within the LTR which contributed to overall level of LTR activity. For example, a negative regulatory element was mapped between -340 and -278 that appeared to regulate both basal and mitogen-induced expression of the HIV-LTR. Further deletions within the LTR revealed a moderate but not complete loss of PHA-induced activity associated with removal of a region that shares significant homology with two imperfect repeats present within the 5' regulatory regions of both the IL-2 and IL-2 receptor genes. As noted earlier, each of these cellular genes are inducible with PHA and PMA and the segment containing these imperfect repeats has been determined by Fujita et al. (1986) to be important for mitogen-induced activation of the IL-2 promoter.

Analysis of *tat* I-induced activation of the HIV-LTR revealed a very different pattern of stimulation than that obtained with PHA and PMA. Changes induced by *tat* I were quite small until the LTR was deleted to -117. With this deletion, marked stimulatory effects of this viral *trans*-activator protein were evident. These findings suggest the possible presence of a second negative regulatory element located between -278 and -117 that impairs activation by *tat* I but not by PHA and PMA. A second difference in *tat* I-induced activation versus stimulation by PHA and PMA was found in studies using combinations of these inducers. The *tat* I protein consistently produced additive effects when combined with either PHA or PMA. In contrast, combinations of PHA and PMA produced effects that were no greater than with either agent alone.

Further deletion of the LTR revealed that removal of the direct repeat or enhancer element of the HIV-LTR was associated with a near complete loss of both mitogen-induced and basal activity of the LTR. To further explore the role of the enhancer element in this response, an oligonucleotide corresponding to the enhancer was synthesized and ligated to the inactive -65 deletion of the HIV-LTR in both the sense and antisense orientation. Mitogen activation of Jurkat T cells transfected with these plasmids revealed marked increases in HIV-LTR-CAT activity. Thus, this enhancer element appears to play an important role in mitogen activation of the LTR and is consistent with the recent report of Nabel and Baltimore (1987) who have shown that mutation of the direct repeat blocks the interaction of a mitogen-inducible nuclear protein (NF-κB) with the HIV-LTR as assayed by gel retardation. However, production of the NF-κB protein does not appear to be cell-type specific. For example, PMA stimulation of HeLa cells leads to its production (Sen and Baltimore 1986). It is possible that the additional positive and negative regulatory elements located further 5mp within the LTR not only importantly contribute to regulating the overall level of expression of the HIV-LTR but also control expression in a cell-specific or tissue-specific manner.

The effects of cyclosporin A on mitogenic activation of the HIV-LTR were also investigated, as this agent has been proposed as a possible treatment for AIDS patients based on its capacity to block many steps in T-cell activation, including IL-2 and γ-interferon gene expression. Cyclosporin A was shown to effectively inhibit PHA-induced activation of the LTR. In sharp contrast, however, LTR activation mediated by PMA or *tat* I was entirely resistant to the inhibitory effects of this immunosuppressive agent.

CONCLUSION

In summary, our results indicate that four different classes of T-cell mitogens are capable of activating the LTR of the human immunodeficiency virus. These findings suggest that immunological stimulation of T cells latently infected with HIV may play an important role in the initiation of viral replication. The multiplicative effects obtained with combinations of mitogen and the *tat* protein of HIV underscore the potential amplifying effects produced by mitogen activation of *tat* III protein synthesis. The capacity of the *tat* I protein to stimulate the HIV-LTR is also of potential clinical importance in view of the recently recognized high incidence of dual HIV and HTLV-I infection in some populations of intravenous drug abuse patients (Robert-Guroff et al. 1986). Earlier studies have also revealed the capacity of other *trans*-activating proteins from many DNA viruses to stimulate the HIV-LTR (Gendelman et al. 1986; Mosca et al. 1987). These findings highlight the potential cofactor role played by intercurrent viral infection in the regulation of HIV gene expression.

REFERENCES

Arima, N., Y. Daitoku, S. Ohgaki, J. Fukumore, H. Tanaka, Y. Yamamoto, K. Funimoto, and K. Onoue. 1986. Autocrine growth of interleukin 2-producing leukemic cells in a patient with adult T cell leukemia. *Blood* **68:** 779.

Arya, S.K., F. Wong-Staal, and R.C. Gallo. 1984. T-cell growth factor gene: Lack of expression on human T-cell leukemia lymphoma virus infected cells. *Science* **223**: 1086.

Broder, S. and R.C. Gallo. 1984. A pathogenic retrovirus (HTLV-III) linked to AIDS. *N. Engl. J. Med.* **311**: 1292.

Cross, S.L., M.B. Geinberg, J.B. Wolf, N.J. Holbrook, F. Wong-Staal, and W.J. Leonard. 1987. Regulation of the human interleukin-2 receptor α chain promoter: Activation of a nonfunctional promoter by the transactivator gene of HTLV-I. *Cell* **49**: 47.

Dalgleish, A.G., P.C.L. Beverly, P.R. Clapham, D.H. Crawford, M.F. Greaves, and R.A. Weiss. 1984. The CD4 (T4) antigen is an essential component of the receptor for the AIDS retrovirus. *Nature* **312**: 763.

Depper, J.M., W.J. Leonard, M. Krönke, T.A. Waldmann, and W.C. Greene. 1984. Augmented T-cell growth-factor receptor expression in HTLV-I infected human leukemia T-cells. *J. Immunol.* **133**: 1691.

Folks, T., D.M. Powell, M.M. Lightfoote, S. Benn, M.A. Martin, and A.S. Fauci. 1986. Induction of HTLV-III/LAV from a nonvirus-producing T-cell line: Implications for latency. *Science* **231**: 600.

Fujita, T., H. Shibuya, T. Ohishi, K. Yamanishi, and T. Taniguchi. 1986. Regulation of human interleukin-2 gene: Functional DNA sequences in the 5′ flanking region for the gene expression in activated T-lymphocytes. *Cell* **46**: 401.

Gendelman, H.E., W. Phelps, L. Feigenbaum, J. Ostrove, A. Adachi, P.M. Hawley, G. Khoury, H.S. Ginsberg, and M.A. Martin. 1986. Transactivation of the human immunodeficiency virus long terminal repeat sequence by DNA viruses. *Proc. Natl. Acad. Sci.* **83**: 9759.

Haseltine, W.A., J. Sodroski, R. Patarca, D. Briggs, D. Perkins, and F. Wong-Staal. 1984. Structure of the 3′ terminal repeat region of type-II human T-lymphotropic virus: Evidence for a new coding region. *Science* **225**: 419.

Hoxie, J.A., B.S. Haggarty, J.L. Rackowski, N. Pillsbury, and J.A. Levy. 1985. Persistent noncytopathic infection of normal human T lymphocytes with AIDS-associated retrovirus. *Science* **229**: 1400.

Inoue, J., M. Seiki, T. Taniguchi, S. Tsuru, and M. Yoshida. 1985. Induction of interleukin-2 receptor gene expression by p40^x encoded by human T cell leukemia virus type 1. *EMBO J.* **5**: 2883.

Kiyokawa, T., M. Seiki, K. Imagawa, F. Shimizu, and J. Yoshida. 1984. Identification of a protein (p40^x) encoded by a unique sequence pX of human T-cell leukemia-virus type-I. *Gann* **75**: 747.

Klatzmann, D., E. Champagne, S. Chamaret, J. Gruest, D. Guetard, T. Hercend, J.C. Gluckman, and L. Montagnier. 1984. T-lymphocyte T4 molecule behaves as the receptor for the human retrovirus LAV. *Nature* **312**: 767.

Krönke, M., W.J. Leonard, J.M. Depper, and W.C. Greene. 1985. Deregulation of interleukin receptor gene expression in HTLV-I induced adult T cell leukemia. *Science* **228**: 1215.

Maruyama, M., H. Shibuya, H. Harada, M. Hatakeyama, M. Seiki, T. Fujita, J. Inoue, M. Yoshida, and T. Taniguchi. 1987. Evidence for aberrant activation of the interleukin-2 autocrine loop by HTLV-I encoded p40^x and Tc/Ti complex triggering. *Cell* **48**: 343.

McDougal, J.S., M.S. Kennedy, J.M. Sligh, S.P. Cort, A. Mawle, and K.A. Nicholson. 1986. Binding of HTLV-III/LAV to T4$^+$T cells by a complex of the 110 K viral protein and the T4 molecule. *Science* **231**: 382.

Mosca, J.D., D.P. Bednarik, N.B.K. Raj, C.A. Rosen, J.F. Sodroski, W.A. Haseltine, and P.M. Pitha. 1987. Herpes simplex virus type-1 can reactivate transcription of latent human immunodeficiency virus. *Nature* **325**: 67.

Nabel, G. and D. Baltimore. 1987. An inducible transcription factor activates expression of human immunodeficiency virus in T cells. *Nature* **326**: 711.

Poiesz, B.J., F.W. Ruscetti, M.S. Reitz, V.S. Kalyanaraman, and R.C. Gallo. 1981. Isolation of a new type-C retrovirus (HTLV) in primary uncultured cells of a patient with Sezary T-cell leukemia. *Nature* **294**: 268.

Poiesz, B.J., F.W. Ruscetti, A.F. Gazdar, P.A. Bunn, J.D. Minna, and R.C. Gallo. 1980. Detection and isolation of type-C retrovirus particles from fresh and cultured lymphocytes of a patient with cutaneous T-cell lymphoma. *Proc. Natl. Acad. Sci.* **77**: 7415.

Robert-Guroff, M., S.H. Weiss, J.A. Giron, A.M. Jennings, H.M. Ginzburg, I.B. Margolis, W.A. Blattner, and R.C. Gallo. 1986. Prevalence of antibodies to HTLV-I, -II, and -III in intravenous drug abusers from an AIDS endemic region. *J. Am. Med. Assoc.* **255**: 3133.

Seiki, M., R. Eddy, T. Shows, and M. Yoshida. 1984. Nonspecific integration of HTLV provirus genome into adult T-cell leukemia cells. *Nature* **309**: 640.

Seiki, M., S. Hattori, Y. Hirayama, and M. Yoshida. 1983. Human adult T-cell leukemia virus: Complete nucleotide sequence of the provirus genome integrated in leukemia cell DNA. *Proc. Natl. Acad. Sci.* **80**: 3618.

Seiki, M., A. Hikihoshi, T. Taniguchi, and M. Yoshida. 1985. Expression of the pX gene of HTLV-I: General splicing mechanism in the HTLV family. *Science* **228**: 1532.

Sen, R. and D. Baltimore. 1986. Inducibilty of κ immunoglobulin enhancer-binding protein NF-κB by a posttranslational mechanism. *Cell* **47**: 921.

Shimotohno, K., M. Miwa, D.J. Slamon, I.S.Y. Chen, H.O. Hoshino, M. Takano, M. Fujino, and T. Sugimura. 1985. Identification of new gene-products encoded from X-regions of human T-cell leukemia viruses. *Proc. Natl. Acad. Sci.* **82**: 302.

Siekevitz, M., M.B. Feinberg, N. Holbrook, E. Wong-Staal, and W.C. Greene. 1987. Activation of interleukin-2 and interleukin-2 receptor (Tac) promoter expression by the transactivator (*tat*) gene product of human T cell leukemia virus, type 1. *Proc. Natl. Acad. Sci.* **84**: 5389.

Sodroski, J.G., C.A. Rosen, and W.A. Haseltine. 1984. Trans-acting transcriptional activation of the long terminal repeat of human T lymphotropic viruses in infected cells. *Science* **225**: 381.

Waldmann, T.A., W.C. Greene, P.S. Sarin, C. Saxinger, D.W. Blayney, W.A. Blattner, C.K. Goldman, K. Bongiovanni, S. Sharrow, J.M. Depper, W.J. Leonard, and R.C. Gallo. 1984. Functional and phenotypic comparison of human T cell leukemia/lymphoma virus positive adult T cell leukemia with human T cell leukemia/lymphoma virus negative Sezary leukemia. *J. Clin. Invest.* **73**: 1711.

Wong-Staal, F. and R.C. Gallo. 1985. Human T lymphotropic retroviruses. *Nature* **317**: 395.

Yoshida, M., I. Miyoshi, and Y. Hinuma. 1982. Isolation and characterization of retrovirus from cell lines of human adult T cell leukemia and its implication in disease. *Proc. Natl. Acad. Sci.* **78:** 6476.

Zagury, D., J. Bernard, R. Leonard, R. Cheynier, M. Feldman, P.S. Sarin, and R.C. Gallo. 1986. Long-term cultures of HTLV-III infected T cells: A model of cytopathology of T-cell depletion in AIDS. *Science* **231:** 850.

Mobile Genetic Elements, Spontaneous Mutations, and the Assessment of Genetic Radiation Hazards in Man

KRISHNASWAMY SANKARANARAYANAN
Department of Radiation, Genetics, and Chemical Mutagenesis
State University of Leiden, Sylvius Laboratories, 2333 AL Leiden and
The J.A. Cohen Interuniversity Institute for
Radiopathology and Radiation Protection
The Netherlands

OVERVIEW

Studies during the past 20 years have amply documented the existence of mobile genetic elements (MGEs) and the significant role they play in the genesis of "spontaneous" mutations in a number of eukaryotes. Furthermore, some of the dispersed repetitive DNA sequences present in mammalian (including human) genomes are thought to represent putative MGEs on the basis of their molecular structural similarities to the authentic MGEs in other organisms. However, at present, data that bear on spontaneous mutational effects of insertion of MGEs and on the question of whether MGEs can be induced to become mobile by radiation exposures are very limited in these species.

If a major proportion of spontaneous mutations in man that lead to heritable disease is due to insertion of MGEs, and if the mobility of these elements is unaltered by radiation exposures, then some of the principal assumptions of a currently used method (the so-called doubling dose method) for radiation genetic risk evaluation in man would become invalid. It is argued that, at the present state of knowledge, there is no need to abandon the use of the above method for genetic risk evaluation. Possible reasons for our current inability to identify MGE-mediated spontaneous mutations in man are discussed, and it is suggested that chromosomally normal spontaneous abortuses and congenitally abnormal births may be useful "hunting grounds" in the search for MGE-mediated spontaneous mutations in man.

INTRODUCTION

The pioneering genetic studies of McClintock in maize over three decades ago laid the conceptual framework for studies on what are now known as MGEs, i.e., DNA sequences that move from one genomic location to another. MGEs have now been discovered in a wide variety of prokaryotes and eukaryotes and several of these elements have been characterized at the molecular level.

There is substantial evidence from bacteria, yeast, maize, and *Drosophila*, which demonstrates that a significant proportion of spontaneous mutations originates from insertion of MGEs into known genes. Furthermore, several families of dispersed repetitive DNA sequences occurring in the genomes of mammals, including that of man, have been postulated to represent putative MGEs on the basis of similarities in molecular structure to authentic MGEs known in other systems. However, there are, as yet, no demonstrated instances of insertion mutations in man which lead to heritable genetic disease.

The aims of this paper are (1) to briefly review some of the important properties of eukaryotic MGEs, including genetic effects caused by these and the effects of mutagens; (2) to examine the possible consequences of mobility of MGEs in the human genome and the probable reasons for our inability to detect MGE-mediated spontaneous mutations in man; (3) to consider the relevance of this analysis in the context of the estimation of genetic risks to man associated with exposures to ionizing radiation; and finally, (4) to present some views on where to seek evidence for MGE-mediated mutational effects in man.

DISCUSSION

Properties of Authentic Eukaryotic MGEs

The authentic MGEs thus far identified in eukaryotes include among others, the transposons in the yeast, *Saccharomyces cerevisiae*, (the Tys; Roeder and Fink 1983; Boeke et al. 1985; Clare and Farabaugh 1985; Boeke et al., this volume), the autonomous (e.g., *Ac*) and the nonautonomous (e.g., *Ds*) elements in maize (Fedoroff 1983; Fedoroff et al. 1983; Döring and Starlinger 1986; Fedoroff et al.; Chandler; both this volume) and several families of dispersed repetitive sequences in *Drosophila* such as the copia-like elements; P elements, I factors, etc. (Spradling and Rubin 1981; Finnegan et al. 1982; Engels 1983; Rubin 1983; Fawcett et al. 1986; Finnegan et al.; Kidwell; both this volume). The yeast Tys and the *Drosophila* copia-like elements have in common an internal coding domain of one to several kilobase pairs, large terminal direct repeat units of several hundred base pairs, usually with short inverted terminal repeats, and short flanking direct repeats of host DNA duplicated during insertion. Some of these features distinguish this class from other groups of MGEs, such as the classes that include the P elements, foldback elements, and I factors in *Drosophila* and the *Ac*, *Ds*, and *Mu1* in maize.

The evidence for their mobility comes from (1) observations of polymorphisms in the number and position of several dispersed repetitive DNA sequences between different strains, cell lines, etc. (*Drosophila*, yeast); (2) the genetic effects they have been found to cause, insertion mutations, deletions, duplications, inversions, etc. (yeast, maize, *Drosophila*); and (3) molecular analysis of insertion mutations and the strong correlations between the genetic behavior of the insertion mutations and the structure of the responsible mobile elements (yeast, maize, *Drosophila*).

In yeast, mutants that arose as a result of transposition of Ty have been isolated at the histidine-4, alcohol dehydrogenase (*ADH*) and cytochrome-c loci (for reviews, see Roeder and Fink 1983; Paquin and Williamson 1986; Morawetz 1987). Cloning of these genes, restriction mapping, and sequence analysis indicated that these mutations resulted from the insertion of an intact Ty element into the 5' noncoding region of the affected gene(s) (Roeder and Fink 1983). McClanahan and McEntee (1984) showed that DNA damage induced by UV or 4-nitroquinoline-1-oxide stimulates transcription of specific genes (the authors call them DNA-damage responsive, Ddr, genes) which share homology with the Ty*1* element, the inference being that one or more members of the Ty*1* elements are regulated transcriptionally by DNA damage. More recently, Morawetz (1987) showed that Ty insertion mutations at the *ADH*2 locus (antimycin-A-resistant mutations) could be increased by UV- or gamma-irradiation or by treatment with ethylmethane sulfonate (EMS) in a dose-dependent manner, EMS being the strongest agent in this regard.

The fact that in maize, MGEs are capable of causing unstable mutations as well as a wide variety of rearrangements has long been known from genetic studies (e.g., McClintock 1951, 1957, 1965). Many distinct families of such elements as well as mutants at a number of loci (e.g., *Wx, Sh, bz, Adh*) have now been studied at the molecular level (see Table 1 in Fedoroff 1983 and Tables 1 and 2 in Döring and Starlinger 1986). Insertions have been found to occur in introns, intron-exon borders, and exons, as well as in the 5' noncoding regions of these genes; these insertions affect gene structure and expression in a variety of ways.

In *Drosophila*, the genetic effects caused by MGEs include high spontaneous mutability, instability of these mutations as indicated by high spontaneous reversions to wild type at known loci, male recombination, sterility, distorted segregation ratios, chromosome breakage at specific sites, generation of additional variation for quantitative traits, etc. (e.g., Hiraizumi 1971; Green 1977, 1980, 1984; Kidwell et al. 1977; Engels and Preston 1981; Bregliano and Kidwell 1983; Engels 1983; Crow 1984; Mackay 1986; J.C.J. Eeken, unpubl.). Spontaneous mutations at a number of gene loci were first inferred, on the basis of genetic evidence, to result from insertions of MGEs (e.g., Green 1977; Golubovsky et al. 1977; Thompson and Woodruff 1978; Eeken 1982; Gerasimova et al. 1984). Molecular and in situ hybridization studies have clearly demonstrated that a high proportion of such spontaneous mutations, at over ten visible loci examined in this regard (Table 1), is in fact

Table 1
Proportion of Insertion Mutations among Spontaneous Mutations in *Drosophila*

Locus or region	Size (kb)	n/N	Insert[a] no.	size (kb)/nature	localization	Reference
y (yellow)	~4.6	7/9	4	3.5 to 7	N.T.R.	Chia et al. (1986)
			1	7.3 (gypsy)	N.T.R.	
			2	?	N.T.R.	
v (vermilion)	~2.0	4/5	3	7.5 (412)	T.R.	Searles and Voelker (1986)
			1	9.0 (B104)	T.R.	
ct (cut)	~200					Jack (1975)
Lethal group		14/14[b]	14	7.3 (gypsy)	T.R. (probably)	
Wing group		14/14[c]	13	7.3 (gypsy)	T.R. (probably)	
			1	5.0 (copia)		
ry (rosy)	~7.3	3/5[d]	1	7.6 (calypso)	T.R.	Coté et al. (1986)
			2	8.5 (B104)		
w (white)	~6.0	7/10	1[e]	5.0 (copia)	T.R. (intron)	Levis et al. (1982); Zachar and Bingham (1982)
			3	5.7 to 10	T.R.	
			1		N.T.R.	
			1		—[f]	
			1	13 to 14	—[g]	
f (forked)	?	4/4	1	7.3 (gypsy)	T.R.	Parkhurst and Corces (1985);
			1	2 × 7.3 (gypsy)	T.R.	

Locus	~bp	n/N	n	Size (kb)	T.R./N.T.R.	Reference
su(s) (suppressor of sable)	?	7/7	1	3.0	T.R.	McLachlan (1986)
			1	≥6.0	N.K.	Chang et al. (1986)
			2	3.8 to 6	N.T.R.	
			5	5 to 7.3 (gypsy)	N.T.R.	
Bx (Beadex)	?	4/4	1	2 × 7.3 (gypsy)	N.T.R.	Mattox and Davidson (1984)
			1	8.5 (B104)	N.T.R.	
			1	10.3 (B104)	N.T.R.	
			1	12 to 16	N.T.R.	
bx (bithorax)	~19	9/9	7	7.3 (gypsy)	T.R.	Peifer and Bender (1986)
			1	7.9 (B104)	T.R.	
			1	5.4 (I)	T.R.	
sc (scute)	~75	2/2	2	7.3 (gypsy)	N.K.	Campuzano et al. (1985)
Antp (Antennapedia)	~107	3/5	1	8.0	T.R.	Scott et al. (1983)
			1	8.0 (B104)	T.R.	
			1	3.0	T.R.	
N (Notch)	~37	4/7	4	3.5 to 8	T.R.	Kidd et al. (1983)

Abbreviations: (kb) kilo basepairs; (n) number of insertion mutations; (N) total number of spontaneous mutations studied; (T.R.) in the transcribed region; (N.T.R.) in the nontranscribed region; (N.K.) not known if the site of insertion is in T.R. or N.T.R.; (?) not known.

[a] The sizes of the inserts given in this table are those given in the respective papers. Recent sequence data (see Parkhurst and Corces 1986 and references cited therein) indicate that the *gypsy* element is slightly longer than 7.3 kb i.e., 7.469 kb.
[b] All insertion mutations except one arose in *Uc* mutable flies (Lim 1979).
[c] 10 out of the 14 arose in *Uc* mutable flies.
[d] 4 were spontaneous and 1 was recovered in an EMS-mutagenesis study which the authors consider as probably a spontaneous event.
[e] w^a and its derivatives.
[f] 200-bp deletion from the *w* region plus an insertion.
[g] Outside the *w* locus.

associated with insertions of MGEs (see also Simmons and Lim 1980; Rubin et al. 1982; Engels 1983; Rubin 1983; Eeken et al. 1985). Worthy of note in Table 1 are (1) the preponderance of gypsy insertions, but it should be mentioned that the cut wing (*ct*) mutations were from *Uc* mutable flies, and *Uc* system is known to mobilize the gypsy elements, and (2) insertion of the MGEs occur in the coding region and noncoding region, as well as in regions of genes yet to be determined.

Again in *Drosophila*, Sobels and Eeken (1981) found that X-irradiated males carrying P elements responded with higher frequencies of X-linked recessive lethals, but not of translocations. In another study, Eeken and Sobels (1983) showed that gonial treatment of male larvae with ethylnitrosourea (ENU) or methyl methanesulfonate (MMS) was either ineffective (1 mM, ENU; 5 and 10 mM, MMS) or led to a twofold increase (2 mM, ENU) in the reversion rate of P element-induced singed (*sn*) mutations. In an extension of the above work, Eeken and Sobels (1986) found that, in the presence of active P elements, there was no measurable increase in the rate of X-ray-induced reversions from *sn* to sn^+ and, after formaldehyde treatment, a 50% reduction (see also Sobels and Eeken, this volume). In similar experiments involving treatment of male larvae with some known carcinogens, Fahmy and Fahmy (1983) recorded an enhancement of reversions from white-crimson (w^c) to wild type (w^+) or to other *w* alleles. However, since w^c is known to be due to an insertion of a 10-kb mobile foldback element into the white-ivory (w^i) allele (and w^i is due to a tandem 3-kb duplication within the *w* locus) (see Rubin 1983), the increase in reversion rate is most probably due to recombination.

It is thus clear that (1) a significant proportion of spontaneous mutations at specific loci in eukaryotes well-studied in this respect owe their origin to the mobility of MGEs; (2) the number of loci at which this problem has been studied is still relatively limited, and the proportion of insertion mutations among all spontaneous mutations is not known for any species; (3) the mutations analyzed are probably not a random sample and were selected by virtue of their unusual phenotypes; and (4) quantitative data that bear on the rates of mobility of MGEs, either spontaneous or under the influence of mutagens, are very limited at present.

Putative MGEs in Mammals

Information on mammalian interspersed repetitive DNA sequences and why they are considered to be (or were at one time) MGEs have been the subject of a number of recent reviews (e.g., Jelinek and Schmid 1982; Singer 1982; Schmid and Paulson 1984; Sun et al. 1984; Rogers 1985; Singer and Skowronski 1985; Deka et al. 1986; Sakaki et al. 1986; Skowronski and Singer 1986; Singer et al., this volume). These putative MGEs in mammals are currently divided into two classes: *s*hort *in*terspersed *e*lements (SINEs), which are typically less than 500-bp *l*ong, and *l*ong *in*terspersed *e*lements (LINEs), which range in length from a few hundred bps up to 7 kb. Examples of SINEs include the *Alu* in primates, the B1 and B2 in rodents, and *A* and *C* in artiodactyls (Jelinek and Schmid 1982; Singer 1982; Rogers 1985) and of LINEs, the LINE-1 (L1; formerly *Kpn1*) and THE-1 in primates and the L1 in rodents (e.g., L1Md, for L1 in *Mus domesticus*; formerly M1F-1) (Burton et al. 1985; Paulson et al. 1985; Shyam and Weaver 1985; Skowronski and Singer 1986). All these elements share common properties that have suggested that they are generated from RNA intermediates by a common mechanism involving reverse transcription; they are therefore called "retroposons." It is worth noting here that the primate L1 elements and the *Drosophila* I factors have a number of structural features in common (Fawcett et al. 1986).

Another group of DNA sequences that fall under the heading of MGEs in mammals is constituted by the murine-reiterated elements having retroviral structures (Shell et al. 1987 and references cited therein); of these intracisternal A-particles (IAPs), the early transposon-like (E.Tn) sequences and the ecotropic murine leukemia viruses (MLV) are pertinent in the present context. The IAPs are proretroviral-like elements that are found budding from the endoplasmic reticulum of many transformed mouse cells and during very early stages of normal embryogenesis; they are present in about 1000 copies per haploid genome and appear to have no infectious or recognized extracellular phase (Cole et al. 1981; Kuff et al. 1983; Kuff, this volume). The E.Tn sequences form a family of long moderately repeated sequences that are abundantly transcribed in the pluripotent cell lineage between days 3.5 and 7.5 of early mouse embryogenesis; the element is present in approximately 1000 copies in the mouse genome (Brûlet et al. 1983; Kaghad et al. 1985). Several high leukemic strains of the mouse carry multiple (3–5) copies of the ecotropic viral sequences, whereas many low leukemic strains contain a single copy per haploid genome.

At present, with the exception of IAPs, E.Tns, and the endogenous MLVs (see below), there is no direct evidence that the putative mammalian MGEs move from one genomic location to another or that they are associated with any phenotypic effect. However, there are several suggestive indications. With a cloned segment of human DNA initially present in a cluster of *Alu* repeat sequences, Calabretta et al. (1982) recorded DNA rearrangements involving restriction fragment polymorphism and variation in copy number (between tissues of the same donor and between lymphocytes of normal and leukemic donors); the rearrangements involved both extrachromosomal circular duplex DNAs (presumed intermediates in transposition) and integrated sequences. In the work with THE-1, Paulson et al. (1985) found members of this family in extrachromosomal circular DNA molecules and that discrete-length transcripts from HeLa cells are homologous to THE-1.

Skowronski and Singer (1986) have argued that the dispersed positions of L1 sequences between genes, in introns, and in interrupting tandem arrays of species-specific satellites, as well as the target-site duplications associated with many L1 sequences, suggest that they were mobile in the past (in an evolutionary sense). Furthermore, these authors have compiled a list of five instances of mammalian (e.g., dog, rat, and mouse) polymorphic alleles that differ by the presence or absence of L1 units. On the basis of all these (and some unpublished data on the insertion of a L1 unit in a *myc* allele in DNA from a human breast adenocarcinoma), they suggest that the L1s are still capable of being inserted into new genomic loci.

Hawley et al. (1982) and Kuff et al. (1983; see also Kuff, this volume) have shown that the IAPs are responsible, by insertion, for mutations involving the κ-immunoglobulin light-chain gene in mouse hybridoma cells (see Canaani et al. 1983; Cohen et al. 1983; Greenberg et al. 1985; Ymer et al. 1985 for effects of IAP insertions in other genes). Recently, Shell et al. (1987) reported that the interruption of two immunoglobulin heavy-chain switch regions in murine plasmacytoma was due to E.Tn insertions. Again in the mouse, Jenkins et al. (1981) and Copeland et al. (1983) (see also Strobel et al., this volume) showed that the dilute (d) coat color mutation is due to the insertion of the ecotropic murine provirus designated as *Emv-3*. Revertants that are left with only one of the LTRs have normal coat color (see also Favor 1986).

I do not imply that the putative MGEs are not involved in mutant phenotypes in man; there are some known instances, but these do not appear to be associated with the mobility of the MGEs. One form of α-thalassemia involves a deletion of the α-globin genes which extends into an *Alu* repeat (Orkin and Michelson 1980). In both the hereditary persistence of fetal hemoglobin and δ°-β°-thalassemia, a deletion of globin genes extends into two inverted, repeated *Alu* family members (Jagadeeswaran et al. 1982; Ottolenghi and Giglioni 1982). The precise role of *Alu* repeats in these different deletions is unknown. Recently, Russell and colleagues (1986) reported that several naturally occurring deletions in the low-density-lipoprotein (LDL)-receptor mutations can be explained as the consequence of unequal but homologous exchanges between *Alu* family repeats that lie within the LDL receptor gene. Furthermore, the same group of authors (Lehrman et al. 1987) found that, in a child with familial hypercholesterolemia, the mutant LDL-receptor gene contains a 14-kb duplication that encompasses exons 2 through 8, which presumably arose through homologous recombination between *Alu* sequences in intron 1 and 8.

Probable Reasons for Our Inability to Detect MGE-insertion-mediated Mutations in Man

There are several reasons why we have not so far detected MGE-mediated insertion mutations in man. First, the organization of the human genome is quite complex and one of interrupted heterogeneity, made up of unique, moderately repetitive, and highly repetitive sequences. The functional genes are split and have intervening sequences. The coding sequences themselves are but a small subset of the functional gene sequences and there are pseudogenes. Thus, unless the MGEs specifically "seek out" functional structural genes as we know them and insert themselves in or in close proximity to them, resulting in an identifiable phenotypic change consistent with viability, such insertion mutations may not be detected.

Second, there is the problem of genetic heteogeneity, i.e., the same clinical phenotype can be produced by a variety of genetic defects as is now well documented in a number of genetic disorders in man (Rimoin and Rotter 1982; Forget et al. 1983; McKusick 1986). This effectively precludes the assessment of the contribution of insertion mutations to the total of spontaneous mutations.

Third, there may be suppressor mutations, i.e., mutations at specific loci, which are capable of reversing the effect of different mutations located elsewhere in the genome, partially or completely, restoring the wild-type phenotype. Evidence from yeast, *Drosophila*, and mouse (for review, see Parkhurst and Corces 1986; Geyer et al., this volume) permits one to conclude that suppressible mutations are caused by the insertion of retroposons. The questions of whether suppressors exist in man and, if so, how widespread remain to be answered.

So far the difficulties listed assume that the mutational effect caused by insertion of MGEs is consistent with viability. However, several genomic changes mediated by MGEs may be incompatible with viability and can therefore be lost to view. The pregnancy wastage in the human species is considerable. Large-scale studies suggest that the rate of unsuspected pregnancy loss, i.e., early postimplantation losses prior to the clinical recognition of pregnancy, may be 30% or more (Miller et al. 1980). There must also be considerable preimplantation loss of conceptuses; if these were true, the addition of clinically diagnosed spontaneous abortions, estimated at around 15–20% (Hassold and Jacobs 1984), would mean that the rate of human reproductive failure is at least of the order of 50%. Here then is a big "basket" in which a sizeable proportion of the MGEs may end up. An argument strengthening the above premise comes from studies aimed at a direct analysis of human spermatozoa (e.g., Brandiff et al. 1985; Martin 1985; Kamiguchi and Mikamo 1986). In the work of the Japanese investigators, the incidence of structural anomalies of chromosomes of spermatozoa was over 13%, about half of which were breaks. The majority of eggs fertilized by these abnormal spermatozoa would be incapable of development beyond the early stages.

Relevance of MGE-mediated Spontaneous Mutations in the Context of Genetic Risk Assessment for Radiation Exposures of Man

Owing to the paucity of human data, the estimation of genetic risks to man from exposure to ionizing radiation (i.e., those expressed in the first and subsequent generations of progeny following exposure in the parental generation) is still based on extrapolation of rates of induction from animal data, with appropriate corrections to make them applicable to man (for review, see Sankaranarayanan 1982; UNSCEAR 1986). One of the methods used for this purpose is the so-called doubling dose method for which knowledge about spontaneous mutation rates and "natural" prevalence of mutation-maintained diseases in man is essential. The doubling dose is the amount of radiation required to produce as many mutations as those occurring naturally in a generation; this is obtained by dividing the average spontaneous rate by the average rate of induction per unit dose. Thus, for instance, if the average spontaneous rate is m_1 per locus and the rate of induction, m_2 per locus, then the doubling dose, $c=m_1/m_2$. The reciprocal of c (i.e., $1/c$) is the relative mutation risk. It is easy to see that the lower the doubling dose the higher the relative mutation risk and vice versa. The risk to a population at equilibrium (under conditions of continuous exposure with a dose of d/generation) is the product of natural prevalence, $1/c$ and d. In practice, the doubling dose is estimated from animal data and used in conjunction with those on the natural prevalence of genetic disorders in man.

At present, it is not known whether MGEs are involved in any significant manner in the genesis of spontaneous mutations in man and, if so, whether the rates of insertion mutations can be enhanced by radiation exposures. Thus, the question that we address here, namely, the relevance of MGE-mediated mutations in the context of genetic risk estimation using the doubling dose method, is a theoretical one (Sankaranarayanan 1986). At least three possible situations can be envisaged:

Situation 1: All spontaneous mutations arise as a result of errors made during DNA replication and/or repair (mechanism A) and radiation-induced mutations are qualitatively similar to those arising spontaneously.

Situation 2: Some spontaneous mutations arise through mechanism A (mentioned above) while others through mechanism B, (i.e., the B-type mutations are due to insertions of MGEs); the frequencies of both types of mutations can be enhanced by irradiation. If the assumptions postulated for each of these situations are valid, then one can estimate the doubling dose in the manner outlined earlier and use it in risk estimation.

Situation 3: This is similar to situation 2 except that here it is assumed that only the A-type mutations are susceptible to induction by radiation, whereas the B-type mutations are not. If these assumptions are true, the doubling dose calculation and its use are valid only for A-type mutations. The predictions for situation 3 are illustrated in Table 2 using a hypothetical example in which the relative proportions of the natural prevalence of genetic disorders in man (P_A: disorders due mutations arising by mechanism A and P_B; disorders due to mutations arising by mechanism B) are allowed to vary, and the doubling doses are estimated only for A-type mutations and used to make risk estimates. It is clear that as P_A decreases (and P_B increases), the doubling doses also decrease since the induction rate is assumed to be applicable only to A-type mutations and therefore held constant. Consequently, the relative mutation risks (i.e., $1/c$) increase progressively, but the expected increases at equilibrium (last column in Table 2) stay constant (since $E = P_A \times 1/c \times 1$).

The principal conclusions from this exercise are that (1) in the absence of evidence for a significant contribution of MGE-insertion mutations in man that lead to disease states, there is no need to abandon the use of the doubling dose method for genetic risk estimation, and (2) even if MGE-insertion mutations were to occur at a few gene loci, since only average rates are used, the overall effect on the magnitude of risks estimated is likely to be small.

Table 2
Expected Increases in the Frequency of Genetic Disorders at Equilibrium in a Human Population Continuously Exposed to Low LET (X- or γ-ray) Irradiation at a Rate of 1 rad (0.01 Gy) per Generation, Estimated Using the Doubling Dose Method

P_A	P_B	m_1	m_2	c	1/c	E
10,000	0	10^{-5}	10^{-7}	100 rad	1/100	100
7500	2500	0.75×10^{-5}	10^{-7}	75 rad	1/75	100
5000	5000	0.50×10^{-5}	10^{-7}	50 rad	1/50	100
2500	7500	0.25×10^{-5}	10^{-7}	25 rad	1/25	100

Abbreviations: (P_A) prevalence of naturally-occurring genetic disorders (number of cases per 10^6 livebirths) which are due to mutational events not associated with MGEs; (P_B) prevalence of naturally occurring genetic disorders (number of cases per 10^6 livebirths) which are due to MGE-insertion mutations; (m_1) average spontaneous mutation rate of A-type mutations; (m_2) average rate of induction of A-type mutations; (c) doubling dose in rad; (1/c) relative mutation risk per rad; and (E) expected increase in the number of cases of genetic disorders (per 10^6 livebirths) at equilibrium. Note that $c = m_1/m_2$ and $E = P_A \times 1/c \times 1$.

Future Prospects: Where Does One Look for Evidence for Mobility of MGEs in Man?

As was mentioned earlier, 15–20% of clinically diagnosed pregnancies end in spontaneous abortions and about half of these are chromosomally abnormal, predominantly trisomies (Hook 1982; Hassold and Jacobs 1984; Boué et al. 1985). It would seem that chromosomally normal abortuses would provide one useful starting place to look for MGEs.

A second and perhaps a more fruitful hunting ground would be the poignant world of imperfect births, namely, congenital abnormalities (CAs) in man. These have a high-birth prevalence, of the order of about 60/1000 livebirths (Czeizel and Sankaranarayanan 1984; UNSCEAR 1986). The inheritance patterns of a small proportion of these are consistent with classical Mendelian expectations. A further small proportion of them is associated with one or other kind of chromosomal anomalies. In some cases, the disorder is part of a recognizable syndrome, e.g., cleft lip and palate in trisomy 13. Finally, a small proportion of some CAs may be associated with known environmental causes, e.g., patent ductus arteriosus is one feature of rubella embryopathy. The relative proportions of CAs due to these, i.e., mutations in major genes, chromosomal anomalies, and environmental (including maternal) factors, have been estimated to be about 6%, 5%, and 6%, respectively (UNSCEAR 1986).

One is thus left with about 80% of CAs whose familial distributions neither neatly fit Mendelian expectations nor can be explained away as due to common familial environments. For nearly two thirds of these, twin and family studies suggest a complex etiology depending on polygenic predisposition and environmental factors that might also be multiple (Carter 1976; Czeizel and Tusnády 1984). It is tempting to hypothesize that a significant proportion of these CAs may in fact be associated with the mobility of MGEs in man during embryogenesis. Although at present, in view of the high copy numbers of putative MGEs in man, it might appear difficult to envisage methods for testing this hypothesis, given the ingenuity of molecular biologists, one would hope that the hypothesis does become amenable for testing in the not too distant future.

ACKNOWLEDGMENTS

I wish to thank my colleagues, Dr. A. Pastink for placing at my disposal a number of publications in *Drosophila* that are pertinent to this review and for checking the accuracy of the information presented in Table 1 of this paper, Dr. J.C.J. Eeken and Dr. A. Pastink for many profitable discussions, and Professors F.H. Sobels and P.H.M. Lohman for encouragement. I am grateful to Dr. V.G. Corces (Johns Hopkins University, Baltimore) for his useful comments on Table 1, Dr. D.J. Finnegan (University of Edinburgh) for his useful comments on some *Drosophila* data, and to Dr. D.W. Russell (University of Texas, Dallas) for providing me with preprints of his papers in press. I wish to dedicate this paper to my good friend Dr. A.G. Searle, MRC Radiobiology Unit, Chilton, England, on the occasion of his formal retirement. The writing of this paper was supported by EURATOM contract no. BI0-E-406-81-NL with the University of Leiden.

REFERENCES

Boeke, J.D., D.J. Garfinkel, C.A. Styles, and G.R. Fink. 1985. Ty elements transpose through an RNA intermediate. *Cell* **40**: 491.

Boué, A., A. Gropp, and J. Boué. 1985. Cytogenetics of pregnancy wastage. *Adv. Hum. Genet.* **14**: 1.

Brandriff, B., L. Gordon, L. Ashworth, G. Wachtmaker, D. Moore II, A.J. Wyrobek, and A.V. Carrano. 1985. Chromosomes of human spermatozoa: Variability among individuals. *Hum. Genet.* **70**: 18.

Bregliano, J.C. and M.G. Kidwell. 1983. Hybrid dysgenesis determinants. In *Mobile genetic elements* (ed. J.A. Shapiro), p. 363. Academic Press, New York.

Brûlet, P., M. Kaghad, Y.S. Xu, O. Croissant, and F. Jacob. 1983. Early differential tissue expression of transposon-like repetitive DNA sequences of the mouse. *Proc. Natl. Acad. Sci.* **80**: 5641.

Burton, F.H., D.D. Loeb, S.F. Chao, C.A. Hutchison, and M.H. Edgell. 1985. Transposition of a long member of the L1 major interspersed DNA family into the mouse beta globin locus. *Nucleic Acids Res.* **13**: 5071.

Calabretta, B., D.L. Robbertson, H.A. Barrera-Saldana, T.P. Lambrou, and G.F. Saunders. 1982. Genomic instability in a region of human DNA enriched in *Alu* repeat sequences. *Nature* **296**: 219.

Campuzano, S., L. Carramolino, C.V. Cabrera, M.R. Gomez, R. Villares, A. Boronat, and J. Modolell. 1985. Molecular genetics of the achaete-scute gene complex of *Drosophila melanogaster*. *Cell* **40**: 327.

Canaani, E., O. Dreazen, A. Klar, G. Rechavi, D. Ram, J.B. Cohen, and D. Givol. 1983. Activation of the c-*mos* oncogene in a mouse plasmacytoma by insertion of an endogenous intracisternal A-particle genome. *Proc. Natl. Acad. Sci.* **80**: 7118.

Carter, C.O. 1976. Genetics of common single malformations. *Br. Med. Bull.* **32**: 21.

Chang, D.Y., B. Wisely, S.M. Huang, and R.A. Voelker. 1986. Molecular cloning of suppressor of sable, a *Drosophila melanogaster* transposon-mediated suppressor. *Mol. Cell. Biol.* **6**: 1520.

Chia, W., G. Howes, M. Martin, Y.B. Meng, K. Moses, and S. Tsubota. 1986. Molecular analysis of the yellow locus of *Drosophila melanogaster*. *EMBO J.* **5**: 3597.

Clare, J. and P.J. Farabaugh. 1985. Nucleotide sequence of a Ty1 element: Evidence for a novel mechanism of gene expression. *Proc. Natl. Acad. Sci.* **82**: 2829.

Cohen, J.B., T. Unger, G. Rechavi, E. Canaani, and D. Givol. 1983. Rearrangement of the oncogene c-mos in mouse myeloma NS1 and hybridomas. *Nature* **306**: 797.

Cole, M.D., M. Ono, and R.C.C. Huang. 1981. Terminally redundant sequences in cellular intracisternal A-particle genes. *J. Virol.* **38**: 680.

Copeland, N.G., K.W. Hutchison, and N.A. Jenkins. 1983. Excision of the DBA ecotropic provirus in dilute coat color revertants of mice occurs by homologous recombination involving viral LTRs. *Cell* **33**: 379.

Coté, B., W. Bender, D. Curtis, and A. Chovnick. 1986. Molecular mapping of the rosy locus in *Drosophila melanogaster*. *Genetics* **112**: 769.

Crow, J.F. 1984. The P factor: A transposable element in *Drosophila*. In *Mutation, cancer and malformation* (ed. E.H.Y. Chu and W.M. Generoso), p. 257. Plenum Publishing, New York.

Czeizel, A. and K. Sankaranarayanan. 1984. The load of genetic and partially genetic disorders in man. I. Congenital anomalies. Estimates of detriment in terms of years of life lost and years of impaired life. *Mutat. Res.* **128**: 73.

Czeizel, A. and G. Tusnády. 1984. *Aetiological studies of isolated common congenital abnormalities in Hungary*. Akadémiai Kiadó, Budapest.

Deka, N., K.E. Paulson, C. Willard, and C.W. Schmidt. 1986. Repetitive human DNA sequences. II. Properties of a transposon-like-human element. *Cold Spring Harbor Symp. Quant. Biol.* **51(1)**: 473.

Döring, H.P. and P. Starlinger. 1986. Molecular genetics of transposable elements in plants. *Annu. Rev. Genet.* **20**: 175.

Eeken, J.C.J. 1982. The stability of mutator (MR)-induced X-chromosome recessive visible mutations in *Drosophila melanogaster*. *Mutat. Res.* **96**: 213.

Eeken, J.C.J. and F.H. Sobels. 1983. The effect of two chemical mutagens, ENU and MMS on MR-mediated reversion of an insertion sequence mutation in *Drosophila melanogaster*. *Mutat. Res.* **110**: 297.

———. 1986. The effect of X-irradiation and formaldehyde treatment of spermatogonia on the reversion of an unstable, P-element insertion mutation in *Drosophila melanogaster*. *Mutat. Res.* **175**: 61.

Eeken, J.C.J., F.H. Sobels, V. Hyland, and A.P. Schalet. 1985. Distribution of MR-induced sex-linked recessive lethal mutations in *Drosophila melanogaster*. *Mutat. Res.* **150**: 261.

Engels, W.R. 1983. The P family of transposable elements in *Drosophila*. *Annu. Rev. Genet.* **17**: 315.

Engels, W.R. and C.R. Preston. 1981. Identifying P factors in *Drosophila* by means of chromosome breakage hot spots. *Cell* **26**: 421.

Fahmy, M.J. and O.G. Fahmy. 1983. Differential induction of altered gene expression by carcinogens at mutant alleles of a *Drosophila* locus with a transposable element. *Cancer Res.* **43**: 801.

Favor, J. 1986. Molecular characterization of a radiation-induced reverse mutation at the dilute locus in the mouse. In *Abstracts from the 16th Annual Meeting of the European Environmental Mutagen Society*, August 1986. Brussels, p. 64.

Fawcett, D.H., C.K. Lister, E. Kellet, and D.J. Finnegan. 1986. Transposable elements controlling I-R hybrid dysgenesis in *Drosophila melanogaster* are similar to mammalian LINEs. *Cell* **47**: 1007.

Fedoroff, N.V. 1983. Controlling elements in maize. In *Mobile genetic elements* (ed. J.A. Shapiro), p. 1. Academic Press, New York.

Fedoroff, N.V., S. Wesler, and M. Sure. 1983. Isolation of the transposable controlling elements *Ac* and *Ds*. *Cell* **35:** 235.

Finnegan, D.J., B.H. Will, A.A. Bayev, A.M. Bowcock, and L. Brown. 1982. Transposable DNA sequences in eukaryotes. In *Genome evolution* (ed. G.A. Dover and R.B. Flavoll), p. 29. Academic Press, New York.

Forget, B.G., E.J. Benz, Jr., and S.M. Weissman. 1983. Normal human globin gene structure and mutations causing β-thalassemia syndromes. *Banbury Rep.* **14:** 3.

Gerasimova, T.I., L.J. Mizrokhi, and G.P. Georgiev. 1984. Transposition bursts in genetically unstable *Drosophila melanogaster*. *Nature* **309:** 714.

Golubovsky, M.D., Y.N. Ivanov, and M.M. Green. 1977. Genetic instability in *Drosophila melanogaster*: Putative multiple insertion mutants at the singed bristle locus. *Proc. Natl. Acad. Sci.* **74:** 2973.

Green, M.M. 1977. Genetic instability in *Drosophila melanogaster*. De novo induction of putative insertion mutations. *Proc. Natl. Acad. Sci.* **74:** 3490.

―――. 1980. Transposable elements in *Drosophila* and other Diptera. *Annu. Rev. Genet.* **14:** 109.

―――. 1984. Genetic instability in *Drosophila melanogaster*. On the identity of the MR and P-M mutator systems. *Biol. Zentralbl.* **103:** 1.

Greenberg, R., R. Hawley, and K.B. Marcu. 1985. Acquisition of an intracisternal A-particle element by a translocated c-myc gene in a murine plasma cell tumour. *Mol. Cell. Biol.* **5:** 3625.

Hassold, T.J. and P.A. Jacobs. 1984. Trisomy in man. *Annu. Rev. Genet.* **18:** 69.

Hawley, R.G., M.J. Shulman, H. Murialdo, D.M. Gibson, and N. Hozumi. 1982. Mutant immunoglobulin genes have repetitive DNA elements inserted into their intervening sequences. *Proc. Natl. Acad. Sci.* **79:** 7425.

Hiraizumi, Y. 1971. Spontaneous recombination in *Drosophila melanogaster* males. *Proc. Natl. Acad. Sci.* **68;** 268.

Hook, E.B. 1982. Contribution of chromosomal abnormalities to human morbidity and some comments upon surveillance of chromosome mutation rates. *Prog. Mutat. Res.* **3:** 9.

Jack, J.W. 1985. Molecular organization of the cut locus of *Drosophila melanogaster*. *Cell* **42:** 869.

Jagadeeswaran, P., D. Tuan, B.G. Forget, and S.M. Weissman. 1982. A gene deletion ending in the midpoint of a repetitive DNA sequence in one form of hereditary persistence of fetal globin. *Nature* **296:** 770.

Jelinek, W.R. and C.W. Schmid. 1982. Repetitive sequences in eukaryotic DNA and their expression. *Annu. Rev. Biochem.* **51:** 813.

Jenkins, N.A., N.G. Copeland, B.A. Taylor, and B.K. Lee. 1981. Dilute(d) coat color mutation of DBA/2J mice is associated with the site of integration of an ecotropic MuLV genome. *Nature* **293:** 370.

Kaghad, M., L. Maillet, and P. Brûlet. 1985. Retroviral characteristics of the long terminal repeat of E.Tn sequences. *EMBO J.* **4:** 2911.

Kamiguchi, Y. and K. Mikamo. 1986. An improved efficient method for analysing human sperm chromosomes using zona-free hamster ova. *Am. J. Hum. Genet.* **38:** 724.

Kidd, S., T.J. Lockett, and M.W. Young. 1983. The Notch locus of *Drosophila melanogaster*. *Cell* **34:** 421.

Kidwell, M.G., J.F. Kidwell, and J.A. Sved. 1977. Hybrid dysgenesis in *Drosophila melanogaster*. A syndrome of aberrant traits including mutations, sterility and male recombination. *Genetics* **86:** 813.

Kuff, E.L., A. Feenstra, K. Lueders, L. Smith, R. Hawley, N. Hozumi, and M. Shulman. 1983. Intracisternal A-particle genes as movable elements in the mouse genome. *Proc. Natl. Acad. Sci.* **80:** 1992.

Lehrman, M.A., J.L. Goldstein, D.W. Russell, and M.S. Brown. 1987. Duplication of seven exons in LDL-receptor gene caused by Alu-Alu recombination in a subject with familial hypercholesterolemia. *Cell* **48:** 827.

Levis, R., P.M. Bingham, and G.M. Rubin. 1982. Physical map of the white locus of *Drosophila melanogaster*. *Proc. Natl. Acad. Sci.* **79:** 564.

Lim, J.K. 1979. Site-specific instability in *Drosophila melanogaster*: The origin of the mutation and cytogenetic evidence for site specificity. *Genetics* **93:** 681.

Mackay, T.F.C. 1986. Transposable element-induced fitness mutations in *Drosophila melanogaster*. *Genet. Res.* **48:** 77.

Martin, R. 1985. Chromosomal abnormalities in human sperm. In *Aneuploidy, etiology and mechanisms* (ed. V.L. Dellarco et al.), p. 91. Plenum Press, New York.

Mattox, W.W. and N. Davidson. 1984. Isolation and characterization of the Beadex locus of *Drosophila melanogaster:* A putative cis-acting negative regulatory element for the heldup-a gene. *Mol. Cell. Biol.* **4:** 1343.

McClanahan, T. and K. McEntee. 1984. Specific transcripts are elevated in *Saccharomyces cerevisiae* in response to DNA damage. *Mol. Cell. Biol.* **4:** 2356.

McClintock, B. 1951. Mutable loci in maize. *Carnegie Inst. Wash. Year Book* **50:** 174.

———. 1957. Controlling elements and the gene. *Cold Spring Harbor Symp. Quant. Biol.* **21:** 197.

———. 1965. The control of gene action in maize. *Brookhaven Symp. Biol.* **18:** 162.

McKusick, V.A. 1986. *Mendelian inheritance in man, catalogs of autosomal dominant, autosomal recessive and X-linked phenotypes*, 7th edition. Johns Hopkins University Press, Baltimore, Maryland.

McLachlan, A. 1986. The *Drosophila* forked locus. *Mol. Cell. Biol.* **6:** 1.

Miller, J.F., E. Williamson, J. Glue, Y.B. Gordon, J.G. Grudzinskas, and A. Sykes. 1980. Fetal loss after implantation. *Lancet* **II:** 554.

Morawetz, C. 1987. Effect of irradiation and mutagenic chemicals on the generation of ADH-2 constitutive mutants in yeast. Significance for the inducibility of Ty transposition. *Mutat. Res.* **177:** 53.

Orkin, S.H. and A. Michelson. 1980. Partial deletion of the alpha globin structural gene in human alpha-thalassemia. *Nature* **286:** 538.

Ottolenghi, S. and B. Giglioni. 1982. The deletion in a type of $\delta°$-$\beta°$-thalassemia begins in an inverted Alu repeat. *Nature* **300:** 770.

Paquin, C.E. and V.M. Williamson. 1986. Ty insertions at two loci account for most of the spontaneous antimycin A resistance mutations during growth at 15°C of *Saccharomyces cerevisiae* strains lacking ADH1. *Mol. Cell. Biol.* **6:** 70.

Parkhurst, S.M. and V.G. Corces. 1985. Forked, gypsys and suppressors in *Drosophila. Cell* **41:** 429.
———. 1986. Retroviral elements and suppressor genes in *Drosophila. Bioessays* **5:** 52.
Paulson, K.E., N. Deka, C.W. Schmid, R. Misra, C.W. Schindler, M.G. Rush, L. Kadyk, and L. Leinwand. 1985. A transposon-like element in human DNA. *Nature* **316:** 359.
Peifer, M. and W. Bender. 1986. The anterobithorax and bithorax mutations of the bithorax complex. *EMBO J.* **5:** 2293.
Rimoin, D.L. and J.E. Rotter. 1982. Genetic heterogeneity in common disease. *Prog. Clin. Biol. Res.* **103:** 97.
Roeder, G.S. and G.R. Fink. 1983. Transposable genetic elements in yeast. In *Mobile genetic elements* (ed. J.A. Shapiro), p. 300. Academic Press, New York.
Rogers, J.H. 1985. The origin and evolution of retroposons. *Int. Rev. Cytol.* **93:** 187.
Rubin, G.R. 1983. Dispersed repetitive DNAs in *Drosophila*. In *Mobile genetic elements* (ed. J.A. Shapiro), p. 329. Academic Press, New York.
Rubin, G.M., M.G. Kidwell, and P.M. Bingham. 1982. The molecular basis of P-M hybrid dysgenesis: The nature of induced mutations. *Cell* **29:** 987.
Russell, D.W., M.A. Lehrman, T.C. Südhof, T. Yamamoto, C.G. Davis, H.H. Hobbs, M.S. Brown, and J.L. Goldstein. 1986. The LDL-receptor in familial hypercholesterolemia: Use of human mutations to dissect a membrane protein. *Cold Spring Harbor Symp. Quant. Biol.* **51(2):** 811.
Sakaki, Y., M. Hattori, A. Fujita, K. Yoshioka, S. Kuhara, and O. Takenaka. 1986. The LINE-1 family of primates may encode a reverse transcriptase-like protein. *Cold Spring Harbor Symp. Quant. Biol.* **51(1):** 465.
Sankaranarayanan, K. 1982. *Genetic effects of ionizing radiation in multicellular eukaryotes and the evaluation of genetic radiation hazards in man*. Elsevier Biomedical Press, Amsterdam.
———. 1986. Transposable genetic elements, spontaneous mutations and the doubling-dose method of radiation genetic risk evaluation in man. *Mutat. Res.* **160:** 73.
Schmid, C.W. and K.E. Paulson. 1984. Interspersed repeats in mammalian DNAs. In *Genetics, new frontiers: Proceedings XV international congress of genetics* (ed. V.L. Chopra et al.), vol. I, p. 255. Oxford and IBH Publishing, New Delhi.
Scott, M.P., A.J. Weiner, T.I. Hazelrigg, B.A. Polisky, V. Pirotta, F. Scalenghe, and T.C. Kaufman. 1983. The molecular organization of the Antennapedia locus of *Drosophila melanogaster. Cell* **35:** 763.
Searles, L.L. and R.A. Voelker. 1986. Molecular characterization of the *Drosophila* vermilion locus and its suppressible alleles. *Proc. Natl. Acad. Sci.* **83:** 404.
Shell, B., P. Szurek, and W. Dunnick. 1987. Interruption of two immunoglobulin heavy-chain switch regions in murine plasmacytoma P3.26Bu4 by insertion of retrovirus-like element ETn. *Mol. Cell. Biol.* **7:** 1364.
Shyam, S. and S. Weaver. 1985. Chromosomal rearrangements associated with LINE elements in the mouse genome. *Nucleic Acids Res.* **13:** 5093.
Simmons, M.J. and J.K. Lim. 1980. Site-specificity of mutations arising in dysgenic hybrids of *Drosophila melanogaster. Proc. Natl. Acad. Sci.* **77:** 6042.

Singer, M.F. 1982. SINEs and LINEs: Highly repeated short and long interspersed sequences in mammalian genomes. *Cell* **28:** 433.

Singer, M.F. and J. Skowronski. 1985. Making sense out of LINEs: Long interspersed repeat sequences in mammalian genomes. *Trends Biochem. Sci.* **10:** 119.

Skowronski, J. and M.F. Singer. 1986. The abundant LINE-1 family of repeated DNA sequences in mammals: Genes and pseudogenes. *Cold Spring Harbor Symp. Quant. Biol.* **51(1):** 457.

Sobels, F.H. and J.C.J. Eeken. 1981. Influence of the MR (mutator) factor on X-ray-induced genetic damage. *Mutat. Res.* **83:** 201.

Spradling, A.C. and G.M. Rubin. 1981. *Drosophila* genome organization: Conserved and dynamic aspects. *Annu. Rev. Genet.* **15:** 219.

Sun, L., K.E. Paulson, C.W. Schmid, L. Kadyk, and L. Leinwand. 1984. Non-Alu family interspersed repeats in human DNA and their transcriptional activity. *Nucleic Acids Res.* **12:** 2669.

Thompson, J.N. and R.C. Woodruff. 1978. Mutator genes—pacemakers of evolution. *Nature* **274:** 317.

UNSCEAR. 1986. *Genetic and somatic effects of ionizing radiation. United Nations scientific committee on the effects of atomic radiation 1986 report to the general assembly, with annexes.* United Nations, New York.

Ymer, S., Q.J. Tucker, C.J. Sanderson, A.J. Hapel, H.D. Campbell, and I.G. Young. 1985. Constitutive synthesis of interleukin-3 by leukemia cell line WEHI-3B is due to retroviral insertion near the gene. *Nature* **317:** 255.

Zachar, Z. and P.M. Bingham. 1982. Regulation of the white locus expression: The structure of mutant alleles at the white locus of *Drosophila melanogaster*. *Cell* **30:** 529.

Author Index

Anderson, G.R., 265
Arthur, W.L., 155

Banks, J., 63
Begemann, M., 289
Bingham, P.M., 115
Boeke, J., 169
Bradshaw, V.A., 245
Brown, M.R., 219
Bucheton, A., 209
Busseau, I., 209

Callahan, R., 91
Chandler, V.L., 275
Chou, T.-B., 115
Copeland, N.G., 51
Corces, V.G., 123
Csink, A.K., 219

Dragani, T.A., 289
Dukovich, M., 309

Eeken, J.C.J., 195
Eichinger, D., 169

Fanning, T.G., 71
Farkas, B.K., 265
Fawcett, D.H., 209
Fedoroff, N., 63
Fink, G.R., 169
Finnegan, D.J., 209
Fredholm, M., 131

Garrels, J.I., 255
Gateff, E., 299

Georgiev, P.G., 103
Georgieva, S.G., 103
Gerasimova, T.I., 103
Geyer, P.K., 123
Green, M.M., 41, 123
Greene, W.C., 309

Hsiao, W.L., 289
Hsieh, L.-L., 289

Ilyin, Y.V., 103

Jenkins, N.A., 51

Kidwell, M.G., 183
Kleckner, N., 17
Kuff, E.L., 79

Lambert, M.E., 255
Lister, C.K., 209
Lynch, M., 209

Masson, P., 63
McDonald, J.F., 219
McEntee, K., 245
Mizrokhi, L.J., 103
Mongkolsuk, S., 71
Moore, K.J., 51
Morisato, D., 17

Paquin, C.E., 235
Paskewitz, S.M., 219
Pélisson, A., 209

Peraino, C., 289
Policastro, P.F., 131

Roberts, D., 17
Roegner-Maniscalco, V., 29
Ronne, H., 155
Rothstein, R.J., 155

Sang, H.M., 209
Sankaranarayanan, K., 319
Scott, J.P., 265
Seperack, P.K., 51
Shapiro, J.A., 3
Siekevitz, M., 309
Singer, M.F., 71
Skowronski, J., 71
Sobels, F.H., 195
Stoler, D.L., 265
Strand, D.J., 219
Strobel, M.C., 51

Tchurikov, N.A., 103
Thomas, B.J., 155

Voss, S.H., 219

Wallis, J.W., Jr., 155
Wano, Y., 309
Weinstein, I.B., 289
Williamson, V.M., 235
Wilson, M.C., 131
Winston, F., 145
Witkin, E.M., 29

Zachar, Z., 115

Subject Index

a gene, 64
AIDS, 309
Alcohol dehydrogenase (ADH) genes
  novel gene product ADHIV, 237, 240
  Ty transposition to
    effect of, 235–243, 246
    NQO stimulation, 249–251
Alu family, 326
Anoxic stress
  induction of multiple proteins, 267
  induction of VL30 RNA, 266
Antimycin-A resistance, 237–243, 249–250, 321

β-Glucoside promoter (*bgl*), 3, 6
Bacterial differentiation, 10–13
bgl. See β-Glucoside promoter
Blood cells, tumor mutants in, 300–306
BS-1 element of maize, 48

Cancer
  presence of 34-kD anoxic stress protein/ LDH_k in, 271–273
  presence of VL30 elements in, 266, 271
Carcinogenesis
  hepatic, 289–297
  transcription of endogenous retrovirus-like elements and, 291
Carcinogen-induced protein-1, 257
Chloramphenicol acetyltransferase (CAT), 221–224, 295, 311–314
Chromosome inversion, during I-R dysgenesis, 211–213
Chromosome rearrangements
  in *Drosophila*, 209
  Ty-induced, 290
  of *ADH2* or *ADH4*, 250
c-*myc*, 293, 311
Copia element
  capsid protein, 226
  distribution, 221, 225
  enhancement of transcription
    in neuroblast tumor cells, 302
  by stress, 292
  expression during development, 224
  germ line excision, 59
  in RLPs, 299, 301
  stress drives transcription from LTR, 223
Cryptic elements, 275, 283
c-*src* expression, 299, 304, 306
ct locus. See *Drosophila*, *ct* locus
Cyclin, 255–257
Cycloheximide, 294
Cyclosporin A, 315

dam methylase, 18. See also DNA methylation
Deletion mutations in *w* locus of *Drosophila*, 109–111
Dilute mutation (*d*)
  lethal, 56
  phenotypes, 52
  physical map, 54
  revertants
    germ line d^+, 53
    somatic d^+, 55
  specific transcripts, 56
  viral d^v, 53
Dilute suppressor, 51–59
Disposable transposables, 76
DNA-binding proteins, 17, 23–26
DNA damage
  agents of, 257
  protein markers for, 255–257
  responsive genes
    NQO-induced transcription, 247
    structure of Ddr78A, 246
DNA methylation, 17–22
  effect on IAP expression, 84
  in G-C rich region of Spm, 67
  hypomethylation of endogenous retrovirus sequences, 294
  potential CpG sites in viral LTR, 140
DNA modification. See also DNA methylation
  effect on *Mu* transposition, 275–283

339

DNA rearrangements, 243
DNA recombination
 copy choice model, 172
 direct repeat, 156–157
 genetic control, 155
 heteroduplex intermediates, 156
 homologous, 134
 retrotransposition and, 137
 Ty assays, 164
DNA repair
 in *E. coli*, 29–33
 and transposition of P elements, 200
DNA replication, translesional, 30
DNA topology, 5
Double mutants containing *recA718*, UV sensitivity in, 31
Doubling dose method, 196, 328–329
Drosophila melanogaster
 ct locus, 103–106
  cloning of, 107
  mutation hot spot, 109
  transposition events in, 108
 effect of I factor, 209
 gypsy element, 123–129
  insertion site in yellow gene, 125
 hybrid dysgenesis in (I-R), 209–214
 malignant transformation in, 299–307
 mass mutation in, 47
 mgd4, 104–106. See also Gypsy element
 mobile genetic elements in, 43, 320,
  spontaneous mutation and, 41, 321
 mutation induction in, 195–204
 P elements in, 183–192, 195–204
  mutagenesis on tumor genes, 305
 recessive-lethal genes, tumorigenesis in, 299–300
 retrotransposons of, 115
 retroviral-like particles, 224, 300
  copia transcripts in, 300–301
 spontaneous mutation in, 41–45
 su(wa) suppressor
  regulation of, 117–121
  transcription of, 116
 transposition explosions in, 103–113
 viral DNA, effect on mutation rate, 47–48
 white gene, 211–214
  A30 mutation, 213
  deletion mutations and reversions, 109–111
  I factor in mutagenesis, 209–211
  mutation changes in, 103–113
  suppressor of, 115–121
 yellow gene, 123–129
  in development, 124
  insertion of I factors, 211–213
  mutations, 211–213

regulation of, 126
revertants, 127
splice pattern, 125
structure of, 124–125
Drosophila simulans, spontaneous mutation rate, 45

Early transposon-like sequences (E.Tn), 325
 in mutation of immunoglobulin gene, 326
ebgA promoter, silencing by Tn*1000*, 4–5
Endogenous retroviral genomes. See Retroviral genomes, endogenous
Endogenous type-C leukemia virus, 269–270
Enhancer sequence
 in ecotropic and xenotropic virus, 139
 in Ty, 235
env, 220
Envelope protein
 lack of in IAP, 85
 in retrovirus classification, 92, 95
Envelope sequences
 deletions in, 136
 in murine retroviral transcripts, 133–134
Epigenetic to genetic change, 291
Escherichia coli
 differentiation in, 10–14
 in regulation of Tn*10* transposition, 17–26
 SOS response, 29–36
 UV mutagenesis, 29–33
Ethylmethanesulfonate (EMS), 321
Ethylnitrosourea, (ENU), 195–200, 324
Eukaryotic mobile genetic elements, 320

F element, 43
Formaldehyde, 195–200

gag
 function, 219
 gene product of IAP, 84
gag-pol sequences, deletion of in retroviral elements, 135
GAL1/Ty fusion, 170
Gene conversion, 159, 178
Genetic mapping, mouse, 51
Genetic risk
 assessment of, 196, 204
 of ionizing radiation, 328–329
Genome stability, 155
G_{IX} system regulation, 132–139
Globin genes, 326

β-Glucoside promoter (*bgl*), 3, 6
  activation by IS*1* and IS*5*, 6
G proteins, 272
Gypsy element. *See also* Drosophila
    melanogaster
  mechanism of mutagenesis, 123–129

Hemagglutinating unit (HU), 17, 23–26
Hepatocarcinogenesis, 289, 292–296
Hepatocellular adenomas and carcinomas,
    292–293
Herv. *See* Retroviral genomes
Heterochromatin, 79
  constituative, 83
Heteroduplex DNA, 156
High-copy elements
  intracisternal A particles, 81
  KPN-1 family, 93, 95
  LINE-1, 76
High-resolution two-dimensional protein
    gel electrophoresis, to detect
    induced proteins after DNA
    damage, 255–262
his4-917δ, 146–148, 162
Histone mRNA synthesis, 247
Hobo element, 46
Homologous recombination,
    intrachromosomal, 55, 58
HU. *See* Hemagglutinating unit
Human
  endogenous retroviral genomes (Herv),
      91–98
  genetic risk assesment, 328–330
  LINE-1, 71–77
  mobile genetic elements and, 330
  neoplasias, 92
Human immunodeficiency virus (HIV)
  expression of interleukin-2 and IL-2
      receptor, 312–313
  latent vs. productive infection, 313–314
  trans-acting transcriptional activator, 221
Human T-cell leukemia virus (HTLV-I),
    309–312
Hybrid coding sequence, role of *Mu*, 9
Hybrid dysgenesis, in Drosophila
  I-R system, 209–214
    structure of I-factor, 210
  other mutations, 211–214
  P-M syndrome, 184

IAP. *See* Intracisternal A-particle
I Factors, 209–214
  mutations associated with, 211–213
    by deletion, 213
    by inversion, 212
  structure, 210

IHF. *See* Integration host factor
Immunoglobin gene, mutations in, 326
Insertional mutation
  in Drosophila, proportion among
      spontaneous mutation, 322–323
  by exogenous retrovirus, 290
  in man, lack of detection, 327
  in mouse hybridoma cells, 326
In situ hybridization, 104
  in detection of transposition explosions,
      106
Integration host factor (IHF), 17, 23–26
Interleukin-2 and receptor, effect of HIV
    infection, 309–311
Interleukin-3, 290
Interspersed repeats, 71–77
Intracisternal A-particle (IAP)
  chromosomal effects, 83
  insertion into immunoglobulin gene, 290
  insertions and deletions, 82
  -LTR-CAT construct, 295
  properties, 84–86
  pseudogenes, 83
  related retrovirus MMTV, 91, 94
  transpositions, 79–81
Ionizing radiation, genetic risks to man,
    328–329
Ionomycin, 314
I-R-induced mutations. *See* Hybrid
    dysgenesis
IRR factor
  double mutants and, 31
  factor, 31–36
  SOS mutagenesis and, 34
IS elements
  IS*1*, enhancer of transcription, 3, 6–7
  IS*2*, creation of novel promoter, 5–6
  IS*5*, enhancer of transcription, 3, 6–7
  IS*10*, regulation by methylation, 19–21

Kirsten and Harvey sarcoma viruses
  possession of VL30 element, 265
  role of VL30 in transforming activity,
      271–273
KP elements, 187
KPN-1 repetitive cellular DNA, 93–97

L1 elements. *See* LINE-1 elements
Lactate dehydrogenase-K, 268
LDL-receptor mutation, 326
Lethal mutations, MR-induced, 196–198
Leukemia, 309
LexA protein, 29–30
LINE-1 elements
  cDNA structure, 72–75

LINE-1 elements (*continued*)
  definition, 325
  ORF1, 73
  ORF2, 74
  presence in IAP genes, 83
  sequence similarity in I factor, 211
lmbn blood cells, 299–305
Long terminal repeat sequences (LTRs)
  demethylation, 86
  Herv class I, 93
  Herv class II, 95
  in homologous recombination, 58
  of human immunodeficiency virus
    (HIV), 309, 314–316
  initiation of *cat* mRNAs
    levels of transcription, 222
    posttranscriptional regulation, 222–229
  LTR-CAT fusions, 221
  number of in genome, 290
  proviral *Emv-3*, 53, 58
  solitary, 93, 96
  solo
    δ, 145
    in *Drosophila ct* locus, 106, 108
    in mouse genome, 137, 139
  target for *trans*-regulation, 142

Maize
  BS-1 element, 48
  insertion of mobile genetic elements, 42
  Mu elements and mutator stocks, 275–283
  spontaneous mutation in, 42
  suppressor-mutant element, 63–69
Mammalian mobile genetic elements, 324–328. *See also* Early transposon-like sequences; Intracisternal A-particle
Mass mutation, 47. *See also* Transposition explosions
mdg4 element, 104–106. *See also* *Drosophila* gypsy element
Meiotic recombination
  effect of mutagenic agents (ENU, formaldehyde, MMS and X-rays), 195, 198–202
  lethal mutants in, 197–198
  role of P elements, 195–207
Methylation. *See also* DNA methylation
  role in mutator activity, 275–283
Methylmethanesulfonate (MMS), 195–200, 324
N-Methyl-*N*-nitro-*N*-nitrosoguanidine (MNNG), 255–258
Mitotic recombination (MR), 46, 195–200
MMS. *See* Methylmethanesulfonate

MMTV. *See* Mouse mammary tumor virus
Mobile DNA elements. *See also* Transposable elements
  mobility of, 46
Mobilization of P elements, 187
Moloney murine leukemia virus (Mo-MLV), 292
Mouse
  dilute locus, 51–60
  endogenous retroviral genomes, 131–142
  intracisternal A-particle, 79
Mouse mammary tumor virus (MMTV), 91, 94
Movable elements. *See* Transposable elements
MR element, 46
mucAB, IRR activity and, 35–36
Mu element, 3, 8–10, 14, 275–283
Murine leukemia virus
  as mobile genetic element, 325
  regulatory sequence, 139–140
  related cellular sequences, 93
  related endogenous retroviral elements, 131–141
Mus musculus. *See* Mouse
Mutagenesis
  I-R type, 46
  mechanism of inversion, 212
  of P elements, 185–187
Mutation
  to antimycin-A resistance, 237
  associated with I factor activity, 209–214
  deletion, 159
  induction
    effect of mutagens on, 198–204
    by ethylnitrosourea (ENU), 195, 198–200
    by formaldehyde, 195
    by methylmethanesulfonate, 195, 198–200
    by P elements, 195–205
    by X-irradiation, 195
  insertion, 145
    proportion among spontaneous mutations, 322–323
  knock out, 169
  lethal. *See* Lethal mutations
  mass. *See* Mass mutation
  rate, 196–198
    effect of viral DNA on, 47
    of P element-induced mutations, 185–187
  spontaneous. *See* Spontaneous mutation
  sporulation defects, 149
  temperature-sensitive, 146
  of white gene, *Drosophila*, 209–214
Mutator stock, 275–283

c-*myc*, 293, 311
Myelomas, 86

Natural selection, 41
Neuroblasts, 300–301
4-Nitro-quinoline-1-oxide, 247–251, 321
Novel promoters, 5–6
Nucleic-acid-binding proteins, 74
Nucleocapsid, 219

P elements, 183–192, 195–205, 324
  correlation between number and
    potency, 203
  evolution of, 188–190
  expression of, 184–187
  frequency of mobilization, 187–188
  induction of mutation by, 195–205
  insertion, 185–187
  mobility of, 183–194
    expression and control, 185
  mutagenesis of, 185–187
    in lethal tumor genes, 305
  potency of reverse mutation, 203
  regulation, 188–189
    by KP elements, 187
  role of DNA repair enzymes, 200
  transposition and X-irradiation, 200–202
Phorbol esters, 311–312
pKM101, 29, 34
pol, 219
pol genes, in retrovirus classification,
    92–93, 97
Polyadenylation signal, 75
Primer tRNA binding site
  glutamic acid, 93
  glutamine tRNA, 140
  in IAP transposition, 83
  lysine, 95
Prokaryotes
  regulation of transposition in, 17–26
  transposable elements in, 3–14
Proliferating cell nuclear antigen (PCNA),
    induction following DNA
    damage, 255–259
Prophage induction, 22
Protein gel electrophoresis, 2-D, 255–262
Protein kinase pp60^{c-src}, 304. *See also* c-*src*
    expression
Protein induction by DNA damage,
    255–262
Proto-oncogenes, activation by insertional
    mutagens, 131
Proviral genomes. *See also* Retroviral
    genomes
  amplification, 97

of ectotropic and xenotropic viruses, 137
intracisternal A-particle, 79
murine leukemia virus related, 93
species distribution, 97
Provirus
  integrational excision, 51–59
  latency period, 220
  murine leukemia, *Emv-3*, 53–54

ras oncogene, 271
RecA
  double mutants in, 31
  model for eukaryotic response, 255
  protease, 22–23
  UV mutagenesis and, 29
Recessive tumor mutants, 300–305
Repetitive sequence elements. *See also*
    High-copy elements
  IAPs located in, 83
  solo δ and Ty, 155
Replisome reactivation, 29
Retroposons, 112, 211
Retrotransposition
  by endogenous proviral genomes,
    131–141
  limiting factors for intracisternal A-
    particle, 81
  mediation by coexpressed elements, 86
Retrotransposons. *See also Drosophila*,
    mobile elements; Copia elements;
    Ty elements
  class II lacking terminal repeats, 75–76
  in *Drosophila*, 123–129
  endogenous, 81
  gypsy, 151
  mutagenic effects, 246
  reversion frequency in dilute mutation,
    55
  suppressor-of-white-apricot, *su(wa)*,
    115–122
Retroviral elements. *See also*
    Retrotransposons; Retrovirus
  comparative analysis in cultured cells,
    220
  definition, 219
  host-mediated regulation of expression,
    219–229
  stress-response effect, 226
Retroviral genomes
  ecotropic and xenotropic, 139
  human endogenous (Herv)
    class I, 93
    class II, 94–95
    expression, 96
    species distribution, 97
  murine endogenous

Subject Index / 343

Retroviral genomes (*continued*)
  distribution of, 133
  enhancer region, 139
  promoter region, 140
  transcriptional regulation, 132–141
Retroviral-like particles (RLPs)
  copia element in, 299, 301
  in *Drosophila* cell nuclei, 224, 300
  increase in tumor cells, 300, 302, 305
  in malignant lethal blood cell neoplasm (*lmbn*), 299
  tumor gene mutation and, 305
  in wild-type and mutant cells, 303–304
Retroviral-like sequences, endogenous
  carcinogenesis and, 294
  expression of, 293
Retroviral transcripts, 133
Retrovirus
  classification, 91–92
  germ line integration, 92
  HIV, 221
  RSV, 221
  similarity to retroposons, 211
  similarity to Ty and copia elements, 289–290
Reverse mutation
  effect of P-element number, 203
  study mechanism of, 204
Reverse transcriptase
  IAP Mg-dependent, 85
  from LINE-1 ORF2, 74–75
  role of during carcinogenesis, 291–292
Reverse transcription, in IAP transposition, 79
RNA processing in regulation of white suppressor, *su(wa)*, 115–120
Rous sarcoma virus (RSV), oncogenic potential, 221

Saccaromyces cerevisiae
  mobile genetic elements in, 320–321
  NQO-induced transcription, 247–248
  recombination between repeated sequences, 155–165
  sporulation and tetrad dissection, 238
  Ty element transposition
    effect of temperature, 235–243
    genes affecting, 145–152
    regulation, 169–180
Sectioned colony, 156–157
SINE elements, 325
Sister chromatid, 160–161
Solo δ LTR
  definition, 145
  enhanced recombination at, 161
  genetic control of recombination, 158
  rearrangements, 158, 162
  transcription initiation from, 148
SOS response, 17, 23
  analogous response in eukaryotes, 255–263
  phenotype, 31–36
  repressor of, 29
Splicing of *su(wa)* transcript, 118–120
Spontaneous mutation
  definition of, 41
  in *D. simulans*, 45
  intergenic and intragenic events, 42
  mobile element insertion and, 42
  molecular characterization of, 43–44
  rate of, 196–197
    in sibling species, 45
  relationship to mobile element insertion, 44
SPT3, 147–149, 171, 177. *See also* Suppressor of Ty
SPT6, 7, and 8, 149–152
c-src expression, 299, 304, 306
  pp60 protein kinase activity in *lmbn* cells, 304
Suppressor of forked, 151
Suppressor of hairy wing, 151
Suppressor-mutator elements
  cryptic, 66
  inactive, 65
  insertion of defective, 63–64
Suppressor of Ty
  mutations
    isolation and analysis, 146
    sporulation defects, 149, 151
    transcriptional activation, 171
Suppressor of white-apricot, *su(wa)*
  autogenous regulation, 118–120
  transcription, 116–117

tat-1. *See* Tyrosine aminotransferase-1
T-cell lines, in study of HIV and HTLV infection, 309–315
Thalassemia, 326
THE-1, 326
Tn*3*
  consensus sequence, 35
  replicative mechanism of, 7
Tn*10*
  IS*10* transposition, 17–22
  dam methylase in, 19
  role of HU in, 23–26
  role of IHF in, 23–26
Tn*1000*
  consensus sequence, 35
  as transcriptional silencer, 3–4, 13
Trans-acting elements

Gu-1, 132–142
suppressors of Ty and δ, 145–152
Transcription
 control and expression of RLPs, 295
 NQO induction in yeast, 247–248
Transcriptional activation, 158
Transcriptional signals, 145
Transferrin, 74
Transposable elements. *See also* Copia elements; Intracisternal A-particle; P elements; Retrotransposons; Ty element; VL30
 of bacteria,
  DNA structure and, 7
  effects on expression, 4–7
  genome restructuring and, 3–14
  physiological change and, 7
  plasmid inheritance and, 10–13
  regulation by host cell, 17–26
  role of, 4–10
 in carcinogenesis, 289–297
 class II lacking terminal repeats, 75
 of *Drosophila*, gypsy, 123–129
 high copy number, 76
 LINE-1, 71–77
 maize suppressor-mutator (*Spm*), 63–69
Transposition
 cellular response to, 22
 deletion mutations in, 109–112
 explosions, 103–112
  in *ct* locus, 108–113
 mechanism for, by transcription of IAP, 291
 of *Mu* elements, 275–283
 NQO stimulation of Ty, 249–251
 prophage induction and, 22
 rate of, 103
  effect of temperature on, 235–243
 reverse-transcription-mediated, 71–77
 reversion of deletion mutations in, 109–112
 trans-acting function, 63–68
Transposition memory, 104
 mechanism of, 106
Tumor formation in liver, 289
Tumor genes, recessive, 300–305
Tumor viruses, 91
Ty element. *See also* Suppressor of Ty
 genetic control of recombination, 158
 homology with retroviruses, 235
 host transcriptional activator (*SPT3*), 171
 insertion into ADH genes, 236–237

NQO-induced transcripts, 248
NQO-induced transposition, 249–251
 overexpression effects, 86
 recombination enhancer mutation, 161
 regulation of transposition, 169–179
  defective, 175, 177
  promoter shuffling model, 178–179
  RNA overproduction, 174
 role in yeast gene expression, 145
 transcription, 148
  enhancer, 292
Tyrosine aminotransferase-1 (*tat-1*), 309–310
 activation of interleukin-2 receptor, 311

UmuDC, 29, 33
$URA3^+$ gene, 156
UVC light, effects on fibroblasts and keratinocytes, 258–262
UV-irradiation
 DNA damage by, 321
 double mutants and, 31–33
 induction of proteins by
  major histocompatibility class I protein (MHC-I), 258–262
  PCNA, 255–258
 mechanism of mutagenesis, 34–36
 mutagenic effects of, 29–33
 transient gene expression by, 256
UV mutagenesis
 mechanism of, 29–31
 resistance to by plasmids, 34–36
 role of RecA, 29–31
 role of UmuDC, 29–31

Viral DNA/RNA as mutagens, 48
Viral infection, trigger for latent to productive infection, 313
Viruses, mutagenicity of, 48
Virus-like particles, 170
VL30 element
 definition and distribution, 266
 RNA
  expression in tumors, 292–293
  induction by anoxic stress, 267
  presence in exercised muscle, 270

White gene. *See Drosophila*, white gene

Yeast. *See Saccharomyces cerevisiae*
Yellow locus. *See Drosophila*, yellow gene